Valuing Crop Biodiversity

On-farm Genetic Resources and Economic Change

The International Food Policy Research Institute (IFPRI) was established in 1975. IFPRI's mission is: to identify and analyse alternative national and international strategies and policies for meeting food needs of the developing world on a sustainable basis, with particular emphasis on low-income countries, poor people and sound management of the natural resource base that supports agriculture; to make the results of its research available to all those in a position to use them; and to help strengthen institutions conducting research and applying research results in developing countries. IFPRI is one of 15 Future Harvest agricultural research centres and receives its principal funding from governments, private foundations and international and regional organizations, most of whom are members of the Consultative Group on International Agricultural Research.

IFPRI
2033 K Street NW
Washington, DC 20006
USA

The International Plant Genetic Resources Institute (IPGRI) is an independent international scientific organization that seeks to improve the well-being of present and future generations of people by enhancing conservation and the deployment of agricultural biodiversity on farms and in forests. It is one of 15 Future Harvest Centres supported by the Consultative Group on International Agricultural Research (CGIAR), an association of public and private members who support efforts to mobilize cutting-edge science to reduce hunger and poverty, improve human nutrition and health, and protect the environment. IPGRI has its headquarters in Maccarese, near Rome, Italy, with offices in more than 20 other countries worldwide. The Institute operates through four programmes: Diversity for Livelihoods, Understanding and Managing Biodiversity, Global Partnerships, and Improving Livelihoods in Commodity-based Systems.

IPGRI
Via dei Tre Denari 472/a
00057 Maccarese
Rome, Italy

The geographical designations employed and the presentation of material in this publication do not imply the expression of any opinion whatsoever on the part of IFPRI, IPGRI or the CGIAR concerning the legal status of any country, territory, city or area or its authorities, or concerning the delimitation of its frontiers or boundaries. Similarly, the views expressed are those of the authors and do not necessarily reflect the views of these organizations. Mention of a proprietary name does not constitute endorsement of the product and is given only for information.

Valuing Crop Biodiversity

On-farm Genetic Resources and Economic Change

Edited by

Melinda Smale

Research Fellow
International Food Policy Research Institute
Washington DC
USA
and
Senior Economist
International Plant Genetic Resources Institute
Rome
Italy

CABI Publishing

In association with
International Food Policy Research Institute
International Plant Genetic Resources Institute
Food and Agriculture Organization of the United Nations

CABI Publishing is a division of CAB International

CABI Publishing
CAB International
Wallingford
Oxfordshire OX10 8DE
UK

CABI Publishing
875 Massachusetts Avenue
7th Floor
Cambridge, MA 02139
USA

Tel: +44 (0)1491 832111
Fax: +44 (0)1491 833508
E-mail: cabi@cabi.org
Website: www.cabi-publishing.org

Tel: +1 617 395 4056
Fax: +1 617 354 6875
E-mail: cabi-nao@cabi.org

©CAB International 2006. All rights reserved. No part of this publication may be reproduced in any form or by any means, electronically, mechanically, by photocopying, recording or otherwise, without the prior permission of the copyright owners.

A catalogue record for this book is available from the British Library, London, UK.

Library of Congress Cataloging-in-Publication Data
Valuing crop biodiversity : on-farm genetic resources and economic change / edited by Melinda Smale.
 p. cm.
 Includes bibliographical references and index.
 ISBN 0-85199-083-5 (alk. paper)
 1. Crops--Germplasm resources. 2. Crops--Germplasm resources--Economic aspects. 3. Plant innovations. 6. Farms, Small. I. Smale, Melinda. II. Title.
 SB123.3.V35 2005
 631.5'23--dc22
                                                                    2005002725

ISBN   0 85199 083 5
        978 0 85199 083 5

Typeset by SPI Publisher Services, Pondicherry, India
Printed and bound in the UK by Biddles Ltd, King's Lynn

Contents

About the Authors   vii

Foreword *Joachim von Braun and Emile Frison*   xiv

Acknowledgements   xvi

PART I INTRODUCTION

1. **Concepts, Metrics and Plan of the Book**   1
   M. Smale

PART II PRIVATE VALUE: STATED PREFERENCES OF FARMERS

2. **Crop Valuation and Farmer Response to Change: Implications for *In Situ* Conservation of Maize in Mexico**   17
   G.A. Dyer

3. **Farmer Demand for Agricultural Biodiversity in Hungary's Transition Economy: A Choice Experiment Approach**   32
   E. Birol, A. Kontoleon and M. Smale

4. **An Attribute-based Index of Coffee Diversity and Implications for On-farm Conservation in Ethiopia**   48
   E. Wale and J. Mburu

PART III PRIVATE VALUE: REVEALED PREFERENCES OF FARMERS

5. **Missing Markets, Migration and Crop Biodiversity in the *Milpa* System of Mexico: A Household-farm Model**   63
   M.E. Van Dusen

6. **Explaining the Diversity of Cereal Crops and Varieties Grown on Household Farms in the Highlands of Northern Ethiopia**   78
   S. Benin, M. Smale and J. Pender

7. **Demand for Cultivar Attributes and the Biodiversity of Bananas on Farms in Uganda** 97
   S. Edmeades, M. Smale and D. Karamura

8. **Explaining Farmer Demand for Agricultural Biodiversity in Hungary's Transition Economy** 119
   E. Birol, M. Smale and Á. Gyovai

9. **Rural Development and the Diversity of Potatoes on Farms in Cajamarca, Peru** 146
   P. Winters, L.H. Hintze and O. Ortiz

PART IV PUBLIC VALUES, VILLAGES AND INSTITUTIONS

10. **Managing Rice Biodiversity on Farms: The Choices of Farmers and Breeders in Nepal** 162
    D. Gauchan, M. Smale, N. Maxted and M. Cole

11. **Determinants of Cereal Diversity in Villages of Northern Ethiopia** 177
    B. Gebremedhin, M. Smale and J. Pender

12. **Social Institutions and Seed Systems: The Diversity of Fruits and Nuts in Uzbekistan** 192
    M.E. Van Dusen, E. Dennis, J. Ilyasov, M. Lee, S. Treshkin and M. Smale

13. **Community Seed Systems and the Biodiversity of Millet Crops in Southern India** 211
    L. Nagarajan and M. Smale

14. **Seed Supply and the On-farm Demand for Diversity: A Case Study from Eastern Ethiopia** 233
    L. Lipper, R. Cavatassi and P. Winters

15. **Institutions, Stakeholders and the Management of Crop Biodiversity on Hungarian Family Farms** 251
    G. Bela, B. Balázs and G. Pataki

16. **Cooperatives, Wheat Diversity and the Crop Productivity in Southern Italy** 270
    S. Di Falco and C. Perrings

PART V CONCLUSIONS

17. **Scope, Limitations and Future Directions** 280
    M. Smale, L. Lipper and P. Koundouri

18. **An Annotated Bibliography of Applied Economics Studies about Crop Biodiversity *In Situ* (On Farms)** 296
    P. Zambrano and M. Smale

Index 310

About the Authors

Bálint Balázs is a sociologist, a social historian and a PhD student of the Institute of Environmental and Landscape Management at St István University, Gödöllő, Hungary. He has an MA in Sociology from the Faculty of Social Sciences, Eötvös Loránd University, Budapest, and an MA in Modern History from the Central European University, Budapest. Currently, his research focuses on social aspects of agrobiodiversity, exploring the interface and dynamics between social, economic and ecological systems. He is also working on the application of participatory social science methodology in several research projects ranging from environmental sociology to organic farming policy and sustainable rural development.

Györgyi Bela is an economist and a researcher of the Institute of Environmental and Landscape Management at St István University, Gödöllő, Hungary. She is specialized in social and economic valuation of nature and in environmental decision support tools. She is responsible for coordinating various research projects on environmental valuation of biodiversity, wetlands and the impact assessment of multifunctional agriculture.

Ekin Birol is a Research Associate of the Department of Land Economy, University of Cambridge; a Fellow of Homerton College; and an Affiliate Lecturer at the Department of Economics, University of Reading. She has been working as an economics consultant for the International Plant Genetic Resources Institute and the International Food Policy Research Institute for several years, focusing on economics methods for analysis of agricultural biodiversity on farms. She was awarded a PhD in Economics (2004), an MPhil in Economics with a concentration in econometrics and development economics (2001) and an MSc in Environmental and Resource Economics (1999), all from University College London. Her main research interests include sustainable use and management of agrobiodiversity, and conservation and sustainable management of wetlands.

Samuel Benin is an Agricultural Economist/Research Fellow with the International Food Policy Research Institute, currently based in Kampala, Uganda. He leads a research project to inform the design and implementation of programmes supporting the United States Agency for International Development Expanded Sustainable Economic Opportunities for Rural Sector Growth in Uganda, as well as the Government of Uganda's plan for the modernization of agriculture and eradication of poverty. Between 1999 and 2003, he worked with the International Livestock Research Institute, Addis Ababa, Ethiopia, first as Postdoctoral Scientist and then Scientist leading research projects to identify policy, institutional and technological strategies for sustainable land management and enhancing technology adoption and increasing returns on investment by smallholders in the highlands of Amhara Region, Ethiopia. He obtained his PhD from the University of California at Davis (1999) where he carried out his doctoral research on the efficiency and distribution implications of traditional land inheritance institutions in Ghana.

About the Authors

Romina Cavatassi is an economist with the Agricultural Sector in Economic Development Service of the Food and Agriculture Organization in Rome. She obtained an MSc from the London School of Economics in Environmental Assessment and Evaluation and a *Laurea* in Economics (MA equivalent) from the University of Bologna. She has worked on research projects in Ethiopia and Costa Rica that involved survey design, data collection and data analysis. Her research interests are agricultural, environmental and development economics, GIS application and water management.

Matthew Cole is a Senior Lecturer in Environmental Economics in the Department of Economics, University of Birmingham, UK.

Evan Dennis received his BA in Political Science from Yale University and is currently pursuing a doctorate in Environmental Anthropology at Indiana University. He has conducted research on plant genetic resources in Uzbekistan and Turkmenistan for the International Plant Genetic Resources Institute in Rome, Italy, and for the International Food Policy Research Institute in Washington, DC.

Salvatore Di Falco is a Research Fellow at the Agricultural and Resource Economics Department, University of Maryland. He earned his PhD at the University of York, UK, working with Charles Perrings. He has worked on research projects in Italy and Ireland. His research interests are in applied econometrics, agricultural and resource economics and development economics.

George A. Dyer is an economist at El Colegio de Mexico, Morelia Campus. He has a degree in Biology from the Universidad Nacional Autonoma de Mexico (1991), a Master's in Economics from the El Colegio de México (1994) and a PhD in Agricultural and Resource Economics from the University of California at Davis (2002). He has worked on the economics of maize cultivation and conservation in Mexico for a number of years. His current research focuses on seed and grain flows and the spread of transgenics to Mexican maize landraces.

Svetlana Edmeades is a Natural Resource Economist working as a Postdoctoral Fellow at the International Food Policy Research Institute, Washington, DC. She earned her PhD in Economics from North Carolina State University (2003), specializing in the fields of Development and Natural Resources/Environmental Economics. Her dissertation focused on the development of a conceptual framework for the analysis of variety choice and variety demand in semi-subsistence agriculture, where markets are imperfect and intrinsic variety characteristics are important factors in farmers' growing decisions for staple crops. Her dissertation work was part of a project about assessing the impacts of improved banana varieties in Uganda and Tanzania. She also holds a Master's degree in Economics from the University of Waikato in New Zealand (1998) and a Bachelor's degree in International Relations from the University of the Americas in Mexico City (1995).

Devendra Gauchan is an Agricultural Economist with Nepal Agricultural Research Council, based in Kathmandu. He conducts socio-economic and policy research work for the National Agricultural Research System in Nepal. Since 1997, he has worked closely with the International Plant Genetic Resources Institute in the Nepal country component of a global project entitled 'Strengthening the Scientific Basis for *In Situ* Conservation of Agrobiodiversity On-farm'. He has recently participated in an International Plant Genetic Resources Institute-led global project about genetic resources policies. He obtained his PhD from the University of Birmingham, UK, where his field research work focused on economic and policy incentives for conserving rice genetic diversity on-farm in Nepal. Between 1992 and 1997, he worked in farming systems, participatory technology development and technology adoption, and diffusion studies. Before 1992, he conducted plant breeding and seed production research on maize in the National Maize Research Programme in Rampur, Chitwan (lowland) and seed research on temperate vegetables in the Marpha and Mustang (high mountain) regions of Nepal.

Berhanu Gebremedhin is a scientist at the International Livestock Research Institute in Addis Ababa, Ethiopia. He earned his MA in Economics (1994) and PhD in Agricultural Economics

(1998) from Michigan State University, USA, and his BSc in Agricultural Economics with distinction (1983) from the Alemaya College of Agriculture, Ethiopia. He has taught and worked for many years in the area of sustainable resource management and agriculture, both in Ethiopia and in the USA. Major research experiences in Ethiopia include economic evaluation of soil and water conservation technologies; investigation of the adoption of agricultural technologies; analysis of the determinants of effective collective action for community natural resource management; policy and institutional analysis for irrigation development; policy and institutional analysis of sustainable agricultural development and land management; and community and farm level conservation of cereal crops diversity.

Ágnes Gyovai works at the Institute of Agrobotany, Tápiószele, Hungary, where she has been involved in the global project of 'Strengthening the Scientific Basis of *In Situ* Conservation of Agricultural Biodiversity'. She is a third-year PhD student in the training programme at St István University, Gödöllő, investigating the social and economic relevance and means of conserving agrobiodiversity in Hungary. She holds an MSc in Agricultural Environment Management, specializing in environmental protection and nature conservation. She is also an agroenvironmental adviser.

L. Hernando Hintze is an Agricultural Economist who has been a consultant at the Inter-American Development Bank, Washington, DC since 2002, working in the preparation and design of sustainable development programmes in Central America and in rural development issues. He has also worked as a member of the Cabinet of Advisers for the Peruvian Ministry of Economy and Finance in 2001 and as an economic and policy analyst at Apoyo, a consultant group in Peru, from 1991 to 1995. He obtained a PhD in Economics from North Carolina State University with a dissertation on the adoption of maize varieties among small farmers in Honduras.

Jarilkasin Ilyasov earned his BSc in Agriculture (general) from Aberdeen University, Scotland (2002). He attended the university as a recipient of a scholarship from the '*Umid* Foundation of the President of Uzbekistan to Support Study of the Talented Youth Abroad'. Since May 2003, he has worked for the International Plant Genetic Resources Institute in Tashkent as a Consultant on Participatory Research Approaches under the System-Wide Program on Collective Action and Property Rights project 'Strengthening Community Institutions to Support the Conservation and Use of PGR in Uzbekistan and Turkmenistan'. He also holds a position as an Economist with the district municipal authority, where his major responsibility is to monitor implementation of national economic reform policies in and among agricultural and rural enterprises at the district level.

Deborah Karamura joined IPGRI-INIBAP (International Network for the Improvement of Banana and Plantain) in 2000 as a Musa Genetic Resource specialist, responsible for INIBAP's *in situ* banana conservation project in East and South Africa, based in Kampala, Uganda. Bringing her wide experience in banana taxonomy and nomenclature, particularly of highland bananas, she has developed and tested methodologies for the characterization and conservation of Musa diversity and created innovative approaches for mobilizing and supporting farming communities in East Africa in the quest to advance banana diversity conservation on-farm. Her research, which focuses on the determination and documentation of cultivar diversity in traditional banana-based systems; the identification of genetic erosion factors; the documentation of traditional management approaches with the aim of integrating them with research-derived technologies through scientific methodologies; and on the use of banana diversity to improve rural livelihoods as an on-farm conservation strategy, has helped turn a predominantly staple food into a crop with commercial prospects in the region.

Phoebe Koundouri is a Senior Lecturer in the Department of Economics, University of Reading, UK. She obtained her PhD in 2000 from the Department of Economics, Faculty of Economics and Politics, University of Cambridge. She has previously taught at the Department of Economics of the University of Cambridge and the Department of Economics, University College London. She was a Research Fellow at the Department of Applied Economics of the University of Cambridge and at the Centre for Economic Forecasting of the London Business School. She is also a Senior Research Fellow in the Department of Economics and the Centre for Socio-economic Research on the Global Environment of University College London, a member of the World Bank Groundwater Management Advisory Team and a member of the World Bank Water Resource Management Group on Economic Incentives. She is also a member of Peterhouse College of the University of Cambridge and an honorary fellow of Cambridge Commonwealth Trust. She has coordinated projects for, and acted as an economic adviser to, various international organizations, such as the World Bank, World Health Organization, European Commission, International Institute for Environment and Development, World Wildlife Fund, UK Treasury, UK Department of Water Affairs, Ministry of Agriculture, Water and Rural Development, United States Environmental Protection Agency, and various governments of other developed and developing countries.

Andreas Kontoleon is a Research Fellow at the Department of Economics and Centre for Social and Economic Research for the Global Environment, University College London. He is also a university lecturer in Environment Economics in the Department of Land Economy, University of Cambridge. He earned his PhD in Economics from University College London (2002).

Marina Lee works at Uzbek Research Institute of Market Reforms as a junior scientist carrying out the monitoring of legislative basis in Uzbekistan and processing of methodological and policy recommendations for legislative basis improvement from 1999 to 2002. She is also a postgraduate student majoring in the economic and ecological aspects of agricultural production. She has a state degree in Business and Management, having specialized as an expert translator in Djizak Center of Foreign Languages from 1996 to 1998. She graduated with honours from Djizak Polytechnic Institute, Faculty of Economics (1998). Since 2002, she has worked at the International Plant Genetic Resources Institute in Tashkent as a research assistant.

Leslie Lipper is a Staff Economist in the Agricultural and Development Economics Division of the Food and Agriculture Organization of the United Nations since 2000. She has a PhD in Agricultural and Resource Economics from the University of California at Berkeley, and an MSc in International Agricultural Development from the University of California at Davis. The focus of her professional work is the empirical economic analysis of relationships between rural poverty and environmental management. The two main topics of her current research are the potential for environmental service payments to contribute to poverty alleviation and the impact of seed and commodity markets on the sustainable utilization of crop genetic resources. Before coming to the Food and Agriculture Organization of the United Nations, she spent one year in northeast Brazil as a Fulbright Fellow, conducting research on the impact of agrarian reform policy on environmental management. She worked for several years as a consultant for several international development agencies in the design and evaluation of development projects in China, Vietnam and Bhutan. She taught English in Beijing, China, from 1980 to 1982, as a participant in the Volunteers in Asia Program.

Nigel Maxted has directed international research projects addressing *in situ* and *ex situ* conservation of plant genetic resources in Europe, Asia and Africa, as well as studying the taxonomy and ecogeography of legume diversity worldwide. He has coordinated two successful EC-funded projects: International Solanaceae Information Network and European Crop Wild Relative Diversity Assessment & Conservation Forum (PGR Forum). He regularly works as a consultant for the leading international conservation agencies and is Senior Scientific Adviser for the GEF/World Bank (Plant Genetic Resources Conservation) in the Middle East; Chair of the ECP/GR *In Situ* and On-Farm Network; Chair of the IUCN Crop Wild Relative Specialist Group; and a member of the UK Plant Genetic Resources Group.

About the Authors

John Mburu is, since 2002, a Senior Research Fellow and the head of the Biodiversity Research Subgroup of the Department of Economic and Technological Change at the Center for Development Research, University of Bonn. He holds a PhD in agricultural and resource economics (2002) and an MSc in socio-economics of rural development, both from the University of Goettingen, Germany. Currently, he is specializing in economic valuation of forests, animal and crop genetic resources; analysis of efficient and sustainable approaches of conservation of natural resources; and cost–benefit analysis and assessment of incentives for conservation of natural resources. Besides lecturing in the Center for Development Research in doctoral courses and within the lecture programme of the University of Bonn, he coordinates a number of projects that are involved in biodiversity research in Ethiopia, India and Kenya.

Latha Nagarajan recently completed her PhD in the Department of Applied Economics at the University of Minnesota. Her dissertation was about 'Managing Millet Diversity: Farmers' Choices, Seed Systems and Genetic Resource Policies'. During the early part of 2002, she worked on a rice technology evaluation project in Egypt with the International Food Policy Research Institute, Washington, DC. She obtained her BSc and MSc in agricultural science from Tamil Nadu Agricultural University, specializing in plant sciences and economics. Upon graduation she worked on sustainable agriculture and rural development issues at the M.S. Swaminathan Research Foundation in Chennai, India, from 1993 to 1998. Her fields of interest are international trade and development, resources and environmental economics.

Oscar Ortiz is an agronomist with specialization in knowledge systems and participatory research. Born in Cajamarca, the northern highlands of Peru, he received his BSc in Agronomy at the local university (1986). After working with the National Agricultural Research Institute, National Program of Andean Crops, he was employed by the Nestlé Company in Cajamarca in 1988, where he was responsible for the extension service related to industrial crops. He obtained his MSc in Crop Production and Agricultural Extension at La Molina University (1991). From 1992, he worked in the Social Sciences Department at the International Potato Center in a number of projects in Latin America and the Caribbean. He earned his PhD from the Agricultural Extension and Rural Development Department at the University of Reading, UK (1998). For the International Potato Center, he has coordinated special projects in Latin America, Africa and Asia, and implemented projects on participatory research for Integrated Pest Management and Integrated Disease Management. Interim Project Leader for Integrated Pest Management in 2003, he is currently Division Leader for Integrated Crop Management at the International Potato Center and visiting lecturer at the Graduate School of La Molina University in Lima. He is a member of the Latin American Potato Association and the International Society for Horticultural Science.

György Pataki is an Economist/Associate Professor with the Institute of Environmental and Landscape Management at St István University, Gödöllő, Hungary, and also affiliated with the Department of Business Economics, Corvinus University of Budapest. He has a PhD in Management and Organisation Science from the Faculty of Business Administration, Budapest University of Economic Sciences and Public Administration (now called Corvinus University of Budapest). He recently spent six months as a visiting researcher at the Department of Management and Organisation, Helsinki School of Economics and Business Administration, where he has been working on the social theory of corporate greening. In addition, he is experimenting to apply participatory action research techniques, particularly in the context of bottom-up sustainable rural development, in Hungary. He is also committed to the perspective of ecological economics and doing research on biodiversity issues, including plant genetic diversity, the social and cultural value of ecosystem services provided, particularly by forest and wetland ecosystems. As a university lecturer, he frequently applies problem- and project-based learning and teaching techniques that push students and teachers in a less structured and more cooperative learning context. As a concerned citizen, he is also involved with actions of NGOs in Hungary, particularly with 'Protect the Future', a civil political organization.

John Pender leads the International Food Policy Research Institute's research programme on policies for sustainable development of less favoured lands. His research at the International Food Policy Research Institute focuses on the impacts of policies, institutions and technologies on livelihood strategies, land management, agricultural production, poverty and natural resource sustainability in less favoured areas having low agricultural potential or low access to markets and infrastructure. The research also seeks to understand the trade-offs or synergies among these outcomes resulting from different policy and programme interventions. Most of his research has focused on the highlands of East Africa, hillsides of Central America and semi-arid parts of India. He received a Bachelor's degree from California Institute of Technology, a Master's in public policy from the University of California at Berkeley and a PhD in agricultural economics from Stanford University.

Charles Perrings has been Professor of Environmental Economics and Environmental Management at the University of York since 1992. Previous appointments include Professor of Economics at the University of California, Riverside; Director of the Biodiversity Programme of the Beijer Institute, Stockholm; Professor of Economics at the University of Botswana; and Associate Professor of Economics at the University of Auckland. He is editor of the Cambridge University Press journal, *Environment and Development Economics*, and is on the editorial board of several other journals in environmental, resource and ecological economics, and in conservation ecology. He is President of the International Society for Ecological Economics, a society formed to bring together the insights of the ecological and economic sciences to aid understanding and management of environmental problems. His research interests in environmental, resource and ecological economics include the modelling of dynamic ecological-economic systems, the management of environmental public goods under uncertainty, and the environmental implications of economic development. His applied research focuses on the economics of biodiversity change, freshwater and marine resources.

Melinda Smale leads a research programme about economics and genetic resources at the International Food Policy Research Institute in Washington, DC, USA and the International Plant Genetic Resources Institute in Rome, Italy. Her research emphasizes the development of methods to assess the value of crop biodiversity and the identification of policies to enhance the utilization and management of crop genetic resources, particularly in developing economies. From 1994 to 2000, in Mexico, Melinda worked on crop genetic resources and technology adoption issues with the International Maize and Wheat Improvement Center. She conducted research in Malawi from 1989 to 1993 about hybrid maize adoption by smallholder farmers and maize research impacts, also with the International Maize and Wheat Improvement Center. In the 1980s, she worked in Pakistan, Somalia, Mauritania and Niger on short-term assignments for the International Maize and Wheat Improvement Center, Chemonics International, Volunteers in Technical Assistance and the United States Agency for International Development. She obtained her PhD in agricultural economics from the University of Maryland, her MSc from the University of Wisconsin, also in agricultural economics, and her MA from the Johns Hopkins School of Advanced International Studies.

Sergey Treshkin is Regional Specialist in Community Conservation of Plant Genetic Resources from Uzbekistan. He joined the International Plant Genetic Resources Institute in the Tashkent office in June 2002 to work on the project 'Strengthening Community Institutions To Support The Conservation and Use of Plant Genetic Resources in Uzbekistan and Turkmenistan'. He holds a PhD in Biological Sciences in Ecology from the Research Institute of Nature Protection, graduating as an Engineer of Forestry Farms in the Forestry Faculty, Agricultural Institute, Tashkent, Uzbekistan. He worked with the Karakalpak Department of Forestry (1980–1984), joining the Complex Institute of Natural Sciences as a researcher in 1987. In late 1994, he joined the Institute of Bioecology where he held the position of Leading Scientist Researcher. He has participated in various scientific field missions in Germany, Mongolia, Russia, Turkmenistan and Uzbekistan. He has received awards from the John D. and Catherine T. MacArthur Foundation,

USA (1996), National Geographic Society, USA (1998) and the Open Society Institute (1997, 1998). He implements international and local scientific projects, with a research focus on biodiversity conservation and natural resources management.

M. Eric Van Dusen is a Ciriacy-Wantrup Postdoctoral Research Fellow at the University of California at Berkeley. He conducts research in the areas of crop genetic resources, international treaties covering genetic material and intellectual property rights. He received a PhD from the University of California at Davis in Agricultural and Resource Economics. His dissertation was on the conservation of the traditional *milpa* cropping system in a remote area of Puebla, Mexico. He is currently involved in collaborative research through the International Food Policy Research Institute with researchers in Hungary, Nepal, Uzbekistan and India. He is working on developing empirical approaches to understanding farmer seed systems as a way to combine both conservation and development policy objectives.

Edilegnaw Wale is an assistant professor of Agricultural Economics at the Department of Agricultural Economics, College of Agriculture, Alemaya University, Ethiopia. He started his career as a graduate assistant in the Department in 1993. Since then, he has been working at the university in various capacities. He is involved in teaching graduate and undergraduate courses, research and advising Agricultural Economics graduate students on their MSc thesis work. He has recently started working on the economics of genetic resources policy in collaboration with the International Plant Genetic Resource Institute, Nairobi, and Center for Development Research, Germany. He received his PhD in Agricultural Economics from the Department of Economics and Technological Change, University of Bonn, Germany (2003). His field research work focused on incentives, opportunity costs and attribute preferences of farmers in conserving coffee and sorghum genetic resources on-farm in Ethiopia. His research interest is the application of microeconomic theory and econometrics for agricultural development problems.

Paul Winters is an Associate Professor in the Department of Economics at American University in Washington, DC. Previously, he was an Agricultural Economist at the Inter-American Development Bank, Visiting Expert at the Food and Agriculture Organization, Lecturer at the University of New England in Australia, and Rockefeller Foundation Research Fellow at the International Potato Center in Lima, Peru. His current research interests include international migration, project impact evaluation, cash transfer programmes in developing countries, rural non-farm activities, contract farming and agricultural biodiversity. He obtained his PhD in Agricultural and Resource Economics from the University of California at Berkeley.

Patricia Zambrano is a Research Analyst at the International Food Policy Research Institute, Washington, DC. She holds a Master's degree in Economics from the University of California at Davis. She has been involved in different research projects in the fields of intellectual property rights, biotechnology and genetic resources.

Foreword

Societies depend on agricultural innovation processes for food security on local, regional and global scales. Crop genetic resources, embodied in the seed planted by farmers, are the building blocks of these processes. Farmers, plant breeders, gene bank managers and other crop scientists draw on diverse crop genetic resources to innovate, supporting their own livelihoods and benefiting society at large.

Sustainable management of crop genetic resources means assuring their diversity, both in trust collections and on farms. In agricultural systems, crop biodiversity is essential to combat the risks farmers face from plant pests, diseases and climatic shocks. Crop biodiversity also underpins the range of dietary needs and services that consumers demand as economies change.

Crop genetic resources are natural assets that are renewable but vulnerable to losses from either natural or human-made interventions, including the disruptions caused by droughts, floods or wars, as well as the gradual process of social and economic change. Technological changes in agricultural production over the past century, spurred by crop genetic improvement combined with the use of other farm inputs, have transformed rural societies in many parts of the world. Not all of these changes have been positive. Local communities, governments, research organizations and NGOs have expressed growing concern about the potential loss of crop biodiversity associated with social and economic change. The common challenge they now face is to develop strategies that enable crop genetic resources to be managed in ways that satisfy the needs of farmers and consumers at present and in the future.

This book contributes to a better understanding of the challenges involved in maintaining crop biodiversity on farms within a rapidly changing global food system. It is one of the first to assemble a set of empirical case studies conducted in the field with farmers and crop scientists across a range of agricultural economies and income levels, applying economics tools and methods adapted specifically to research about valuing and managing crop biodiversity on farms. All of the case studies were implemented with national and international research partners, most by the International Food Policy Research Institute (IFPRI), the International Plant Genetic Resources Institute (IPGRI) and the Food and Agricultural Organization (FAO) of the United Nations. As a set of studies about the *in situ* (on-site) management of crop genetic resources and their diversity, the findings reported here complement those recently published by CABI about the costs of saving seeds *ex situ* (in gene bank collections), prepared by IFPRI and other Future Harvest Centres for the System-wide Genetic Resources Program.

The collection of studies is intended to illuminate the practical meaning of crop biodiversity to farmers, to specify the sources of its value and to indicate how it might be supported by national policies. It is also intended to be used as a tool kit for applied researchers, particularly those working in national and international research programmes or projects in developing economies. As such, the book extends the dialogue launched in 1992 when the Convention on Biological Diversity (CBD) established international legal norms with respect to biodiversity. The CBD recognizes farmers' contribution to crop improvement and urges the equitable sharing of benefits as an incentive for farmers to conserve their biological resources. This book contributes constructively to these policy debates, and to the development of strategies that can facilitate the sustainable management and conservation of crop genetic diversity for future generations.

Joachim Von Braun
Director General
International Food Policy Research Institute

Emile Frison
Director General
International Plant Genetic Resources Institute

Acknowledgements

Neither man nor his crops have obeyed set rules for a sequence of events or stages of development....Simple solutions simply do not work very well. This is an age of great knowledge and little wisdom, but we have no choice; we must blunder on.

Jack R. Harlan, *Crops and Man*, 1975; 1992

The research reported here was supported first by the men and women farmers who generously shared their knowledge and perceptions in personal interviews. Time to talk is increasingly scarce – whether on the hillsides of Nepal or Ethiopia, in the home gardens of Hungary, in the maize fields of Mexico or in an office. The research was also supported patiently through the tutelage of geneticists, plant breeders and ethnobotanists, in particular Toby Hodgkin, István Már, László Holly, Bhuwon Sthapit, Devra Jarvis, Deborah and Eldad Karamura, W. Tushemereirwe, Mauricio Bellon, Julien Berthaud, Dominique Louette, Richard Jones, Prem Mathur and Muhabbat Turdieva. Insights have been contributed by other economists and social scientists, including Pablo Eyzaguirre, Erika Meng, Michael Morris, Paul Heisey, Ruth Meinzen-Dick, John Pender and Douglas Gollin. Paul Winters, Leslie Lipper, Phoebe Koundouri and IFPRI provided valuable review of the work, as did Amanda King. Amanda, Patricia Zambrano and Maria Meer offered a combination of solid advice and opinions, editorial recommendations and clear thinking. Patty Arce and Annie Huie provided tireless administrative support to enable the research to happen. Funds and resources in kind were provided by a number of institutions, cited in the acknowledgements of each chapter, along with other personal thanks from the authors.

1 Concepts, Metrics and Plan of the Book

M. Smale

A Question of Value

There has been considerable public debate about the economic value of biodiversity and whether economists should attempt to value it at all. Some contend that it is inherently unethical to employ a utilitarian discipline like economics to assess the relative costs and benefits of species survival (Ehrenfeld, 1988); others argue that biodiversity must be priced to ensure that what matters to society is conserved (Randall, 1988). Economists' emphasis on value has often distanced them from natural scientists, especially if the purpose of valuation is to justify rather than to explain human behaviour (Roughgarden, 1995; Dyer, 2002). In recognition of divergent perspectives, the balance of this book is tipped more heavily towards the use of economic concepts to explain and predict human choices than to estimate prices expressed in cardinal terms.

The world's array of crop varieties is a consequence of human choices in close interaction with natural selection processes – on-farm (*in situ*), where crop genetic resources are managed by farmers, and off-farm (*ex situ*), where they are managed by plant breeders or gene banks. Relative to other areas of public policy, economics has contributed relatively little to debates about the value of these resources. In a landmark collection of writings about species preservation (Orians *et al.*, 1990), Brown (1990) explained that 'since most of the genetic resources of interest do not trade in markets, there are no prices'. This is still largely the case. The challenges involved in measuring the value of non-market goods are substantial, despite continued progress in the theory and applications of environmental economics.

Price data remain 'sparse' for crop genetic resources, as is true for many other resources of economic significance to society, because it costs a lot to exclude users. One reason is that crop genetic resources are mixed goods with multiple traits or attributes, some of which are not equally 'visible' to all of the people who manage and exchange them. Such information asymmetries do not contribute to good market performance.

There are signs that new markets for crop genetic resources are being created, and if so, more prices may soon be evident. Private companies supplying crop genetic resources have in recent years sought to strengthen the intellectual property rights over crop varieties, isolated genes and enabling tools such as promoters and markers. Simple economic theory predicts that stronger proprietary regimes will decrease the costs of excluding others from using the same resources, generating incentives for innovation and market formation. Non-governmental organizations and a battery of interest groups have countered with claims over other property rights, ostensibly on behalf of farmers and their communities.

The catch is that as on-farm suppliers of crop genetic resources, farmers, in contrast to plant breeders, also use them as planting material – a reproducible production input. These farmers are different in some respects from those working in fully commercialized agriculture in industrialized economies. Many reside in places that have benefited comparatively less from the green revolutions. For example, although it is now generally accepted that Asia's seed-based

© CAB International 2006. *Valuing Crop Biodiversity: On-farm Genetic Resources and Economic Change* (ed. M. Smale)

green revolution generated substantial benefits beyond (the adopting) farmers in irrigated production environments, large numbers of food-insecure families remain in the less productive lands of that continent. Farmers like these, who manage and supply crop genetic resources, often face unpredictable and undifferentiated markets for their products, relying on their own harvests for at least some of the goods consumed by their families.

The decisions of these farmers are the subject matter of this book. There is a growing recognition that some of them are de facto custodians of socially valuable resources. Acknowledging this role, the International Undertaking on Plant Genetic Resources drew the concept of 'farmers' rights' into the public arena during the 1980s. Ratified by over 40 country signatories, the International Treaty on Plant Genetic Resources for Agriculture became law in 2004. The Treaty establishes a multilateral system for sharing genetic resources for 64 key food crops and 24 forage species through a standard agreement, reducing the costs of bilateral transactions among the many parties exchanging lines and progenitors in the development of improved crop varieties – principally professional scientists.

In 1992, the Convention on Biological Diversity (CBD) established international legal norms (not laws) encouraging nations to manage biodiversity in ways that support already declining levels against greater loss. The CBD recognizes farmers' contribution to crop improvement and urges the equitable sharing of benefits as an incentive for farmers to conserve their biological resources. Though farmers' privilege to save seed from harvests has long been recognized, farmers' rights now specifically refer to the right to claim ownership over their varieties as do plant breeders, and the right to be rewarded for the use of these genetic resources by others. The evolution of the plant variety and farmers' rights legislation in India illustrates the ethical, political and scientific complexity of the issues (Srinivasan, 2003; Ramanna and Smale, 2004). It remains to be seen whether this legislation will be 'effective' as a *sui generis* system under the Agreement on the Trade-Related Aspects of Intellectual Property (TRIPS); a requirement for members of the World Trade Organization (WTO) (Koo *et al.*, 2004a).

A goal of this volume is to advance practical thinking about how levels of crop biodiversity may be sustained in ways that do not conflict with but contribute to sharing benefits from economic change. Arguably, sustaining crop biodiversity on farms makes most sense in locations where both society and the farmers who manage it benefit from the process, i.e. where both the private and public values associated with it are relatively high, taking into account any opportunity costs. Conservation initiatives need also to recognize the dynamic nature of human interactions with crop plants, conforming more to a notion of resource management than to that of preservation or curatorship.

Past economics research has treated related topics that bring much to bear on the methods applied and hypotheses tested in this book. The economic benefits of increasing crop productivity through the diffusion of crop varieties bred by professional plant breeders have been documented comprehensively (Byerlee and Traxler, 1995; Morris and López-Pereira, 1999; Alston *et al.*, 2000; Heisey *et al.*, 2002; Evenson and Gollin, 2003), and state-of-the-art tools developed to assess them (Alston *et al.*, 1998). Surveys discussing the sources of economic value in crop biodiversity are numerous, including Pearce and Moran (1994), Swanson (1996) and Gollin and Smale (1998). The value of diversity in crop or animal species diversity has been modelled theoretically (Brown and Goldstein, 1984; Weitzman, 1993; Polasky and Solow, 1995; Simpson *et al.*, 1996; Rausser and Small, 2000; Brock and Xepapadeas, 2003). Costs and benefits have also been estimated for plant genetic resources conserved in gene banks, destined principally for use by commercial farmers (Evenson and Gollin, 1997; Virchow, 1999; Gollin *et al.*, 2000; Johnson *et al.*, 2003; Koo *et al.*, 2004b). The global values of genetic resources and other ecosystem services (Costanza *et al.*, 1997) as well as the values of plant genetic resources and their diversity in crop breeding (Evenson *et al.*, 1998) have been assessed.

Far less work has investigated the value of increasingly scarce, local varieties to the farmers who grow them. This book is an attempt to address the research gap. The studies assembled here explore the economic incentives farmers and their communities have to maintain crop biodiversity across a range of agricultural economies. The opportunity costs associated with growing diverse crops and varieties depend on the farming system and economic context. Developing economies are represented from Asia, Latin America and Africa, as well as transitional and richer economies in Europe. The structure of crop biodiversity depends also on the crop reproduction system. Authors investigate cash crops and food crops, cereals, tubers and fruits – crops that are self-pollinating, cross-pollinating or vegetatively propagated.

Several features of these studies distinguish them from related economics research. The first is an obvious emphasis on farmers' varieties as compared to modern varieties, sometimes called 'folk varieties' (Cleveland et al., 1994) or 'landraces' (Harlan, 1992; Zeven, 1998). Almost exclusively, data were collected through personal interviews with samples of farm families and other stakeholders. The disciplinary approaches are grounded in microeconomic models of farmer decision making and environmental valuation, although the research has in most cases entailed interdisciplinary work with crop scientists. Values are local rather than global; approaches are often enriched with genetic or taxonomic information.

The collection of studies in this book portrays a glimpse of the relationship between economic change and the determinants of agricultural biodiversity. Much of the research represented was conducted by doctoral students in the context of national projects that were internationally funded and facilitated. Thus, this book is intended to serve as a source for tools and examples that can be further adapted or applied by economists working in national and international research programmes and to provide information of relevance for conservation and development practitioners.

The following sections of this introduction define a common vocabulary of biodiversity and economics concepts as they are invoked throughout the chapters of this book. There are many ways to define these concepts, and some simple conventions are followed. After a discussion of terms and cross-cutting themes, the contribution of earlier economics studies about crop biodiversity on farms is summarized. A roadmap for chapters in this book, and how they interrelate, is then presented.

Common Vocabulary and Concepts

Biodiversity of crop plants

Agricultural biodiversity is a component of biodiversity, referring to all diversity within and among species found in crop and domesticated livestock systems, including wild relatives, interacting species of pollinators, pests, parasites and other organisms (Wood and Lenné, 1999). Since agricultural landscapes are fluid, the term component does not imply that boundaries are firm. Domesticated biodiversity (crops, aquaculture and livestock) is located within agricultural landscapes, complemented outside these systems (*ex situ*) by wild relatives in gene banks, breeders' collections or reserves; it serves as both a component of production and a resource for genetic improvement (Cassman et al., forthcoming). Agricultural landscapes also contain non-domesticated species as weedy or 'casual' elements, or just as a part of natural (non-protected) ecosystems. Species diversity pertains to the diversity among species within which gene flow occurs under natural conditions. Genetic diversity in crops comprises all the variation in the genes of individuals. Some have argued that genetic diversity is the fundamental building block of ecological and organism diversity (Cox and Wood, 1999).

The emphasis of this book is the *in situ* (in place of origin, or source) management of crop diversity by agricultural households and communities, or on farm conservation. Here, on-farm conservation implies the choice by farmers to continue cultivating biologically diverse crops and varieties in their communities in the agricultural ecosystems where the crops have evolved historically through processes of human and natural selection (from Bellon et al., 1997; Jarvis et al., 2000).

In this book, 'crop biodiversity' refers to the biodiversity of crops. The biodiversity of crops encompasses phenotypic as well as genotypic variation, including cultivars recognized as agromorphologically distinct by farmers and varieties recognized as genetically distinct by plant breeders. The terms 'cultivars' and 'varieties' are used here to describe either farmers' varieties or those bred by plant breeders. Typically, farmers' varieties do not satisfy International Union for the

Protection of New Varieties of Plants (UPOV) definitions of variety because they are heterogeneous, exhibit less uniformity and segregate genetically. Where it is necessary to distinguish between varieties selected and managed by farmers and those bred by professional plant breeders, the terms 'landraces' and 'modern varieties' are assigned.

Landraces are understood simply as variants, varieties or populations of crops, with plants that are often highly variable in appearance, whose genetic structure is shaped by farmers' seed selection practices and management, as well as natural selection processes, over generations of cultivation. As Harlan (1992) described them, landraces generally exhibit high degrees of local adaptation, with particular properties or characteristics. Genetic variation in landraces is considerable but not without structure, since their composition is often deliberately manipulated by farmers. Landraces 'usually produce something' (Harlan, 1992: 148; Ceccarelli and Grando, 2002: 305), but they do not have high expected yields like modern varieties.

In this book, an effort has been made to understand the genetic structure of the crop and the units of diversity as managed and understood by farmers, in accordance to the extent possible with recent research by geneticists assessing crop biodiversity levels on farms (Jarvis et al., 2004; Sadiki et al., 2005). For example, banana types grown in the East African highlands (Chapter 7) are classified by genome, use group and phenotype, drawing on primary data elicited from farmers about distinguishing characteristics and published taxonomic work. A similar approach was applied in the research on millet crops presented in Chapter 13, where varieties were also sorted by categories of improvement status hybrids, improved open-pollinated variety and improved pure-line selection. Research by geneticists was the basis of the classification used for rice in Nepal (Chapter 10) and potatoes in Peru (Chapter 9). The distinctiveness of sorghum landraces grown in Eastern Ethiopia was validated by merging information from farmers and geneticists (Chapter 14). Botanical and genetics research supports the classification of crops and varieties for Chapters 3, 8 and 15 about Hungary.

Crop biodiversity on farms has both interspecific (among crops) and intraspecific (within a crop) components (Bellon, 1996). Since the crops studied in this book have variable taxonomic status, the terms 'intercrop' and 'intracrop' designate diversity between and among common crops, respectively.

Economic value

All classes of economic value have a basis in human preference. Total economic value includes current use value, option value and existence value. Current use value derives from the utility gained by an individual from the consumption of a good or service, or from the consumption by others of a good or service. Option value, also a use value, is the value associated with retaining an option to a food or service for which future demand is uncertain. Existence value, a non-use value, derives from human preferences for the existence of the resource as such, unrelated to any use to which the resource may be put.

The global spectrum of genetic variation in crops and livestock has expanded and contracted over the centuries as a direct consequence of human interest. That human interest is practical because crop varieties and livestock races are functional units of food production. The premise of this book is that, compared to an endangered, wild plant or animal species, proportionately more of the economic value in domesticated components of agricultural biodiversity resides in current use and option values, as compared to existence value.

The basic policy dilemma of on-farm conservation stems from the mixed good properties of crop genetic resources. All goods can be situated along two axes defined by the extent of rivalry overuse and ease of exclusion in consumption (Romer, 1993; Fig. 1.1). An impure public good has characteristics of both private and public goods. Seed is highly rival with low cost of exclusion, but the genetic resources embodied in seed are non-rival and the costs of controlling their use can be high. The handful of seed or planting material a farmer places in the ground is a private good that is consumed as a production input. No two farmers can plant the same physical unit of seed. To those same farmers, the genetic resources embodied in the seed and their diversity are public goods. Both can grow the same variety simultaneously, and it is

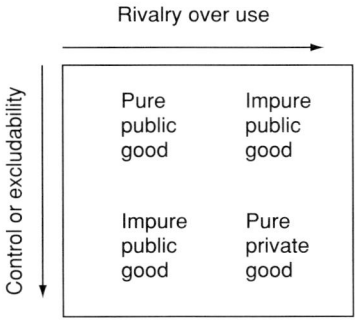

Fig. 1.1. Private, public and impure public goods. Adapted from Romer (1993: 72) and Sandler (1999: 24).

costly to prohibit others in one's community from doing so. Clearly, the costs of exclusion vary by the type of crop genetic resource in question and the institutional context. Controlling the flow of genes among fields is difficult, especially with predominantly cross-pollinating crops as they are managed by farmers in semi-subsistence agriculture. At the same time these are crops for which self-reinforcing forms of intellectual property, such as a professionally bred hybrid, are likely to be profitable for seed companies and encourage private investments (Morris et al., 1998).

The combinations of seed types grown by farmers produce a harvest, which they consume and/or sell and from which they derive private value, but the pattern of genotypes across the landscape contributes to the diversity of the crop genetic resources from which people residing elsewhere and in the future may benefit. The public value of crop biodiversity includes option value for any unforeseen events, such as changes in consumer tastes. Since farmers' decisions on the use and management of crop varieties in their fields can result in smaller plant populations and loss of potentially valuable alleles, their choices have intergenerational and interregional consequences (Sandler, 1999; Fig. 1.1).

Since the diversity of crop genetic resources is never fully apparent to the farmers who provide and use it and is undervalued in markets, farmers are unable to consider the contributions of all other farmers in their community or elsewhere when they make their decisions. Economic theory predicts that, as long as crop biodiversity is a (desirable) 'good', farmers as a group will underproduce it as a group relative to the social optimum and institutional interventions are necessary to close the gap (Cornes and Sandler, 1986; Heisey et al., 1997).

Situations in which individual interests conflict in some way with group interests are called social dilemmas (Sell et al., 2002). Other features of the social dilemma of on-farm conservation are noteworthy. Although some poor farmers in the world depend directly on the biodiversity of the crops they grow, most farmers do not, and most consumers depend on it only indirectly. Those who encourage conservation, and perhaps those who are willing to pay for it, reside largely in other political jurisdictions (Brown, 1990). Sell et al. (2002) classify social dilemmas into public goods problems, in which the individual must decide whether or not to contribute to a common resource, and common property resource problems, in which the individual must decide whether to refrain from taking the resource. They find that individuals are more cooperative when faced with a resource dilemma than a public goods dilemma. On-farm conservation has features of a public goods dilemma.

High benefit–cost ratios for on-farm conservation

Because crop genetic resources are impure public goods, their costs and benefits have both private and public dimensions. Conceptually, the highest benefit–cost ratios for managing crop genetic resources on farms (as compared to their management *ex situ* in breeding programmes or gene banks) will occur where the utility farmers derive from managing them as well as the public value associated with their biodiversity is high (area II of Fig. 1.2). Since farmers are already bearing the costs of maintaining diversity in those areas and they reveal a preference for doing so, the costs of public interventions to support conservation will also be least. Where genetic diversity is assessed as relatively low, no unique traits have been identified in local genetic materials, and farmers derive few benefits from it, there may be no need to invest in any form of conservation (area III of Fig. 1.2). Where the contribution to diversity is great but farmers derive little private value from it, *ex situ* conservation is the only option (area I of Fig. 1.2). Where there is little diversity but farmers care a lot about it, there is no need for public investment at all since no value is associated with conservation (area IV of Fig. 1.2). None the less, some societies might decide to pay farmers to grow certain landraces (examples are found in Tuscany and Ethiopia).

The empirical findings presented in this book and elsewhere demonstrate clearly that in some places in the world, rural people depend on the diversity of their crops and varieties to cope with climatic risk, match them to specific soil and water regimes and meet a range of consumption needs when markets are unreliable. In such environments the opportunity costs of maintaining diversity are likely to be low because development alternatives are limited. Cash-earning opportunities may be few. These locations are often characterized as 'less-favoured', or 'marginalized'. The people who live in them are often considered to be poor on a global scale. Resolving the social dilemma requires some comprehension, however, of how the distribution of costs and benefits from managing crop biodiversity changes with economic changes.

Productivity and diversity trade-offs

Across a crop-producing region diversity is expressed in more distinct genetic types distributed more evenly or equitably. Economics principles suggest that as an economy changes, maintaining intracrop diversity on farms should occur to the extent that trade-offs between productivity and diversity maintenance are consistent with social preferences. Figure 1.3 sketches a hypothetical frontier with points determined by different combinations of biologically distinct cultivated varieties in a reference region with fixed area. The fixity in the area in any growing period or season ensures the concavity of the relationship between the amount that can be produced and the genetic diversity of the crop varieties planted.

Planting all the area in a region to a single variety with the highest expected yield generates the greatest production levels in a given growing season, hypothetically. The short-term costs to farmers of sacrificing expected crop production for the sake of maintaining areas in more numerous but less higher-yielding varieties could be great in the zones with high productivity potential and homogeneous production environments, such as the loci of the Asian green revolutions. Planting all the area to one variety will also augment genetic vulnerability to pests and diseases. Greater equity in the spatial distribution of different varieties, or less genetic uniformity, can improve yield stability over time even in those environments. In more environmentally heterogeneous zones with lower productivity potential, the short-term yield losses to the national economy of growing numerous varieties more equitably distributed across the landscape are likely to be less.

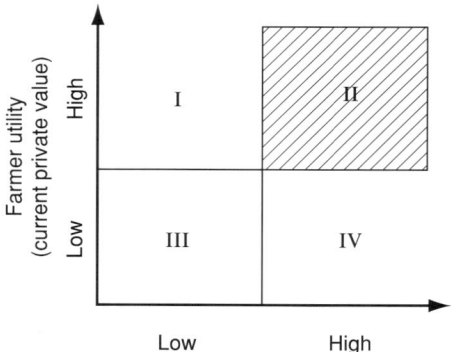

Fig. 1.2. Sites with high benefit–cost ratios for on-farm conservation. Adapted from Smale and Bellon (1999: 395).

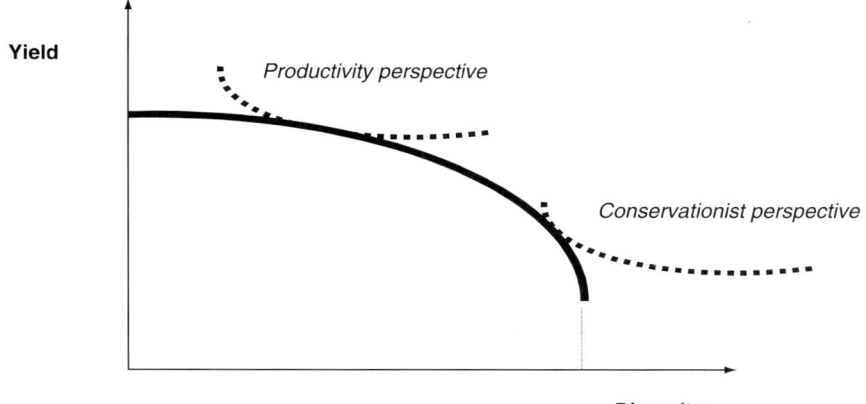

Fig. 1.3. Hypothetical relationship between productivity and crop biodiversity in a reference cropping region with a fixed area. Adapted from Heisey et al. (1997).

The social costs and benefits of crop biodiversity will depend on social preferences, the contour of the yield–diversity relationship and actual production combinations. Although trade-offs are inevitable on the frontier, production does not always occur on the frontier. In some seasons, it is likely that the actual mix of varieties and areas allocated generates both lower overall yields and less diversity than is feasible, so that production points lie within the frontier (Heisey et al., 1997).[1] The social indifference curves depict different preferences for points on the frontier. Some richer societies may be willing to pay for conservation if they produce more than they need; in general, poorer societies are thought to prefer short-term production gains to long-term conservation interests.

Diversity metrics

Metrics for assessing the public value of crop biodiversity (the horizontal axis in Fig. 1.2) can be based on criteria that plant breeders and geneticists employ to identify useful genetic materials for future crop improvement. For example, greater public value might be associated with genes that are locally common but globally rare, on the supposition that these carry both the greatest potential for adaptation and scarcity value. Landraces can be identified for conservation according to rarity, heterogeneity or adaptive traits (as in Chapter 10). Diversity indices can serve as proxies for the public value of a set of crop varieties; these are the dependent variables of the econometric analyses in this book.

To select the appropriate diversity index for an economic analysis several issues must be resolved through interaction with farmers in the study region, knowledgeable crop scientists and geneticists (Meng et al., 1998a; Table 1.1). Different indices represent different diversity concepts; none is universally correct, and more than one may be appropriate in a particular empirical context. For example, the diversity that is 'apparent' to farmers or crop scientists in the physical characteristics of crop populations growing in a field contrasts with the 'latent' diversity revealed though molecular or pedigree analysis. Crop biodiversity can also be differentiated according to its distribution within or among crops or crop varieties; it can express spatial or temporal dimensions.

The diversity concept (latent or apparent; spatial or temporal) is distinguished from the measurement tool that enables the concept to be incorporated into an economic model as a

---

[1] A similar approach to Heisey et al. (1997) has been developed in a game-theoretic framework in Heal et al. (2004).

Table 1.1. Criteria to consider when choosing an index to measure the biodiversity of crop plants.

Crop reproduction	Farming system	Diversity concept	Level or scale	Conservation goal	Data used to construct index
Self Cross Vegetative	Modern Traditional Mixed Microecosystem	Latent/apparent Spatial/temporal Inter/intra[a]	Household Community Region Nation	Rarity Heterogeneity Adaptation	Biochemical Molecular Agromorphological descriptors Pedigree Ecological

[a]Inter/intra could refer to population, variety or species.

diversity index. Diversity indices are scalars constructed from any one of several types of data, and numerous metrics are available in the literature. For example, data may record physical measurements on crop plants grown in controlled experiments. Alternatively, data may document the variation in DNA taken from plant tissue and expressed as patterns on gels.

Fundamental is the definition of the crop population under study. Farmers, plant breeders, molecular biologists and germplasm collectors each have taxonomies or systems for distinguishing among plants. Though a single taxonomy is internally consistent, integrating any pair of taxonomies can be challenging. There is also considerable scientific evidence that the level of diversity identified in a crop plant when measured by one tool, such as molecular markers, may not correspond to the level identified by another, such as morphological descriptors.

This book focuses deliberately on diversity indices that are meaningful to farmers. As a result, the dependent variables in most of the models estimated are diversity indices that are apparent to farmers. Two considerations drove the decision to emphasize apparent over latent diversity concepts. The first reflects the subject matter of human-managed, 'domesticated' as opposed to wild species. Because farmers choose to grow varieties based on the traits and attributes they observe rather than those they cannot see, the relationship between farmers' decisions and molecular or biochemically based indices is far-fetched. Since our dependent variables are derived from the choice variables in models of farmer decision-making, we have adhered to the units that farmers recognize and manage.

Second, molecular and biochemical assays are relatively high cost in terms of laboratory time and materials and the sample sizes that would be required to link them statistically to crop populations as they are managed by farmers. This is especially true for the heterogeneous landraces of open-pollinating crop species. Professionally bred varieties are more uniform and stable genetically across environments than are landraces, though the difference diminishes when they are saved and replanted in successive generations, and applies less to crop species with high rates of outcrossing. Measurement costs of high magnitude are not warranted by the exploratory phase of empirical research conducted in this book, though such investments might be justified once a conservation programme is under implementation.

Named varieties, even when they represent genetic distinctness, are admittedly poor units of analysis for constructing diversity indices because they inform us little about genetic distances. Names can mask genotype redundancies, especially among landraces whose names are linguistic, cultural artefacts. In most of the chapters of this book, variety names have been cross-checked with morphological or genetic evidence to determine genetic distinctness to the extent possible. Distance metrics constructed from taxonomic trees have desirable mathematical properties (Weitzman, 1992; Solow et al., 1993), though for the reasons cited above, chapters in this book have adhered to simpler mathematical constructs. After some experimentation with metrics, the literature concluded that the more sophisticated the construction of the index, the more obscure is its relationship to the decisions of farmers and consequently, the more difficult the interpretation in the context of farm-level data (Meng *et al.*, 1998a, 1999; Van Dusen, 2000). For crop breeding and conservation programmes, on the other hand, understanding latent diversity is of critical importance. During the design of conservation programmes, distance metrics are best handled by scientific experts whose units of analysis are linked to information about farmer-managed units.

With the exception of the attribute-based index used for coffee in Chapter 7, the diversity indices used in this book are adapted from ecological indices, which express spatial diversity concepts for species (Magurran, 1998; Table 1.2). In most cases, the data are compiled from cross-sectional surveys of farm households across villages in sub-national regions. Each represents a unique diversity concept. Richness is measured by a count or Margalef index. The index is constructed from the numbers of crop species, varieties or both encountered per unit of area that is geographically defined, such as the household farm, the village or the region. In the richness index, units are distinct but each has equal importance. Relative abundance, or the distribution of individuals associated with each of the species or varieties, is represented by the Berger–Parker index of dominance (Berger and Parker, 1970). Relative abundance accounts for the frequency that a species or variety is counted. An index that combines both richness and relative abundance (or evenness) concepts is the Shannon index of proportional abundance, sometimes called a heterogeneity index for that reason (Magurran, 1998). The Shannon index, originally used in information theory, embodies no particular assumptions about the shape of the underlying distribution in species abundance, and has been widely used in the agronomic literature to compare diversity within varieties as well as in the ecological literature to evaluate species diversity. The Margalef and Shannon indices have a lower limit of zero if only one variety is grown, while the Berger–Parker index has a lower limit of one when a single variety occupies all of the area.

The proportion of crop area planted to a variety (or area share) is used as a proxy for the number of individual plants encountered in a physical unit of area. Though area shares are not distributed spatially in the same way as plants (since they combine plants of the same crop or variety from several different locations on a farm or in a community), using area shares emphasizes the choice variable that is central to economic analysis. Crop-area shares allocated to modern varieties, as categories, have been choice variables in the constrained optimization models of the adoption literature, representing the 'extent' of adoption (Feder *et al.*, 1985; and later Feder and Umali, 1993). The notion that area shares represent the constrained demand for

Table 1.2. Spatial diversity indices used in this book.

Index	Concept	Construction	Explanation
Count	Richness	$D = S$	S = number of farmer-managed units of diversity
Margalef	Richness	$D = (S-1)/\ln A_j$ $D \geq 0$	A_j = total area planted or total population count over all farmer-managed units of diversity
Shannon	Evenness, equitability, proportional abundance	$D = -\Sigma \alpha_i \ln \alpha_i$ $D \geq 0$	α_i = area share or population share occupied by *i*th farmer-managed unit of diversity
Simpson	Proportional abundance	$D = 1 - \Sigma \alpha_i^2$ $0 \leq D \leq 1$	α_i = area share or population share occupied by *i*th farmer-managed unit of diversity
Berger–Parker	Inverse dominance (relative abundance)	$D = 1/\max(\alpha_i)$ $D \geq 1$	Max (α_i) is the maximum area share planted to any single farmer-managed unit of diversity

Note: As understood here, a farmer-managed unit of diversity is a plant population, cultivated variety, crop, use group or class recognized by farmers as distinct based on observable genetic and/or agromorphological descriptors.

variety traits was described in the analysis of maize landrace diversity in Mexico (Smale et al., 2001) and modern wheat diversity in China and Australia (Smale et al., 2003).

Previous Economic Studies about On-farm Conservation

Earlier applied research by Brush et al. (1992) in Peru, Meng (1997) and Meng et al. (1998b) in Turkey and Van Dusen (2000) and Van Dusen and Taylor (in press) in Mexico developed the approach that serves as a starting point for much of this book. Although the studies are similar to each other, they are not derived from an identical theoretic framework. Like most of the chapters in this book, each involved an econometric estimation accomplished with data collected in household and plot surveys, supplemented by information about the genetics and taxonomy of the crop.

These studies evolved from microeconomic models of crop variety choice, formulated to analyse farmer adoption of high-yielding varieties during the early green revolutions of the 1970s. The models of that period implicitly assumed that the new seed varieties were superior to those grown by farmers. Therefore, the practice of growing both modern varieties and landraces at the same time ('partial adoption') reflected the inefficiencies associated with farmers' learning processes (e.g. Kislev and Shchori-Bachrach, 1973; Hiebert, 1974). In a final equilibrium state, 'efficient' farmers would plant all of their crop area to modern varieties. Later theoretical approaches depicted farmers as efficient but motivated by their attitudes towards risk (e.g. Feder, 1980; Just and Zilberman, 1983). Subsequently, economists argued that partial adoption could be attributed to any one of a number of competing explanations (Smale et al., 1994), including attitudes towards risk, the differential costs that farmers face while transacting in imperfect markets (de Janvry et al., 1991) or by environmental heterogeneity such as soil type differences on farms (Bellon and Taylor, 1993).

Brush et al. (1992) were motivated by what they called the 'displacement hypothesis' – that the rapid of diffusion modern varieties leads inescapably to the loss of potentially valuable landraces and gene complexes, as was observed in Asia during the early phases of the green revolution (Frankel, 1970; Harlan, 1972; Hawkes, 1983). Their model explained the area farmers allocate to improved potato varieties and the effect of adopting these varieties on the number of potato landraces they grow. Whether adoption of improved potato varieties reduced landrace diversity depended on how recently modern varieties had been introduced. Expansion of land per farm into modern varieties displaced landraces among adopters in earlier stages of adoption, but not later. The authors proposed that 'if there are compelling reasons for farmers in cradle areas to retain a minimum level of diversity on their farms, we would expect to find any negative association between the area in improved varieties and diversity to approach zero at late stages of the adoption process...' (Brush et al., 1992, p. 369). In contrast to earlier perspectives, their findings led to the conclusion that the replacement of landraces by modern varieties is not inevitable and cultivation of both types may be optimum for farmers.

Meng's (1997) and Meng et al.'s (1998b) approach reflected both of these perspectives. Her model explained the choice to grow wheat landraces, and conditional on that choice, landrace diversity. She demonstrated that multiple explanations, including missing markets, risk and agroclimatic conditions, influenced the probability that Turkish households grew a wheat landrace on any particular plot. Hence, during processes of economic development and change, a shift in any single factor would be unlikely to cause farmers to cease growing landraces. She found that factors affecting the probability that households grow landraces were independent of those that influence wheat landrace diversity, implying that different policies would be instrumental for maintaining landraces in general as compared to diverse landraces.

Two fundamental aspects of Meng's study distinguished it from that of Brush et al. (1992). First, her economic model was motivated not by the decision to adopt, but by the decision to grow landraces. Variety choice was no longer viewed as equivalent to adoption; nor was diversity viewed as equivalent to growing a landrace. Her policy concern was not whether modern varieties would replace landraces, but how best to target households with genetically diverse wheat landraces in a conservation programme. Second, she used diversity indices calculated with experimental data measuring qualitative traits of wheat landraces that had been sampled from households. One of them, the Shannon index, is among the most widely used in the agronomic literature.

Other hypotheses articulated in these first two studies recur in the chapters of this book. Consistent with biogeographical theory in ecology, Brush et al. (1992) found that greater farm fragmentation was associated with a larger number of landraces per farm in one of the regions. Off-farm employment opportunities for the household head were negatively associated with the number of landraces grown on the household farm in both regions, supporting their hypothesis that maintaining more potato landraces required more labour. Wealth affected the number of landraces cultivated only indirectly, through its influence on adoption in the region with more recent introduction. The rich planted less area in modern varieties, possibly because of the luxury status of some potato landraces; the poor were also less likely to grow them due to imperfect credit and insurance markets.

Meng also found that farmers owning more land or more fertile land, and those with higher wealth indexes (refrigerators, tap water and electricity) had higher probabilities of growing wheat landraces. Variables measuring consumption demand by farm families, the share of wheat output marketed and the distance to market or road quality, all supported the hypothesis that farmers located in areas of less wheat market activity rely on their own production to meet their consumption needs. Some findings for the wheat landrace diversity equation contradicted working hypotheses. Fragmentation was negatively associated with the diversity of wheat landraces, while, unexpectedly, an increase in the percentage of district-level output marketed influenced it positively. A 'striking finding' from the diversity estimations was that 'the effects of explanatory variables differ by diversity index' (Meng, 1997:164). The finding led to the hypothesis that policies influencing an explanatory factor, such as market infrastructure, could have negative effects on one conservation criterion and positive effects on another; in other words, policies may not be neutral to conservation goal. The hypothesis is tested in several chapters of this book.

Building on the last two studies, Van Dusen (2000) developed a model of the household farm with missing markets to explain both species and variety diversity within the *milpa* farming system of Puebla, Mexico. Smale et al. (2001) also analysed the area shares allocated among maize landraces by farmers in the state of Guanajuato, Mexico, considering variety traits, introducing the notion of impure public goods and including a variable to represent the supply of distinct landraces at the community level.

The Plan of this Book

The chapters are grouped according to the approaches used to value crop biodiversity on farms and investigate its determinants. The value of crop biodiversity can be measured with stated or revealed preference approaches. Chapters included in Part II apply stated preference approaches. The applications and case studies found in Part III explore revealed preferences within the modelling framework of farmer decision making with constraints. Approaches used in Parts II and III generate direct or indirect metrics for identifying varieties, farmers or locations that are associated with high private values for managing crop biodiversity on farms situated in the upper segments on the vertical axis of Fig. 1.2. The valuation metrics are partial by construction, since they reflect only use values. In Parts II and III, only the private value to farmers is addressed. The unit of analysis is the farmer. Chapters in Part IV explore aspects of on-farm conservation that are related to public values, the role of institutions and seed systems and conservation within a larger social unit of analysis than the individual farm household.

Part II

Chapters in Part II are applications of stated preference methods. In part, stated preference methods were developed to address the limitations of revealed preference approaches. For example, hedonic pricing methods, a revealed preference approach, have been applied by environmental economists or to estimate trait values in crop or livestock production (Von Oppen and

Rao, 1982; Unnevehr, 1986; Hamath et al., 1997; Scarpa et al., 2003b). Hedonic models relate prices of goods in markets to the attributes that are implicitly traded. There are two major constraints to the use of hedonic pricing methods in valuing agricultural biodiversity during economic change. First, as for many other environmental goods, real or surrogate markets may not exist for the attributes of genetic resources, and even if they do, their value may not be well represented in observed prices due to incomplete markets. Markets for varieties may be thin or lacking in the grades and standards that enable consumers to differentiate quality. Governments may establish uniform pricing. Second, even if markets exist, the market price might not be a good approximation of the value of the environmental resource because, by definition, market values tend to reflect use values only.

By contrast, stated preference approaches have the potential to reveal the total economic value of a change in the provision of a non-marketed good, given that the surveys used to elicit them are properly designed. Contingent valuation is a direct elicitation method. Dyer (2002) applied a contingent behaviour method and computable general equilibrium model to investigate the supply response of maize growers in Mexico to the North American Free Trade Agreement. He presents the method and discusses implications of economic change for farmer valuation of maize landraces in Chapter 2.

Controversy over contingent valuation has led more recently to the development of alternative stated preference methods, including attribute-based choice modelling, and indirect elicitation procedure. These involve rankings or ratings by respondents across alternative options, each of which is associated with a set of attributes, one of which may be a price. Similar to contingent valuation, they are implemented through survey research. Like hedonic pricing methods, they are grounded conceptually on Lancaster's (1966) theory of consumer choice. Two examples from the literature on valuing livestock genetic resources have applied choice experiments. Scarpa et al. (2003a) compared the value of the attributes of creole pigs to those of more productive, but less well adapted, exotic breeds in the Yucatán peninsula of Mexico. Scarpa et al. (2003b) compared, revealed and stated preference methods in valuing cattle traits for indigenous livestock breeds among the Maasai in Kenya.

Birol (2004) developed a choice experiment to estimate the value of home gardens and their agrobiodiversity attributes in Hungary. Birol's modelling framework combines the random utility approach and the Lancaster (1966) theory of consumer choice. In Chapter 3, Birol et al. combine these findings with secondary data on settlements to test hypotheses about economic change and farmer valuation.

Wale and Mburu (Chapter 4) discovered that coffee farmers in Ethiopia do not name varieties, although they distinguish among types according to attributes. They propose an attribute-based index of diversity in their investigation of smallholder production decisions, also drawing on random utility and the Lancaster approaches and incorporating household vulnerability to risk.

Part III

Not all farmers in sites with high benefit–cost ratios today will continue managing diverse crop genetic resources in the future. Encouraging them to do so will have efficiency and equity implications at local, regional and global levels. A first step in designing appropriate policy mechanisms is to identify the factors that increase and decrease the likelihood that farmers will continue to manage crop biodiversity in a given context. Next, farmers with high predicted probabilities of maintaining crop biodiversity in the presence of economic change can be profiled statistically.

To accomplish this, Part III of this book applies econometric models of variety choice derived from the theoretic framework of the agricultural household (Singh et al., 1986; de Janvry et al., 1991). Crop biodiversity levels on individual farms are explained by testing hypotheses about the factors that influence their variation. A combination of microeconomic theory, principles of population genetics and ecology define the set of conceptual explanatory variables that are measured empirically in each case. The incidence, measurement and predicted effects of each variable are location-specific because they depend on the farming system, crop reproduction system and physical features of the environment. Hypotheses test the significance of environmental heterogeneity in the region or on the farm, market infrastructure, human capital, income and assets and the use of improved varieties.

A summary of the theoretical model developed by Van Dusen (2000) is presented in Chapter 5, with an econometric application that focuses on the policy issue of migration. This model provides the analytical basis and format for a number of the chapters in Part III and Part IV. Benin et al. (Chapter 6) analyse the determinants of both intercrop and intracrop diversity of cereals on farms in the northern highlands of Ethiopia. Edmeades (2003) formulated a complete trait-based model of variety demand within the theoretical framework of the household farm. In Chapter 7, Edmeades et al. apply the model to analyse the biodiversity of bananas in the East African highlands. Birol et al. (Chapter 8) explain the revealed preferences of farmers for four attributes of agrobiodiversity in home gardens, based on the Singh et al. (1986) model of the agricultural household. Winters et al. (Chapter 9) re-examine the determinants of potato diversity on farms in Peru, testing hypotheses about rural development interventions and policies.

Part IV

Culled largely from research in progress, the approaches used in Part IV are more disparate than those presented in Parts II and III. Gauchan et al. (Chapter 10) investigate the factors predicting that Nepalese farmers will choose to grow rice landraces that are also of public value for future crop improvement, based on the choice sets identified through interviews with breeders and conservationists. Gebremedhin et al. (Chapter 11) augment the social unit of analysis from the farm household level to the village level in the northern highlands of Ethiopia, expanding on the approach presented in Chapter 6. Seed institutions, including social networks that transmit seed-related information, as well as bazaars and formal seed suppliers, are explored for fruits and nuts in Uzbekistan. (Van Dusen et al., Chapter 12). In Chapter 13, Nagarajan and Smale investigate seed systems for major and minor millet crops in South India, introducing seed system parameters into a village-level estimation of biodiversity determinants. Lipper et al. (Chapter 14) analyse the impact of seed programmes in drought-prone areas of eastern Ethiopia on intercrop diversity. Bela et al. (Chapter 15) use an institutional economics approach to analyse stakeholder interests and strategies for managing crop genetic resources in Hungary, complementing and expanding the findings of the revealed and stated preference approaches reported in Chapters 3 and 8. Di Falco and Perrings (Chapter 16) advance the work by Heisey et al. (1997), estimating diversity–productivity relationships in South Italy, where farmers are organized into producing and marketing cooperatives in a highly articulated, controlled market for durum wheat.

Part V

The concluding chapter presents the combined sense of the authors about the innovative contributions of this book, their limitations and future directions for research that assesses the value of crop biodiversity on farms. An annotated bibliography of related research, focusing explicitly on published articles that apply economics methods and principles, is provided in Chapter 18.

References

Alston, J.M., Norton, G.W. and Pardey, P.G. (1998) *Science Under Scarcity: Principles and Practice for Agricultural Research Evaluation and Priority Setting*, CAB International, Wallingford, UK.

Alston, J.M., Chan-Kang, C., Marra, M.C., Pardey, P. and Wyatt, T.J. (2000) *A Meta-Analysis of Rates of Return to Agricultural R&D: Ex Pede Herculem?* Research Report 113, International Food Policy Research Institute, Washington, DC.

Bellon, M.R. (1996) The dynamics of crop intraspecific diversity: a conceptual framework at the farmer level. *Economic Botany* 50, 26–39.

Bellon, M.R. and Taylor, J.E. (1993) 'Folk' soil taxonomy and the partial adoption of new seed varieties. *Economic Development and Cultural Change* 41, 763–786.

Bellon, M.R., Pham, J.L. and Jackson, M.T. (1997) Genetic conservation: a role for rice farmers. In: Maxted, N., Ford-Lloyd, B. and Hawkes, J.G.

(eds) *Plant Genetic Conservation: The In Situ Approach*. Chapman & Hall, London, New York, pp. 263–289.

Berger, W.H. and Parker, F.L. (1970) Diversity of planktonic Foraminifera in deep sea sediments. *Science* 168, 1345–1347.

Birol, E. (2004) Valuing agricultural biodiversity on home gardens in Hungary: an application of stated and revealed preference methods. PhD thesis, University College London, University of London, UK.

Brock, A.N. and Xepapadeas, A. (2003) Valuing biodiversity from an economic perspective: a unified economic, ecological and genetic approach. *The American Economic Review* 93, 1597–1614.

Brown, G.M. (1990) Valuing genetic resources. In: Orians, G.H., Brown, G.M., Kunin, W.E. and Swierzbinski, J.E. (eds) *Preservation and Valuation of Biological Resources*. University of Washington Press, Seattle, Washington, pp. 203–226.

Brown, G. and Goldstein, J.H. (1984) A model for valuing endangered species. *Journal of Environmental Economics and Management* 11, 303–309.

Brush, S.B., Taylor, J.E. and Bellon, M.R. (1992) Technology adoption and biological diversity in Andean potato agriculture. *Journal of Development Economics* 39, 365–387.

Byerlee, D. and Traxler, G. (1995) National and international wheat improvement research in the post-Green Revolution period: evolution and impacts. *American Journal of Agricultural Economics* 77, 268–278.

Cassman, K.G., Wood, S., Choo, P.S., Cooper, C., Devendra, C., Dixon, J., Gaskell, J., Khan, S., Lal, R., Lipper, L., Pretty, J., Primavera, J., Ramankutty, N., Viglizzo, E., Weibe, K., Kadungure, S., Kanbar, N., Khan, Z., Leakey, R., Porter, S., Sebastian, K. and Tharme, R. (2005) Cultivated systems. In: *Millennium Ecosystem Assessment. Condition Working Group Report*. Island Press, Washington, DC (in press).

Ceccarelli, S. and Grando, S. (2002) Plant breeding with farmers requires testing the assumptions of Conventional Plant Breeding Lessons from the ICARDA Barley Program. In: Cleveland, D.A. and Soleri, D. (eds) *Farmers, Scientists and Plant Breeding: Integrating Knowledge and Practice*. CAB International, Wallingford, UK, pp. 297–332.

Cleveland, C., Solieri, D. and Smith, S.E. (1994) Folk crop varieties: do they have a role in sustainable agriculture? *BioScience* 44, 740–751.

Cornes, R. and Sandler, T. (1986) *The Theory of Externalities, Public Goods, and Club Goods*. Cambridge University Press, Cambridge, UK.

Costanza, R., d. Arge, R., d. Groot, R., Farber, S.C., Grasso, M., Hannon, B., Limburg, K., Naeem, S., O'Neill, R.V., Paruelo, J., Raskin, R.G., Sutton, P. and v. d. Belt, M. (1997) The value of the world's ecosystem services and natural capital. *Nature* 387, 253–260.

Cox, T.S. and Wood, D. (1999) The nature and role of crop biodiversity. In: Wood, P. and Lenné, J.M. (eds) *Agrobiodiversity: Characterization, Utilization and Management*. CAB International, Wallingford, UK, pp. 35–57.

de Janvry, A., Fafchamps, M. and Sadoulet, E. (1991) Peasant household behaviour with missing markets – some paradoxes explained. *Economic Journal* 101, 1400–1417.

Dyer Leal, G.A. (2002) The cost of *in situ* conservation of maize landraces in the Sierra Norte de Puebla, Mexico. PhD dissertation, University of California at Davis, California.

Edmeades, S. (2003) Variety choice and attribute trade-offs within the framework of agricultural household models: the case of bananas in Uganda. PhD dissertation, North Carolina State University, Raleigh, North Carolina.

Ehrenfeld, D. (1988) Why put a value on biodiversity? In: Wilson, E.O. and Peter, F.M. (eds) *Biodiversity*. National Academy Press, Washington, DC, pp. 212–216.

Evenson, R.E. and Gollin, D. (1997) Genetic resources, international organizations, and improvement in rice varieties. *Economic Development and Cultural Change* 471, 471–500.

Evenson, R.E. and Gollin, D. (2003) Assessing the impact of the Green Revolution, 1960 to 2000. *Science* 5620, 758–762.

Evenson, R.E., Gollin, D. and Santaniello, V. (eds) (1998) *Agricultural Values of Plant Genetic Resources*. CAB International, Wallingford, UK.

Feder, G. (1980) Farm size, risk aversion and the adoption of new technology under uncertainty. *Oxford Economic Papers* 3, 263–283.

Feder, G. and Umali, D.L. (1993) The adoption of agricultural innovations: a review. *Technological Forecasting and Social Change* 43, 215–239.

Feder, G., Just, R.E. and Zilberman, D. (1985) Adoption of agricultural innovations in developing countries: a survey. *Economic Development and Cultural Change* 33, 255–298.

Frankel, F.R. (1970) The genetic dangers of the Green Revolution. *World Agriculture* 19, 9–13.

Gollin, D. and Smale, M. (1998) Valuing genetic diversity: crop plants and agroecosystems. In: Collins, W.W. and Qualset, C.O. (eds) *Biodiversity in Agroecosystems*. CRC Press, London, pp. 237–265.

Gollin, D., Smale, M. and Skovmand, B. (2000) Searching an *ex situ* collection of wheat genetic resources. *American Journal of Agricultural Economics* 82, 812–827.

Hamath, A.S., Faminow, M.D., Johnson, G.V. and Crow, G. (1997) Estimating the values of cattle characteristics using an ordered probit model.

American Journal of Agricultural Economics 79, 463–476.

Harlan, J.R. (1972) Genetics of disaster. *Journal of Environmental Quality* 1, 212–215.

Harlan, J.R. (1992) *Crops and Man*. American Society of Agronomy, Crop Science Society of America, Madison, Wisconsin.

Hawkes, J.G. (1983) *The Diversity of Crop Plants*. Harvard University Press, Cambridge, Massachusetts.

Heal, G., Walker, B., Levin, S., Arrow, K., Dasgupta, P., Ehrlich, P., Maler, K.-G., Kautsky, N., Lubchenco, J., Schneider, S. and Starrett, D. (2004) Genetic diversity and interdependent crop choices in agriculture. *Resource and Energy Economics* 26, 175–184.

Heisey, P.W., Smale, M., Byerlee, D. and Souza, E. (1997) Wheat rusts and the costs of genetic diversity in the Punjab of Pakistan. *American Journal of Agricultural Economics* 76, 726–737.

Heisey, P.W., Lantican, M.A. and Dubin, H.J. (2002) *Impacts of International Wheat Breeding Research in Developing Countries, 1966–97*, CIMMYT (International Maize and Wheat Improvement Center), Mexico D.F.

Hiebert, D. (1974) Risk, learning and the adoption of fertilizer responsive varieties. *American Journal of Agricultural Economics* 56, 764–768.

Jarvis, D.I., Myer, L., Klemick, H., Guarino, L., Smale, M., Brown, A.H.D., Sadiki, M., Sthapit, B. and Hodgkin, T. (2000) *A Training Guide for In Situ Conservation On-Farm, Version 1*. International Plant Genetic Resources Institute (IPGRI), Rome.

Jarvis, D.I., Zoes, V., Nares, D. and Hodgkin, T. (2004) On-farm management of crop genetic diversity and the convention on biological diversity programme of work on agricultural biodiversity. *Plant Genetic Resources Newsletter* 138, 5–17.

Johnson, N.L., Pachico, D. and Voysest, O. (2003) The distribution of benefits from public international germplasm banks: the case of beans in Latin America. *Agricultural Economics* 29, 277–286.

Just, R.E. and Zilberman, D. (1983) Stochastic structure, farm size and technology adoption in developing agriculture. *Oxford Economic Papers* 35, 307–328.

Kislev, Y. and Shchori-Bachrach, N. (1973) The process of an innovation cycle. *American Journal of Agricultural Economics* 55, 28–37.

Koo, B., Nottenburg, C. and Pardey, P.G. (2004a) Plants and intellectual property: an international appraisal. *Science* 306, 1295–1297.

Koo, B., Pardey, P.G. and Wright, B.D. (2004b) *Saving Seeds: The Economics of Conserving Crop Genetic Resources* Ex Situ *in the Future Harvest Centres of the CGIAR*. CAB International, Wallingford, UK.

Lancaster, K.J. (1966) A new approach to consumer theory. *Journal of Political Economy* 74, 132–157.

Magurran, A.E. (1998) *Ecological Diversity and its Measurement*. Princeton University Press, Princeton, New Jersey.

Meng, E.C.H. (1997) Land allocation decisions and *in situ* conservation of crop genetic resources: the case of wheat landraces in Turkey. PhD dissertation, University of California at Davis, California.

Meng, E.C.H., Smale, M., Bellon, M.R. and Grimanelli, D. (1998a) Definition and measurements of crop diversity for economic analysis. In: Smale, M. (ed.) *Farmers, Gene Banks and Crop Breeding: Economic Analyses of Diversity in Wheat, Maize, and Rice*. Kluwer Academic Publishers, Boston, pp. 19–31.

Meng, E.C.H., Taylor, J.E. and Brush, S.B. (1998b) Implications for the conservation of wheat landraces in Turkey from a household model of varietal choice. In: Smale, M. (ed.) *Farmers, Gene Banks and Crop Breeding: Economic Analyses of Diversity in Wheat, Maize, and Rice*. Kluwer Academic Publishers, Boston, Massachusetts, pp. 127–143.

Meng, E., Smale, M., Ruifa, H., Brennan, J.P. and Godden, D. (1999) Measurement of crop genetic diversity in economic analysis. Contributed Paper, 43rd Annual Conference of the Australian Agricultural and Resource Economics Society, 18–20 January, Christchurch, New Zealand.

Morris, M.L. and López-Pereira, M.A. (1999) *Impacts of Maize Breeding Research in Latin America, 1966–1997*, International Maize and Wheat Improvement Center (CIMMYT), Mexico D.F.

Morris, M.L., Rusike, J. and Smale, M. (1998) Maize seed industries: a conceptual framework. In: Morris, M.L. (ed.) *Maize Seed Industries in Developing Countries*. Lynne Rienner and CIMMYT, Boulder, Colorado, pp. 35–54.

Orians, G.H., Brown, G.M., Kunin, W.E. and Swierzbinski, J.E (eds) (1990) *Preservation and Valuation of Biological Resources*, University of Washington Press, Seattle, Washington.

Pearce, D. and Moran, D. (1994) *The Economic Value of Biodiversity*. Earthscan, London.

Polasky, S. and Solow, A. (1995) On the value of a collection of species. *Journal of Environmental Economics and Management* 29, 298–303.

Ramanna, A. and Smale, M. (2004) Rights and access to plant genetic resources under India's New Law. *Development Policy Review* 22, 423–442.

Randall, A. (1988) What mainstream economists have to say about the value of biodiversity. In: Wilson, C. (ed.) *Biodiversity*. National Academy Press, Washington, DC, pp. 217–223.

Rausser, G.C. and Small, A.A. (2000) Valuing research leads: bioprospecting and the conservation of genetic resources. *Journal of Political Economy* 108, 173–206.

Romer, P.M. (1993) *Two Strategies for Economic Development: Using Ideas and Producing Ideas.* World Bank, Washington, DC.

Roughgarden, J. (1995) Can economics protect biodiversity? In: Swanson, T.M. (ed.) *The Economics and Ecology of Biodiversity Decline: The Forces Driving Global Change.* Cambridge University Press, Cambridge, UK, pp. 149–155.

Sadiki, M., Jarvis, D., Rijal, D., Bajracharya, J., Hue, N.H., Camacho-Villa, T.C., Burgos-May, L.A., Sawadogo, M., Balma, D., Lope, D., Arias, L., Mar, I., Karamura, D., Williams, D., Chavez-Servia, J.-L., Sthapit, B. and Ramanatna Rao, V. (2005) Variety names: an entry point to crop genetic diversity and distribution in agroecosystems? In: Jarvis, D.I., Padoch, C. and Cooper, D. (eds) *Managing Biodiversity in Agroecosystems.* Columbia University Press, New York.

Sandler, T. (1999) Intergenerational public goods: strategies, efficiency, and institutions. In: Kaul, I., Grunberg, I. and Stein, M.A. (eds) *Global Public Goods.* United Nations Development Programme and Oxford University Press, Oxford, UK pp. 20–50.

Scarpa, R., Drucker, A.G., Anderson, S., Ferraes-Ehuan, N., Gómez, V., Risopatrón, C.R. and Rubio-Leonel, O. (2003a) Valuing genetic resources in peasant economies: the case of 'hairless' creole pigs in Yucatan. *Ecological Economics* 45, 427–443.

Scarpa, R., Ruto, E.S.K., Kristjanson, P., Radeny, M., Drucker, A.G. and Rege, J.E.O. (2003b) Valuing indigenous cattle breeds in Kenya: an empirical comparison of stated and revealed preference value estimates. *Ecological Economics* 45, 409–426.

Sell, J., Chen, Z.-Y., Hunter-Holmes, P. and Johannson, A.C. (2002) A cross-cultural comparison of public good and resource good settings. *Social Psychology Quarterly* 65(3), 285–297.

Simpson, R.D., Sedjo, R.A. and Reid, J.W. (1996) Valuing biodiversity for use in pharmaceutical research. *Journal of Political Economy* 1, 163–185.

Singh, I., Squire, L. and Strauss, J. (eds) (1986) *Agricultural Household Models: Extensions, Applications and Policy.* The World Bank and Johns Hopkins University Press, Washington, DC and Baltimore, Maryland.

Smale, M. and Bellon, M.R. (1999) A conceptual framework for valuing on-farm genetic resources. In: Wood, D. and Lenné, J.M. (eds) *Agrobiodiversity: Characterization, Utilization and Management.* CAB International, Wallingford, UK, pp. 387–408.

Smale, M., Just, R.E. and Leathers, H.D. (1994) Land allocation in HYV adaption models: an investigation of alternative explanations. *American Journal of Agricultural Economics* 76, 535–546.

Smale, M., Bellon, M.R. and Aguirre Gómez, J.A. (2001) Maize diversity, variety attributes, and farmers' choices in southeastern Guanajuato, Mexico. *Economic Development and Cultural Change* 50, 201–225.

Smale, M., Meng, E., Brennan, J.P. and Hu, R. (2003) Determinants of spatial diversity in modern wheat: examples from Australia and China. *Agricultural Economics* 28, 13–26.

Solow, A., Polasky, S. and Broadus, J. (1993) On the measurement of biological diversity. *Journal of Environmental Economics and Management* 24, 60–68.

Srinivasan, C.S. (2003) Exploring the feasibility of farmers' rights. *Development Policy Review* 21, 419–448.

Swanson, T. (1996) Global values of biological diversity: the public interest in the conservation of plant genetic resources for agriculture. *Plant Genetic Resources Newsletter* 105, 1–7.

Unnevehr, L.J. (1986) Consumer demand for rice grain quality and returns to research for quality improvement in Southeast Asia. *American Journal of Agricultural Economics* 68, 634–641.

Van Dusen, E. (2000) *In situ* conservation of crop genetic resources in the Mexican *Milpa* System. PhD thesis, University of California at Davis, California.

Van Dusen, E. and Taylor, J.E. (2005) Missing markets and crop diversity: evidence from Mexico. *Environment and Development Economics* 10, 513–531.

Virchow, D. (1999) *Conservation of Genetic Resources: Costs and Implications for a Sustainable Utilization of Plant Genetic Resources for Food and Agriculture.* Springer, Heidelberg, Germany.

Von Oppen, M. and Rao, P.P. (1982) A market-derived selection index for consumer preferences of evident and cryptic quality characteristics of sorghum. In: ICRISAT (ed.) *Proceedings of the International Symposium on Sorghum Grain Quality*, 28–31 October 1981, Patancheru, Andhra Pradesh, India, pp. 354–364.

Weitzman, M.L. (1992) On diversity. *The Quarterly Journal of Economics* 107, 363–405.

Weitzman, M.L. (1993) What to preserve? An application of diversity theory to crane conservation. *The Quarterly Journal of Economics* 108, 157–183.

Wood, D. and Lenné, J.M. (1999) Why agrobiodiversity? In: Wood, D. and Lenné, J.M. (eds) *Agrobiodiversity: Characterization, Utilization and Management.* CAB International, Wallingford, UK, pp. 1–13.

Zeven, A.C. (1998) Landraces: a review of definitions and classifications. *Euphytica* 104, 127–139.

2 Crop Valuation and Farmer Response to Change: Implications for *In Situ* Conservation of Maize in Mexico

G.A. Dyer

Abstract

The chapter applies a stated preference approach to analyse the management of maize landraces in the Sierra Norte de Puebla, Mexico. Mexico is a centre of origin and diversity for maize. A method is proposed to assess farmer response to changes in income and price expectations associated with the integration of maize markets and compensatory policies through the North American Free Trade Agreement (NAFTA). Given its similarities with the contingent valuation approach used in environmental economics, the method is called 'contingent behaviour analysis'. Differences in responses are evident among households. Commercial growers and better-off households that are able to plant more land are more likely to respond to maize price increases. Worse-off households plant a relatively small amount of land but are willing to expand home production when they have access to additional income. Large growers are less likely to respond to increased income, presumably because non-market benefits of producing maize landraces are lower for them. Unexpectedly, maize supply in Mexico has remained above the 1990 level even in the rainfed areas where maize landraces dominate and semi-subsistence farmers have not benefited from subsidies. Results from this study also offer an explanation for the inelastic aggregate supply response observed following the abrupt decline in the domestic price of maize. This research was conducted in one of the same sites as the revealed preference analysis presented in Chapter 5.

Introduction

The purpose and legitimate use of valuing nature has long been controversial (Wilson, 1988). Scientific experts have questioned whether economics can measure nature's true value and worth, or whether it can only assess the value that society places on it de facto (Hanemann, 1988). At the centre of the controversy are the merits of conservation, and who gets to say what is to be conserved. The emphasis on valuation has had the effect of distancing economists and natural scientists, stifling cooperation at the cost of making progress (Roughgarden, 1995). A point of departure for a better understanding of how to approach conservation drawing on different disciplines is to define a common goal. The United Nations has defined this goal, through the Convention on Biological Diversity, and given a mandate for *in situ* conservation of crop genetic resources. A group of scholars is honouring this mandate. But for many in this group, the 'emerging paradigm' of *in situ* conservation includes surprisingly few contributions by economists (Wood and Lenné, 1997). Economists can and must take a more proactive role in conservation. One way to do so is to develop models that help to estimate the costs of conserving crop biodiversity on farms.

Appraisal of value has always been the point of departure in conservation debates, for reasons that are as much conceptual as practical. As Maxted *et al.* (1997) argue, 'it is difficult to persuade society to meet the cost (of conservation) unless it is seen as being of some value'. Experts have long perceived great strategic value to society in the preservation of crop genetic resources for future use (Brown, 1990; CAST, 1999), but the problems arising from setting priorities have only recently demonstrated the need for a formal assessment of value by economists. Given the importance of setting priorities for conservation, there is still no consensus that economic value is the best criterion to use.

This chapter examines the role of value in understanding farmer responses to changes in the economic context of crop management. Maize is believed to have originated in present-day Mexico more than 6000 years ago and, to date, maize is planted across half of all the arable land in the country. Landraces constitute the major part of this maize area. Wellhausen *et al.* (1952) classified Mexican maize populations into major races using a taxonomy based on morphological traits. Recent genetic analyses suggest that these landraces should not be considered as separate entities, but as an open genetic system. Population differentiation is more visible in analysis of morphological traits over which farmers exert selection pressure than it is with analysis of molecular markers (Pressoir and Berthaud, 2004).

The next section of this chapter discusses valuation issues, expanding on their treatment in Chapter 1. A synopsis of the policy context in Mexico is provided, followed by a description of methods, data and findings. A practical methodology is proposed to analyse farmers' responses to change. In the final section, the responses of farmers are used to predict changes in the management of maize landraces on farms and assess threats to sustaining current levels of maize biodiversity.

Valuation Issues

Crop genetic resources grown in agricultural fields have both private and public value and, ideally, the criterion for conserving their diversity should be total value. In practice, *ex situ* conservation responds to society's interest in preserving the public value of crop genetic resources, and it is indeed this value alone that it preserves: accessions in gene banks rarely become available to farmers that originally grew them but somehow lost them. Accordingly, public funding has borne the cost of *ex situ* conservation either directly or through international agencies (Koo *et al.*, 2004). In contrast, *in situ* conservation preserves the private value of crop genetic resources as much as their public value, largely at the expense of farmers.

The potential use of crop genetic resources in the future and their option value are largely immeasurable. So far, the applied economics literature has focused instead on estimating the value of adding an unspecified seed sample to a particular gene bank, on the marginal value of an accession in crop improvement (Evenson and Lemarié, 1998; Gollin *et al.*, 2000; Zohrabian *et al.*, 2003) or on the costs of conserving it (Koo *et al.*, 2004). The resulting value – the value of conserving an additional seed sample *ex situ* – depends on the size of the collection and the probability distributions of trait more than on the characteristics of the landrace itself. Value estimates obtained this way are therefore useful in determining the optimal size of a collection but useless in selecting among candidate landraces for conservation, unless more is learned about the candidates themselves. The genetic contribution of a particular landrace within a pool of landraces can be approximated using any one of a panoply of diversity of distance indexes or genetic distance metrics (Chapter 1); but this is likely to be an expensive process because of its crop- and location-specificity.

The relevant question for *in situ* conservation is: what are the private costs and benefits of landrace conservation by farmers? Conservation of landrace seed by farmers is generally not an end in itself but a by-product of agricultural activity; with few exceptions, it is not landrace conservation but landrace cultivation that farmers practise. Unravelling the share of value in landrace cultivation that accrues to the genetic resources embodied in the seed seems an insurmountable task, given the numbers of landraces grown in most centres of crop diversity and the challenges of distinguishing them genetically.

If it is taken for granted that farmers act in their own best interest, it can be concluded that the private value of landrace cultivation where it occurs today exceeds all of its costs, including the cost of managing and conserving seed from year to year (Brush and Meng, 1998). The rationale of an *in situ* conservation programme is to ensure that private value continues to exceed private costs well into the future. The programme must ensure that no alternative to landrace conservation is more appealing to farmers; that no opportunity costs emerge. As long as farmers find it optimal to grow and manage local crop genetic resources, they will cover all costs – i.e. the public costs of an *in situ* conservation programme will be minimal (Jarvis *et al.*, 2000; Chapter 1). If comparative advantages vanish and opportunity costs emerge, the programme must produce ways to eliminate them, or absorb the costs. Thus, the public cost of *in situ* conservation can be expressed as the cost necessary to maintain the comparative advantage of landraces above that of competing varieties, crops or off-farm activities (Brush and Meng, 1998).

In fact, a persistent objection to *in situ* conservation has been its cost (e.g. Frankel, 1970). For many years, it was believed that cash payments would be necessary to keep farmers from substituting modern varieties for landraces (Myers, 1979). While it is realized today that there are more efficient ways of reducing the opportunity costs of landrace cultivation, incentives will still be necessary to prevent farmers from ceasing landrace cultivation. Some fear that as communities assimilate into industrial society, there may be no need for traditional markets or germplasm exchange networks (Qualset *et al.*, 1997) or farmers will lose interest in growing diverse crops. Still others believe that declining biodiversity (including agrobiodiversity) is an inevitable by-product of technological advance and economic development. Whatever the reasons, opportunity costs are expected to emerge in landrace cultivation and to grow steadily. Common sense suggests that conservation should take place where opportunity costs are low and diversity high (Hentschel, 1997, Chapter 1).

In seeking ways to minimize the opportunity costs associated with conservation, economic research has tried to gauge the private value of landrace cultivation. One approach to assessing value is based on the principle that households value landraces for their traits, so the value of a variety is some function of the value of its attributes. Transaction costs and other factors largely outside farmers' control (e.g. physical, economic and cultural heterogeneity) influence landrace and trait valuation within the household and condition farmer decisions. Whenever these factors are identified, they can be used to establish targets for conservation, increasing the probability of success and decreasing the cost of conservation. This is the approach explored in several chapters found in Part III of this book.

In situ conservation is a long-term venture. Little is known about how values evolve through time and how farmers respond to economic change. It seems reasonable to expect that in the long run, the value of landrace cultivation will erode slowly as preferences change, substitutes for particular variety traits will emerge in the market and the comparative advantage or relative value of growing landraces will shift vis-à-vis competing varieties, crops or off-farm activities. This line of reasoning provided the basis on which NAFTA's potential consequences for the diversity of Mexican maize landraces were judged when the trade agreement came into effect.

A Policy Enigma

Mexico's total cultivated area in maize peaked in the mid-1960s. The rising cost of inputs and the subsequent removal of subsidies that followed eroded profitability, particularly in rainfed areas, and maize production stagnated for the next 30 years (Hewitt, 1994; CEC, 1999). Surprisingly, maize cultivation began expanding again across Mexico during the 1990s. At the turn of the century, productivity and total output had reached record highs (INEGI, 2001). During the intervening years, maize production spread from its traditional stronghold in small-scale, rainfed areas of South and Central Mexico, to the irrigated, commercial areas in the North. This expansion was a result of commercial growers turning to maize as a last resort after the government ceased to support the price of other crops (Appendini, 1994; Fritscher Mundt, 1996; Yúnez and Barceinas, 2000). It was expected to be a passing response to transitory policy and market conditions.

Fierce protection of the maize sector was a fundamental aspect of Mexican food policy and politics for many years (Austin and Esteva, 1987; Fox, 1992; Ochoa, 2000). Thus, despite its secular decline, the support price of maize remained well above international prices in the early 1990s, benefiting commercial growers. In contrast, subsistence growers were left adrift. Government analysts of all backgrounds had long considered that subsistence agriculture sequestered land and labour, and programmes were conceived over the years to use these resources more efficiently (Montanari, 1987; Levy and van Wijnbergen, 1992). In the early 1990s, the neo-liberal administration in government – weary of the cost and inefficiency of support prices – found an opportunity to discontinue price supports while harnessing peasant labour in non-agricultural sectors. It meant to do so by integrating the liberalization of the maize sector with that taking place in the rest of the economy in association with the NAFTA (Tellez, 1992).

The cornerstone of liberalization in the maize sector was the phase-out of support prices (then paid through the state trading agency, La Compañia Nacional de Subsistencias Popolares (CONASUPO)) and simultaneous removal of trade barriers, which would allow maize imports from the US to fill a growing gap between domestic supply and demand. Liberalization was expected to discourage commercial maize agriculture in irrigated areas that lacked a comparative advantage for growing maize, facilitating a shift to export crops. Maize agriculture in rainfed areas, characterized by subsistence maize production with landraces, was not expected to compete favourably with imports either (Robinson *et al.*, 1993; Levy and van Wijnbergen, 1994). Macroeconomic models predicted that Mexican maize output would decline by up to 20% (depending on the specific rules of the agreement).

Commercial maize growers in rainfed areas would be hurt but only slightly, since lower wages would partially offset lower output prices. Subsistence growers and landless workers would suffer job losses and lower wages, and many were expected to migrate to urban areas and the USA (Levy and van Wijnbergen, 1994). According to forecasts, other agricultural production (namely fruits and vegetables) on irrigated lands would expand, benefiting growers.

The Mexican government attempted to preempt the anticipated adverse welfare effects of agricultural liberalization by compensating growers with cash transfers designed to comply with current international trade agreements, i.e. NAFTA and GATT – General Agreement on Tariffs and Trade (Appendini, 1994). Under the PROCAMPO programme, per-hectare payments were based on the area cultivated by households (immediately before the programme's implementation) in nine different crops. By offering growers a certain income, this programme was intended to create incentives to increase productivity while remaining decoupled from current production (SAGARPA, 2002). PROCAMPO was scheduled to last for the duration of the liberalization process, ending in 2008.

Ten years into NAFTA and PROCAMPO, maize imports have soared, but the expected slump in domestic maize supply has not materialized. In fact, domestic supply of maize has remained above the record 1990 level since the initiation of NAFTA and PROCAMPO and, in 2001, the area cultivated in maize surpassed its 1965 historic high (INEGI, 2001). The increase in maize output can partly be attributed to commercial maize growers in irrigated areas, who were undoubtedly affected by imports but also continued to benefit from subsidies, especially through commercialization programmes. Subsistence growers on rainfed lands, on the other hand, have not benefited from subsidies for commercialization, and many appear to operate with losses (Perales, 1998; Dyer, 2002). Surprisingly, the cultivated area in rainfed maize rose steadily during the 1990s (SAGARPA, 2002). Clearly, farmers in these areas – where landraces are the norm – continue to value maize cultivation above the market price of maize grain alone. The model presented below seeks to explain this behaviour.

Modelling Farmer Response to Change

The physical, cultural and economic environment in which farmers live and work influences their decisions and the value they place on activities. This environment is neither static nor uniform. Most research on the management of crop diversity on farms has focused on farmers' adaptation to physical, cultural and economic heterogeneity (Jarvis *et al.*, 2000; Chapter 1) rather than their adaptation to change. Most of what is known about the role of development in farmer management of crop biodiversity comes from analysis of patterns observable in cross-sectional databases. Lack of longitudinal panels of cross-sectional, time-series data, and the costs of compiling them, have prevented direct scrutiny of farmers' responses to change.

Here, a method is proposed to assess farmers' response to changes in expectations regarding income and maize prices. A convenient term for the method is 'contingent behaviour analysis' because of its similarities with contingent valuation, the non-market valuation method used in environmental economics. The method is relatively inexpensive and less time consuming to make operational, and can be used to assess responses to changes yet to occur.

Contingent valuation is the name given to a collection of methods used in welfare analysis to value goods that are not traded in markets (Cummings *et al.*, 1986; Mitchell and Carson, 1989). These methods are based on the analysis of responses to hypothetical questions or in simulated markets. In a first stage, response data are elicited from a target sample population. Responses are then fitted to an econometric model and used to estimate an individual's willingness to pay for (or accept) changes in non-market goods or services.

The procedures applied in each of these stages have advanced over the last two decades in recognition of various limitations. For instance, the data generating process has evolved from open-ended to dichotomous choice questions, and more recently to the double-bounded dichotomous choice referendum (Bishop and Heberlein, 1979; Hanemann *et al.*, 1991). In parallel, econometric methods have been devised to handle elicited data (e.g. Cameron and Quiggin, 1994). Theoretical developments to justify particular interpretations of data have also appeared (e.g. Hanemann, 1984), and expert panels have since validated and improved the contingent valuation approach (Carson *et al.*, 1998).

The double-bounded dichotomous choice referendum is employed here to elicit data from farmers in a Mexican village about their responses to change; hence the terms 'contingent behaviour'. Data are used to explain changes in demand for land planted to maize (all are landraces) in response to changes in farmer expectations regarding maize prices and income, and not to estimate willingness-to-pay for some good or service. Thus, the theory behind the model

fitted to response data is unrelated to welfare analysis. The model presented below represents the derived demand of a farm household for land to plant maize. The model assumes that only one crop is available to households; so a farmer's need only decides how much land he requires. This assumption is not overly restrictive in the study region, where crops are generally part of a multi-crop system based on maize landraces, known as '*milpa*'. The model further assumes that how much maize area to plant and which maize types to grow are recursive decisions.

Hypothetical scenarios have been used before in developing rural areas (Griffin *et al.*, 1995; Davis and Whittington, 1998; Whittington, 1998). They have been validated with the use of alternative valuation methods and follow-up surveys of actual behaviour (Griffin *et al.*, 1995; Davis and Whittington, 1998). A potential source of error in the data-generating process (and beyond) is survey design and implementation. Care must be taken to avoid certain types of bias (Cummings *et al.*, 1986).

Agricultural production is a relatively lengthy, usually risky and often unprofitable enterprise. Since growers must commit resources long before they know for certain how much they will reap, or its value, they regularly act on their expectations. Those who produce for the market often take calculated risks based on their expectations of future prices. Although losses are not uncommon, commercial growers usually manage to make a profit in the face of price uncertainty. In contrast, financial accounts of subsistence agriculture often reveal persistent losses. These accounts typically do not take into consideration non-market benefits in home production, including the value of goods that are not traded in markets and the qualities of goods that are not reflected in prices.

Market failures are the source of differences between household-specific shadow values and market prices. When markets fail, household production and consumption decisions become intertwined, and factors other than price – such as household preferences and income – influence the value of a crop to the household and affect its production decisions as well as the indirect demand for land, labour and capital (Chapter 4). An extreme case is when there is no market for food staples and the household's budget constrains production. When more than one market is missing, household decisions become intractable using economic theory, and inferences about variables that are not observed directly can be very limited (de Janvry et al., 1991). Thus, the effect of exogenous shocks on the shadow value of a crop can be elusive in a real-life context. While shadow values are unobservable, household responses are observable.

In an imperfect markets scenario where the production and consumption decisions of the household are simultaneous, household i's demand for arable land, D_i, can be expressed as a function of its own characteristics, α_i, and its expectations for the future:

$$D_i = D_i(f_i(\mathbf{p}, \mathbf{y}_{xi}); \alpha_i) \qquad (2.1)$$

where $f_i(\mathbf{p}, \mathbf{y}_{xi})$ is the subjective joint probability density function of prices, $\mathbf{p}$, and household endowments and exogenous income, $\mathbf{y}_{xi}$.[1] Changes in price and income expectations shift the household's demand for land. Shifts can be determined using time-series data whenever it is available. Alternatively, shifts can be determined using response data elicited from farmers facing hypothetical scenarios that represent a change in their expectations. Based on Eq. (2.1), a first order approximation of the change in household i's demand for land in response to a change in exogenous variable v, dD_{vi}, is given by:

$$dD_{vi} = \beta_0 + \beta_1 dv + \beta\alpha_i \qquad (2.2)$$

Farmer responses to changes in the price of the single crop or in household income received as gifts, remittances or transfers can be fitted to a probit model using an index function dD_{vij}^* based on Eq. (2.2). If the change in farmer i's demand for land planted to maize in response to the jth shock in v ($j = 1,2$) is positive, the index function takes a value equal to 1; otherwise it is equal to 0. In other words, $dD_{vij} > 0 \Leftrightarrow dD_{vij}^* = 1$. The complete model is given by:

$$dD_{vij} = \beta_0 + \beta_1 dv_{ij} + \beta\alpha_i + \varepsilon_{ij} \qquad (2.3)$$
$$dD_{vij}^* = 1 \text{ if } dD_{vij} > 0$$
$$dD_{vij}^* = 0 \text{ if } dD_{vij} = 0$$

where ε_{ij} is a random error distributed $N(0,\sigma)$. Farmer responses (dD_{vi1}^*, dD_{vi2}^*) follow a bivariate normal distribution, $BVN(\mu, \mu, \sigma, \sigma, \rho)$, where $\mu = \beta_0 + \beta_1 dv_j + \beta\alpha_i$. Letting $\omega = \mu/\sigma$, and dropping sub-index i, the log-likelihood function is:

$$L = \Sigma \{dD_{v1}^* \, dD_{v2}^* \, \log[\int_{\omega 1} \int_{\omega 2} \phi(w_1, w_2, \rho) \, dw_1 \, dw_2]$$
$$+ (1-dD_{v1}^*) \, dD_{v2}^* \, \log[\int^{\omega 1} \int_{\omega 2} \phi(w_1, w_2, \rho) \, dw_1 \, dw_2]$$
$$+ dD_{v1}^* (1-dD_{v2}^*) \, \log[\int_{\omega 1} \int^{\omega 2} \phi(w_1, w_2, \rho) \, dw_1 \, dw_2]$$
$$+ (1-dD_{v1}^*)(1-dD_{v2}^*) \, \log[\int^{\omega 1} \int^{\omega 2} \phi(w_1, w_2, \rho) \, dw_1 \, dw_2]$$

In this study, Eq. (2.3) was used as a basis for probit models estimating the effects of maize price and income increases on land planted to maize (landraces). Models were estimated using GAUSS 3.2.41 software.[2]

Site and Survey Description

The study site is the village of Zoatecpan, in the Sierra Norte de Puebla. Zoatecpan is one of the communities in the McKnight Milpa Project of which this study was part (Dyer, 2002). It is also one of the communities studied by Van Dusen, who applied a revealed preference approach described in Chapter 5. The location of the Sierra Norte de Puebla is shown in Fig. 2.1.

Zoatecpan is an indigenous community in the rugged Sierra Norte de Puebla. Maize is overwhelmingly the main agricultural product. Nearly all households in Zoatecpan own land, but endowments vary widely in size: 2% of households own 50% of the land (Fig. 2.2), and the average landholding is only 0.4 ha. There is an active land rental market in the village, and nearly half of all households participated in this market in 1999. As many as 46% of households

---

[1] See Dyer (2002) for a more thorough description of the model.
[2] Only 33 out of the original 49 household heads surveyed over the course of a year were available at the time of the interview (see below).

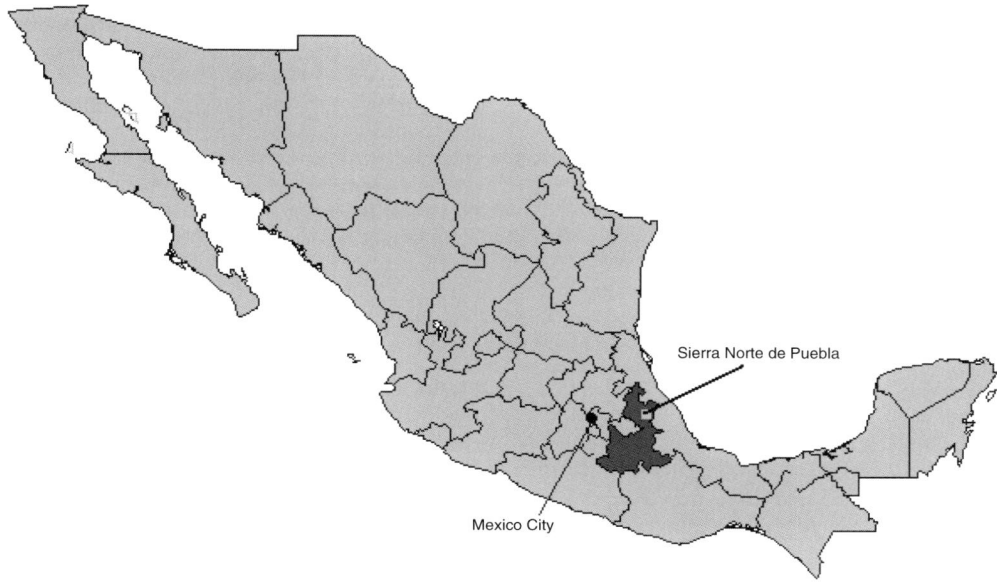

Fig. 2.1. Location of Sierra Norte de Puebla.

rented in land (all for maize) and 5% of households rented out land.

The fact that the few farmers who rented out land for maize are the largest landholders in Zoatecpan reflects a widespread pattern in the region. The rental market fosters a more progressive distribution of arable land among local households and nearly doubles maize area for the average household in this area.

Despite the scope of the rental market and the importance of maize agriculture, Zoatecpan is a net 'importer' of maize, like many Mexican villages. Purchases from outside the village account for approximately three-fourths of total consumption of maize in the village. Two-thirds of local households produce less than 25% of their yearly maize consumption, and most other households produce less than 75% (Fig. 2.3).

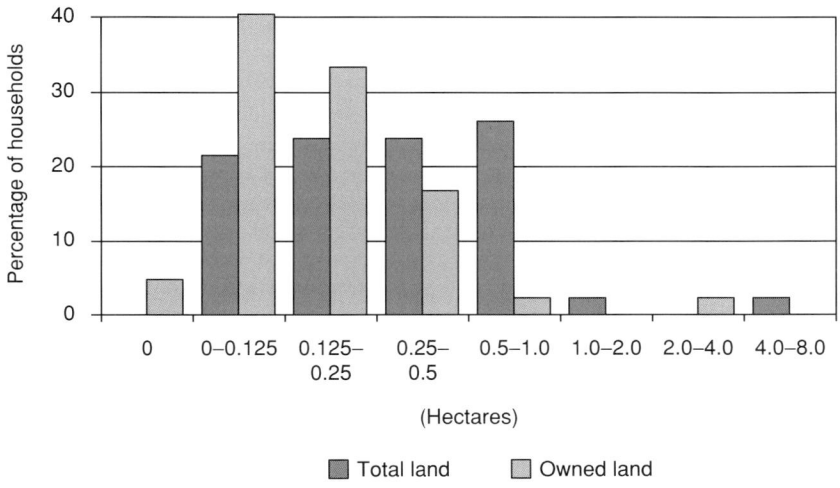

Fig. 2.2. Total land and owned land planted to maize in Zoatecpan, Mexico.

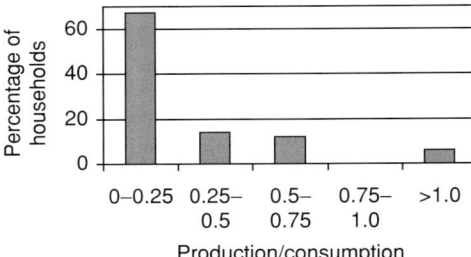

Fig. 2.3. Household maize production to consumption ratio in Zoatecpan, Mexico.

Around 94% of households are by definition subsistence maize growers, producing only for their own consumption and making up the deficit with purchases. Only 4% of the population can be considered commercial growers of maize (households that grow maize with the intention to sell), but all maize grown is landraces.

A random sample of 49 households in Zoatecpan – slightly over 10% of the population – were surveyed quarterly during the 1999–2000 crop cycle. Income and expenditure data gathered in these surveys were used as variables to explain farmer responses to hypothetical scenarios. Responses were then elicited in separate, personal interviews during the 2000–2001 planting season. Only 33 out of the original 49 household-heads surveyed over the course of a year were available. At the time of the contingent survey, all households had made their land use decisions for the season and most were in the process of planting the fields. After recording baseline land use decisions, household-heads were presented with hypothetical changes in maize-price and income expectations. Changes were measured in absolute terms. Farmers were first asked whether they would modify recorded land use decisions in response to a certain change in price or income. In the event of an affirmative answer, they were asked 'how' they would modify it. In the case of income changes, farmers were asked to consider an explicit cash payment delivered the same way as current government transfers. In the case of maize prices, farmers were reminded of the prevailing price for maize at the local store and then asked to consider that an alternative price would prevail throughout the year. In Zoatecpan, as in many other Mexican villages, maize prices in public-sector DICONSA stores are regulated.

Responses were elicited separately for each type of shock (i.e. maize-price and income) using the double bounded referendum (Hanemann et al., 1991; Appendix 2.1). That is, for each type of shock, informants were confronted with two hypothetical changes, one at a time. Based on their answer to the first question, farmers were confronted with a second hypothetical shock in the same variable. If the answer to the first shock was affirmative, the second shock was smaller; if it was negative, the second shock was larger. The magnitude of the second shock was otherwise random and independent of the first. Price shocks ranged from a 33% decrease to a 50% increase in the current maize price (Mex$1.50–3.30/kg). Income transfers ranged from Mex$200 to $1000 (approximately US$20–100).

Results

Sample households in Zoatecpan grow four different maize types that broadly correspond to those identified by previous research in the region (Inzunza, 1988). All of these are local types or landrace complexes that have been present in the region for at least one generation. As in other parts of Mexico, farmers distinguish between types or classes in the first instance according to their colour (see Chapter 4 for further discussion). High yielding varieties are uncommon in the region and were not registered in the village of Zoatecpan. Some local households have different uses for different types, but a majority of households did not express a preference for a particular type except for red maize, which is used ritually to protect the crop in the field (Table 2.1). No land plot is grown entirely in red maize; a few red maize plants are commonly grown interspersed in plots grown in another variety. Most farmers consider yellow, blue and white maize types to be good substitutes.

In all cases, farmer responses consisted of changes in the amount of land rented for maize cultivation. No farmer expressed interest in changing crop choice on land already allocated to agricultural production, which is natural given the lack of crop alternatives in this area (Dyer, 2002). Aggregated responses are based on

Table 2.1. Use-preferences for maize landraces in Zoatecpan, Mexico.

Percentage of households that prefer variety for use as:	White (a)	Yellow (b)	Blue (c)	No preference (d)
Tortillas	18	3	0	79
Atole	38	3	6	53
Tamales	35	4	0	61
Feed	6	47	8	39

farmers' answers to only the first of two questions (Table 2.2). Responses to the second question are used exclusively in the subsequent econometric estimations.

The aggregated response to maize price changes shows a marked asymmetry (Table 2.2, column d). Price increases (averaging 27%) induce 30% of households to increase their demand for land planted to maize landraces, raising the village's aggregate demand by 30%. In contrast, maize price decreases (averaging 19%) induce a response in only 15% of households. Mixed responses to maize price changes resulted in negligible changes in the village's aggregate demand for land in maize landraces. In other words, decreases in demand for land in maize by some farmers were offset by increases by others. For their part, increases in income (averaging Mex$564) prompted 21% of farmers to respond, raising the aggregate demand for land planted to maize by 19%. Only 6% of all farmers responded to increases in both maize prices and income, suggesting that households respond differently to different economic changes. Percentage responses in demand for land in maize are greater in general for white maize landraces, the dominant type. Farmers also respond negatively to price decreases in yellow and blue maize, reducing their demand for land planted to these types.

Regression results confirm this suggestion (Table 2.3). Responses to increases in the maize price (Table 2.3, column a) depend, as expected, on household characteristics as well as on the magnitude of the price increase. Better-off households are more likely to respond to price increases, and surprisingly, so are debtor households. Older household heads are less likely to respond. The likelihood of a positive response also depends on current land use. Growers who are already large-scale are more likely to expand maize production after a price increase. Migration of the household head has a negative but insignificant effect on household response to

Table 2.2. Changes in demand for land planted to maize landraces in Zoatecpan, Mexico.

	White (a)	Yellow (b)	Blue (c)	All maize (d)
Original aggregate land use (ha)	10.1	5.6	0.9	16.6
Changes after price increases				
Aggregate change in land demand	43%	9%	14%	30%
Share of farmers' increasing demand	24%	9%	3%	30%
Changes after price decreases				
Aggregate change in land demand	4%	−7%	−7%	0%
Share of farmers' increasing demand	3%	6%	6%	9%
Share of farmers' decreasing demand	3%	6%	0%	6%
Changes after income increases				
Aggregate change in land demand	27%	8%	0%	19%
Share of farmers' increasing demand	21%	12%	0%	21%

Table 2.3. Price and income elasticity of the demand for land planted to maize landraces in Zoatecpan, Mexico.

	Response to	
Variables	(a) Price increases	(b) Income increases
Constant	−0.29	−1.47
Age of household head	−0.10***	9×10^{-3}
Schooling of household head	0.02	0.06
Debt (= 1 if household indebted)	1.08*	1.82***
Migration (= 1 if head worked away from home)	−0.93	−1.25*
Number of workers in the household	0.42	−0.19
Household income	$9 \times 10^{-4**}$	$-6 \times 10^{-4*}$
Total land used by household	1.08**	0.07
'Shock' variable (see text for definition)	3.05**	7×10^{-4}

*$P \leq 0.10$; **$P \leq 0.05$; ***$P \leq 0.01$.
Note: N = 33

maize price and income changes, while the number of working members in the household has a positive but insignificant effect. The effects of migration on other secondary crops grown with maize in this system are investigated in Chapter 5.

Responses to changes in income depend on household characteristics but not on the size of the change itself (Table 2.3, column b). Although some households indicated that income changes faced were insufficient to finance an increase in maize production, households can spend additional income on non-agricultural goods and services, blurring any association between size of the change and household response. Low-income and debtor households are more likely to respond to an increase in income. Migrant households are less likely to do so. The total amount of land used by the household does not have a significant effect on its response, nor do other household characteristics.

There is little evidence of household-specific transaction costs in the village of Zoatecpan. There is a well-developed local maize market in the region and little household heterogeneity that could affect access to other markets. No household in the village has any of the organizational assets or capital goods associated with access to distant markets (de Janvry et al., 1995). In contrast, non-market benefits in production of maize landraces abound (Dyer, 2002); and they provide a plausible explanation for farmers' responses. Assuming consumption of non-market benefits has decreasing marginal utility, the shadow value of cultivating maize landraces should decrease with scale and approach the market value of maize at the limit. In fact, the marginal value of production for surplus growers should be equal to the market value of grain (net of sales costs) when non-market benefits are embedded in the grain itself (e.g. grain quality). This characterization of the value of maize landraces among Zoatecpan households is supported by cost data. The ratio of total production costs to market value of the maize landraces grown in this village ranges from around 1 to more than 15. It is closest to 1 for surplus growers and attains its highest values among the smallest subsistence growers (Fig. 2.4). Production costs include rents for family land and wages for family labour at market rates.

Under these circumstances, increases in the price of maize should raise the value of production for large growers and surplus producers who should respond by increasing their demand for land in maize. Although price increases could also affect subsistence households, significant non-market benefits could buffer the shadow value of production from price increases. Moreover, since production of non-market benefits is costly and likely to generate monetary losses, response to price changes among subsistence growers should depend on income. Liquidity-constrained growers might forgo expansion after price increases. There is evidence that this is the case in Zoatecpan where better-off households

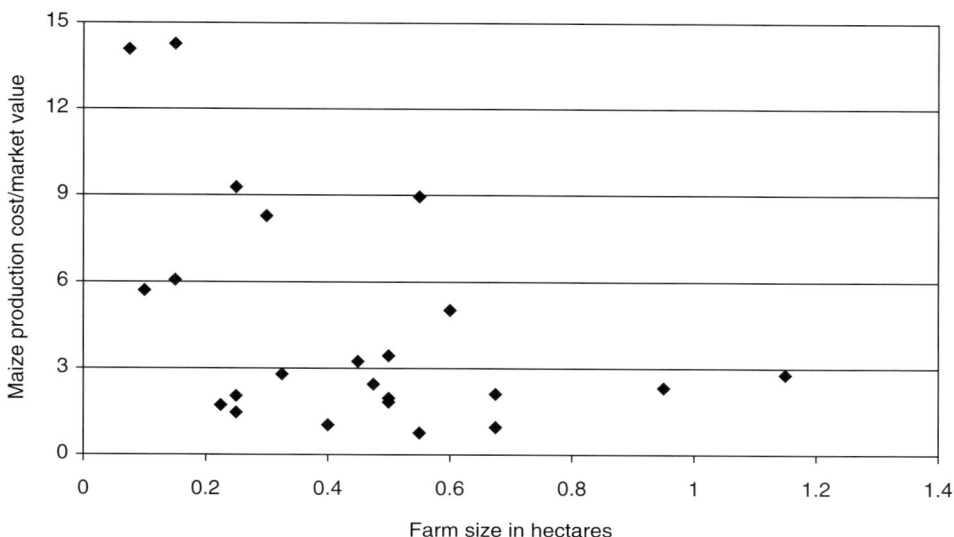

Fig. 2.4. Ratio of production cost to market value for maize in Zoatecpan, Mexico.

are able to plant more land and are more likely to respond to price increases than other households (Table 2.3). Large growers are also more responsive to price increases.

In principle, assuming that demand for the non-market benefits of maize landraces is normal, the elasticity of maize supply with respect to income should be positive among certain groups; specifically, liquidity-constrained households and more generally, households for whom the marginal utility of production is positive. In contrast, surplus growers with sufficient liquidity should not invest additional income in production (assuming that price remains constant) since they produce at a point where the marginal cost of production equals the product price. In Zoatecpan, worse-off households plant a relatively small amount of land, yet they are willing to expand home production when they have access to additional income (Table 2.3). Large growers are less likely to do so, presumably because marginal non-market benefits from home production are low for them.

Conclusions

Evidence collected through formal and informal interviews with Zoatecpan households demonstrates unequivocally that different types of growers respond differently to economic change. Also, changes in the farmer demand for land planted to maize landraces differ by maize types, represented in this study by grain colour. Regardless of the values they ascribe to their local maize landraces, their responses have direct implications for maize management both within the household and at the village level. Household characteristics that explain responses can be statistically associated with particular diversity patterns observed in the study site using a random-utility model (Chapter 5).

Two caveats are in order in interpreting results. First, results can be linked indirectly, but not explicitly, to land use patterns for maize landrace cultivation at either the household or the village level. This is because responses reveal shifts in households' demand for land, while actual changes in land use among different households depend additionally on the supply of land for rent in the market. While the total supply of land in the village is fixed, the amount of land for rent depends on the decisions of a limited number of land-rich households in Zoatecpan. The single landlord in the sample stated clearly that he would not rent out land if the price of maize increased – assuming a constant land-rental rate. However, increased demand for land is bound to raise the rental rate and thus, the supply of rental land in the community.

Second, although tested and validated, contingent methods have important restrictions, such as the number of hypothetical changes that farmers can handle simultaneously in an interview setting. Price or policy changes can also have significant general equilibrium effects on regional economies like that found in Zoatecpan (Taylor and Adelman, 1996), but general equilibrium models are not easily adapted to the approach used in this chapter. Nevertheless, variables estimated through the analysis of contingent behaviour – such as the price and income elasticity of demand – can be used as parameters in village-wide models to investigate the general equilibrium outcome of particular market or policy changes on diversity. For example, Dyer and Taylor (2002) have shown how lump-sum transfers to all households in a village will increase the value of maize production for some households and decrease it for others.

Implications

The idea that de facto conservation of a landrace is a function of its private value has much intuitive appeal, but to develop this idea and its practical implications further value needs to be defined more precisely. Farmers' decisions may be associated with total economic value, including both market and non-market benefits. Even if this is the case, farmers' responses to changes in prices and income, such as a change in the land planted to maize landraces, are based on marginal values.

Use of marginal value to estimate the likelihood of conservation in the face of change suffers an important drawback. In the steady state, alternatives available to farmers all have the same marginal value. That is, a farmer earns no more from cultivating an additional hectare in maize landraces than in the next best alternative. External economic forces tend to disrupt this situation from time to time, prompting farmers to respond and offset value differences that might have surfaced. Once the economy reaches a new equilibrium, marginal values level out again. In summary, steady state marginal values do not determine choices but are jointly determined with them – economic value is endogenous to the decision making process.

When, as in the case of a subsistence producer, the marginal value of a crop landrace is a shadow value that is not strictly determined by the market price of its output, many factors can potentially have an effect on that value. Some are specific to the crop or the household, while others are shared by all households within a region. A single event, such as a generalized increase in income, can have quite distinct direct effects on the value of a crop for different households.

Indirect or general equilibrium effects are likely to complicate the situation. Clearly, variables other than the value of maize will change as a result of income transfers. If the labour market is not perfect, the value of time (the shadow wage) will also be affected in different ways for different persons. As a result, both the value of a crop and its costs can be expected to change in different ways for different households. The magnitude of current opportunity costs alone says very little about the way in which these costs might change with economic change. For economics to contribute to the understanding of crop biodiversity conservation on farms, farmers' response to change needs to be better understood through advances in research methods and applications.

Acknowledgements

This chapter is drawn from the PhD thesis of the author at the University of California at Davis. I am grateful to UC MEXUS-CONACYT and the McKnight Milpa Project for support of this research, and Steve Brush and Melinda Smale for comments on an earlier draft of this chapter.

References

Appendini, K. (1994) Transforming food policy over a decade: the balance for Mexican corn farmers in 1993. In: Hewitt, C. (ed.) *Economic Restructuring and Rural Subsistence in Mexico: Corn and the Crisis of the 1980s.* Transformation of Rural Mexico, No. 2. Center for US–Mexico Studies, University of California at San Diego, California.

Austin, J.E. and Esteva, G. (1987) The path of exploration. In: Austin, J.E. and Esteva, G. (eds) *Food*

Policy in Mexico: The Search for Self-sufficiency. Cornell University Press, Ithaca, New York.

Bishop, R. and Heberlein, T. (1979) Measuring values of extra-market goods: are indirect measures biased? *American Journal of Agricultural Economics* 61, 926–930.

Brown, G.M. (1990) Valuing genetic resources. In: Orians, G.H., Brown, G.M., Kunin, W.E. and Swierzbinski, J.E. (eds) *Preservation and Valuation of Biological Resources*. University of Washington Press, Seattle, Washington, pp. 203–226.

Brush, S.B. and Meng, E. (1998) Farmers' valuation and conservation of crop genetic resources. *Genetic Resources and Crop Evolution* 45, 139–150.

Cameron, T.A. and Quiggin, J. (1994) Estimation using contingent valuation data from a 'dichotomous choice with follow-up' questionnaire. *Journal of Environmental Economics and Management* 27, 218–234.

Carson, R.T., Hanemann, W.M., Kopp, R.J., Krosnick, J.A., Mitchell, R.C., Presser, S., Ruud, P.A., Smith, V., Conaway, M. and Martin, K. (1998) Referendum design and contingent valuation: the NOAA Panel's no-vote recommendation. *The Review of Economics and Statistics* 80, 335–338.

CAST (1999) *Benefits of Biodiversity*. Task Force Report No. 133, Council for Agricultural Science and Technology, Ames, Iowa.

CEC (1999) *Evaluación de los efectos ambientales del Tratado de Libre Comercio de America del Norte*. North American Comission for Environmental Cooperation, Montreal.

Cummings, R.G., Brookshire, D. and Schultze, W.D. (1986) *Valuing Environmental Goods: An Assessment of the Contingent Valuation Methods*. Rowman and Allanheld, Totowa, New Jersey.

Davis, J. and Whittington, D. (1998) Participatory research for development projects: a comparison of the community meeting and household survey techniques. *Economic Development and Cultural Change* 47, 73–94.

de Janvry, A., Fafchamps, M. and Sadoulet, E. (1991) Peasant household behavior with missing markets: some paradoxes explained. *The Economic Journal* 101, 1400–1417.

de Janvry, A., Sadoulet, E. and Gordillo de Anda, G. (1995) NAFTA and Mexico's maize producers. *World Development* 23, 1349–1362.

Dyer, G. (2002) The cost of *in situ* conservation of maize landraces in the Sierra Norte de Puebla, Mexico. PhD dissertation, University of California at Davis, California.

Dyer, G. and Taylor, J.E. (2002) *Rethinking the Supply Response to Market Reforms in Agriculture: Household Heterogeneity in Village General Equilibrium Analysis from Mexico*. Department of Agriculture and Research in Economics, Working Paper, University of California at Davis, California.

Evenson, R.E. and Lemarié, S. (1998) Crop breeding models and implications for valuing genetic resources. In: Smale, M. (ed.) *Farmers, Gene Banks and Crop Breeding: Economic Analysis of Diversity in Wheat, Maize, and Rice*. CIMMYT, Mexico and Kluwer Academic Publishers, Boston, Massachusetts.

Fox, J. (1992) *The Politics of Food in Mexico: State Power and Social Mobilization*. Cornell University Press, Ithaca, New York.

Frankel, O.H. (1970) Genetic conservation of plants useful to man. *Biological Conservation* 2, 162–169.

Fritscher Mundt, M. (1996) El repunte maicero en tiempos de neoliberalismo. In: Lara Flores, S.M. and Chauvet, M. (eds) *La sociedad rural mexicana frente al nuevo milenio. Volumen I. La inserción de la agricultura mexicana en la economía mundial*. INAH, UAM, UNAM, Plaza y Valdez, Mexico, pp. 279–300.

Gollin, D. and Smale, M. (1999) Valuing genetic diversity: crop plants and agroecosystems. In: Collins, W.W. and Qualset, C.O. (eds) *Biodiversity in Agroecosystems*. CRC Press, Boca Raton, Florida.

Gollin, D., Smale, M. and Skovmand, B. (2000) Searching an *ex situ* collection of genetic resources. *American Journal of Agricultural Economics* 82, 812–827.

Griffin, C., Briscoe, J., Singh, B., Ramasubban, R. and Bhatia, R. (1995) Contingent valuation and actual behavior: predicting connections to new water systems in the state of Kerala, India. *The World Bank Economic Review* 9, 373–395.

Hanemann, W.M. (1984) Welfare evaluations in contingent valuation experiments with discrete responses. *American Journal of Agricultural Economics* 66, 332–341.

Hanemann, W.M. (1988) Economics and the preservation of biodiversity. In: Wilson, E.O. (ed.) *Biodiversity*. National Academy Press, Washington, DC, pp. 193–199.

Hanemann, W.M., Loomis, J. and Kanninen, B. (1991) Statistical efficiency of double-bounded dichotomous choice contingent valuation. *American Journal of Agricultural Economics* 73, 1255–1263.

Hentschel, E. (1997) *Opportunity Costs for In Situ Conservation of Plant Genetic Diversity*. Paper presented at the Symposium Building the Basis for Economic Analysis of Genetic Resources in Crop Plants, August 17–19, CIMMYT and Stanford University, Palo Alto, California.

Hewitt de Alcantara, C. (1994) Introduction: economic restructuring and rural subsistence in Mexico. In: Hewitt de Alcantara, C. (ed.) *Economic Restructuring and Rural Subsistence in Mexico: Corn and the Crisis of the 1980s*. Transformation of Rural Mexico, No. 2. Center for US-Mexico Studies, University of California at San Diego, California.

INEGI (2001) Online data base. Instituto Nacional de Estadística, Geografía e Informática, Mexico. Available at: www.inegi.gob.mx

Inzunza, M.F.R. (1988) El proceso de producción agrícola en la porción oriental de la Sierra Norte de Puebla, México. Tebis de Licenciatúra. Universidad Autónoma de Chapingo, Mexico.

Jarvis, D.I., Myer, L., Klemick, H., Guarino, L., Smale, M., Brown, A.H.D., Sadiki, M., Sthapit, B. and Hodgkin, T. (2000) *A Training Guide for* In Situ *Conservation On-farm*. Version 1. IPGRI, Rome.

Koo, B., Pardey, P.G. and Wright, B.D. (2004) *Saving Seeds: The Economics of Conserving Crop Genetic Resources* Ex Situ *in the Future Harvest Centres of the CGIAR*. CAB International, Wallingford, UK.

Levy, S. and van Wijnbergen, S. (1992) Maize and the free trade agreement between Mexico and the United States. *The World Bank Economic Review* 6, 481–502.

Levy, S. and van Wijnbergen, S. (1994) Labor markets, migration and welfare: agriculture in the North American Free Trade Agreement. *Journal of Development* 43, 263–278.

Maxted, N., Ford-Lloyd, B.V. and Hawkes, J.G. (1997) Complementary conservation strategies. In: Maxted, N., Ford-Lloyd, B.V. and Hawkes, J.G. (eds) *Plant Genetic Conservation: The* In Situ *Approach*. Chapman & Hall, London, pp. 15–39.

Mitchell, R.C. and Carson, R.T. (1989) *Using Surveys to Value Public Goods: The Contingent Valuation Method*. Resources for the Future, Washington, DC.

Montanari, M. (1987) The conception of SAM. In: Austin, J.E. and Esteva, G. (eds) *Food Policy in Mexico: The Search for Self-sufficiency*. Cornell University Press, Ithaca, New York, pp. 48–58.

Myers, N. (1979) *The Sinking Ark: A New Look at the Problem of Disappearing Species*. Pergamon Press, Oxford.

Ochoa, E.C. (2000) *Feeding Mexico: The Political Uses of Food Since 1910*. SR Books, Wilmington, Delaware.

Perales, R.H.R. (1998) *Conservation and evolution of maize in the Amecameca and Cuautla Valleys of Mexico*. PhD dissertation, University of California at Davis, California.

Pressoir, G. and Berthaud, J. (2004) Population structure and strong divergent selection-shape phenotypic diversification in maize landraces. *Heredity* 92, 95–101.

Qualset, C.O., Damania, A.B., Zanatta, A.C.A. and Brush, S.B. (1997) Locally based crop plant conservation. In: Maxted, N., Ford-Lloyd, B.V. and Hawkes, J.G. (eds) *Plant Genetic Conservation: The* In Situ *Approach*. Chapman & Hall, London, pp. 160–175.

Robinson, S., Burfisher, M.E., Hinojosa-Ojeda, R. and Thierfelder, K.E. (1993) Agricultural policies and migration in a US-Mexico Free Trade Area: a computable general equilibrium analysis. *Journal of Policy Modeling* 15, 673–701.

Roughgarden, J. (1995) Can economics protect biodiversity? In: Swanson, T.M. (ed.) *The Economics and Ecology Biodiversity Decline: The Forces Driving Global Change*. Cambridge University Press, Cambridge, UK, pp. 149–156.

SAGARPA (2002) Website. Secretaría de Agricultura, Ganadería, Desarrollo Rural, Pesca y Alimentación, Mexico. Available at:www.sagarpa.gob.mx

Taylor, J.E. and Adelman, I. (1996) *Village Economies: The Design, Estimation, and Use of Villagewide Economic Models*. Cambridge University Press, Cambridge, UK.

Tellez, K.L. (1992) Mexican agricultural policy and the nation's modernization process. In: Carter, C. and Carter, H.O. (eds) *North American Free Trade Agreement: Implications for California Agriculture*. Agricultural Issues Center, University of California at Davis, California.

Wellhausen, E., Roberts, J., Roberts, L.M. and Hernández, E. (1952) *Races of Maize in Mexico: Their Origin, Characteristics, and Distribution*. The Bussey Institution, Harvard University, Cambridge, Massachusetts.

Whittington, D. (1998) Administering contingent valuation surveys in developing countries. *World Development* 26, 21–30.

Wilson, E.O. (ed.) (1988) *Biodiversity*. National Academy of the Sciences, National Academy Press, Washington, DC.

Wood, D. and Lenné, J.M. (1997) The conservation of agrobiodiversity on-farm: questioning the emerging paradigm. *Biodiversity and Conservation* 6, 109–129.

Yúnez, N.A. and Barceinas, F. (2000) Efectos de la desaparición de la Conasupo en el comercio y en los precios de los cultivos básicos. *Estudios Económicos* 2, 189–227.

Zohrabian, A., Traxler, G., Caudill, S. and Smale, M. (2003) Valuing pre-commercial genetic resources: a maximum entropy approach. *American Journal of Agricultural Economics* 2, 429–436.

Appendix 2.1. Sample survey form.

Introduction

You have already told me how many plots you are working this season and what you are sowing. I would like you to suppose that things were different from what they are now, and to tell me how you would respond to this new state of things. Your answer will let me understand how you think and how government programmes affect you. Please listen to what I will tell you and think for a moment if you would do things differently; if you would change the amount of land, the number of plots you are working or what you are growing in them.

Price increase

Suppose that the price of maize went 'up' this year, so that maize sold in DICONSA was US$_____ /kg instead of US$2.20, as it is now. Would you work the same amount of *milpa*?

☐ yes (go to second bid) ☐ no
Would you rent more or less land?
How much 'more' land would you rent?
How many plots 'more'?
What would you grow in these plots?

Income increase

Now, please dismiss the previous suppositions and recall that the price of maize today is US$2.20. Suppose instead that the government creates a programme like today's PROGRESA, so that people receive a cash payment to spend as they please and are not required to pay back. In this new programme, each household head receives a one-time payment today totalling US$_____.

Would you work the same amount of *milpa*?
☐ yes (go to second bid) ☐ no
Would you rent more or less land?
How much 'more' land would you rent?
How many plots 'more'?
What would you grow in these plots?

3 Farmer Demand for Agricultural Biodiversity in Hungary's Transition Economy: A Choice Experiment Approach

E. Birol, A. Kontoleon and M. Smale

Abstract

This chapter relates economic development and transition with farmer demand for four components of agricultural biodiversity found on family farms in Hungary using a combination of a stated preference approach and secondary data. Family farms in Hungary are known traditionally as 'home gardens'. Production on these farms is labour-intensive, with few purchased inputs. High levels of crop and variety diversity, and integrated crop and livestock production, are typical of home gardens. It is hypothesized that farmers' demand for home gardens will decrease as Hungary's economic transition proceeds and local, regional and national markets are integrated with European Union (EU) accession. This hypothesis is tested with a choice experiment conducted across 22 settlements in three regions with varying levels of economic development and market integration. Findings indicate that farmers in more economically developed, less isolated settlements will choose to depend less on home gardens for food security and will prefer lower levels of agricultural biodiversity. These results suggest that a vital cultural institution may disappear with EU accession. Data can be used to identify the settlements and farmers who would benefit most by agri-environmental policies that support their maintenance, at least public cost. In some situations, supporting their maintenance is consistent with the multifunctional agriculture approach stated in the EU's reformed Common Agricultural Policy (CAP). The findings of this chapter complement those of the revealed preference analysis presented in Chapter 8 and the institutional analysis shown in Chapter 15, conducted in the same sites.

Introduction

Much of the agricultural biodiversity remaining *in situ* today is found on the semi-subsistence farms of poorer countries and in the small-scale farms or backyard gardens of more industrialized nations (Brookfield, 2001; Brookfield *et al.*, 2002; IPGRI, 2003). The small-scale farms of Hungary, termed 'home gardens', are an example found in a higher income country. These homestead fields were privately owned and cultivated throughout the period of collectivized state farming and the subsequent change to market-oriented, large-scale farming (Kovách, 1999; Swain, 2000; Meurs, 2001). Farmed with family labour and scant use of purchased inputs, many are rich in crop and livestock species, varieties and breeds, relative to the large-scale mechanized farms that now dominate the nation's cropland (Már, 2002). Home gardens also contribute to food security in Hungarian society, supplying farm produce that contributes colour, flavour and nutrients to the diets of both rural and urban people in time periods and locations when markets or state institutions do not.

A nation with an economy in transition, Hungary joined the EU in May 2004. The agricultural and environmental policies and programmes now being developed to comply with the *acquis communautaire* of the EU appear not to recognize the private and public values of agricultural biodiversity found in home gardens. To investigate some of these values, the Institute of Agrobotany, Institute of Environmental and Landscape Management and International Plant Genetic Resources Institute implemented an integrated set of pilot studies in three environmentally sensitive areas (ESAs) of Hungary. In each site, agri-environmental programmes consistent with EU policies have already been envisaged. The purpose of the research was to generate information regarding the prospects for efficient integration of home gardens into these programmes. Findings from three of these studies are reported in this chapter and Chapters 8 and 15.

In this chapter, a combination of the choice experiment method and secondary data is applied to test the effects of economic development and transition on farmer demand for agricultural biodiversity in home gardens and food self-sufficiency. The role of home gardens in Hungary and its policy context are presented in the next section. The choice experiment approach is then summarized briefly. The regions in which the data were collected are described. The design of the choice experiment and econometric analysis are discussed. Conclusions are drawn and implications stated in the final section.

Background

Home gardens in Hungary

Hungarian agriculture today has a dual structure consisting of large-scale, mechanized farms alongside semi-subsistence, small-scale farms operated with traditional farming practices. Dualism has persisted in some form throughout Hungarian history, and most recently during the socialist period of collectivized agriculture from 1955 to 1989 (Szelényi, 1998; Kovách, 1999; Swain, 2000; Szép, 2000; Meurs, 2001). Even today semi-subsistence agricultural production is a significant component of economic activity in Hungary. Of the about 10 million people populating Hungary today, nearly 2 million Hungarians produce agricultural goods for their own consumption and as a source of additional income on 697,336 home gardens with an average size of 591 m^2 (Hungarian Central Statistical Office (HCSO), 2001; Már, 2002). The 1996 Microcensus implemented by the HCSO reported that 33% of people aged 14 and over were engaged in auxiliary agricultural work, although few relied on agriculture as a main occupation.

Although there is extensive variation among home gardens, it is generally the case that production in home gardens is still accomplished with family labour, ancestral crop varieties and livestock races, traditional farming practices and limited use of purchased inputs or machinery. As a consequence, Hungarian home gardens are known as national 'repositories of agricultural biodiversity', described by agricultural scientists as micro-agroecosystems that are relatively rich in numbers of crop species, varieties and livestock races, as well as other soil microorganisms (Már, 2002; Csizmadia, 2004). Bela *et al.* (2003) remark that the few extant Hungarian crop and livestock genetic resources, many of which originated in the Bronze age and Roman period, can only be found in the country's home gardens.

Home gardens played a critical role in food security during the socialist period when markets were run by the state and families were permitted to cultivate privately the small plots located adjacent to dwellings (Szelényi, 1998; Kovách, 1999; Swain, 2000; Szép, 2000; Meurs, 2001). Historically, food market formation was discouraged. Several factors explain the persistence of thin food markets in many rural communities since the transition to a market economy began in 1989. Feick *et al.* (1993) found that along with high inflation and unemployment rates, Hungarian consumers have faced difficulties in obtaining reliable product information and in predicting product availability. Transaction costs are high, including search costs and transport costs to the nearest food market. The number of hypermarkets in Hungary has grown from only 5 in 1996 to 63 in 2003 (HCSO, 2003), but a study by WHO (2000) found that these have contributed to the disappearance of the existing local shops and markets. Today, rural households still rely on their home gardens for the production of some of the foods they consume, enhancing the breadth and quality of their diet.

Only a few studies have documented the economic importance of home gardens in the livelihoods of rural families during economic transition. Szép (2000) found that income in kind generated by part-time agricultural production in Hungarian home gardens amounted to 14% of total income of the households. During the early stages of economic transition in Russia, Seeth *et al.* (1998) found that households engaging in subsistence agriculture on garden plots had higher levels of real income and food consumption than others. Both were crucial to combating poverty during an era when risky food prices prevailed and real incomes declined dramatically. Studying experiences across several transitional economies, Wyzan (1996) likened family survival strategies to those found in developing economies, where families rely on their own production to meet their subsistence requirements when markets are unreliable. Home garden production subsidizes rural settlements and lifestyles, enabling people to remain in the countryside (Seeth *et al.*, 1998; Juhász *et al.*, 2000).

Policy context

This stylized depiction of Hungarian home gardens is consistent with the notion of multifunctional agriculture, which views agriculture as providing a bundle of public goods in addition to private goods of food and fibre. Public goods supplied by agriculture include rural settlement and economic activity, food security, safety and quality, biodiversity, agricultural biodiversity, cultural heritage, amenity and recreational values (Lankoski, 2000; Romstad *et al.*, 2000). The concept of multifunctional agriculture is embraced by the EU's reformed CAP and is stated in the EU's 2078/92 agri-environmental regulation. Regulations stipulate that each EU member country, including those preparing to become full members, is expected to encourage production of agricultural public goods through the development of a National Agri-Environmental Programme (NAEP).

Hungary's NAEP was accepted by the Ministry of Agriculture and Regional Development in 2000 and launched experimentally in 2002 (Juhász *et al.*, 2000). NAEP proposes that the intensity of agricultural production in a region should depend on its natural and human resource endowments. Several areas of Hungary with low agricultural productivity and high environmental value have been designated as ESAs. NAEP seeks to protect these areas as habitats for endangered plant and animal species. Direct payments, training programmes and technical assistance are provided to the farmers who are willing to participate in agri-environmental schemes that promote the use of specified farming methods. The agri-environmental measures of NAEP were integrated into the National Rural Development Plan (NRDP) in 2004.

The Hungarian NAEP recognizes that extensive agricultural methods are most suitable for conserving biodiversity of endangered wildlife and providing other agricultural public goods, but the role of home gardens in the programme has not yet been elucidated. Proposed EU agricultural policies designed for accession states also fail to recognize the possibility of provision of public goods through home garden production. The Special Accession Programme for Agriculture and Rural Development (SAPARD), which provides assistance to the new member states in the period 2000–2006, considers the dual structure of agriculture that exists in several of these countries as inefficient. SAPARD proposes either: (i) subsidies for transformation of semi-subsistence small farms to commercial farms; or (ii) direct payments to landholdings larger than 0.3 ha on the condition that the land is managed in a way compatible with protection of the environment, as suggested by the NAEP of the member country (Commission of the European Communities, 2002).

The expected loss of home gardens has been cited by many experts as one of the risks of EU accession, economic transition and development (Vajda, 2003; Weingarten *et al.*, 2004). Risky food prices and uneven quality, high transaction costs and the low wages that led to dependence on home-grown food are expected to decrease with greater market access. EU accession could lead to improved rural infrastructure through SAPARD, along with rural development and the growth of employment opportunities outside agriculture (Weingarten *et al.*, 2004).

In addition to Hungary's EU level obligations to promote public values of agricultural production, such as conservation of agricultural biodiversity, Hungary is also a signatory to several international agreements whose aim is to generate incentive mechanisms that encourage farmers to sustain levels of agricultural biodiversity remaining *in situ*, on farms. Relevant international agreements include the International Convention on Biological Diversity (CBD, 1992), the Global Plan of Action for the Conservation and Sustainable Use of Plant Genetic Resources for Food and Agriculture (GPA, 1997) and the International Treaty on Plant Genetic Resources for Food and Agriculture (IT, 2001). Like other signatories Hungary is obliged to develop policies that address the commitments stemming from these agreements (Bela *et al.*, 2003).

Development and agricultural biodiversity

Two strands of applied economics literature motivate the hypothesis tested in this chapter. The first analyses the relationship between market development and farmers' choice of production technology, which influences agricultural biodiversity (Fafchamps, 1992; Goeschl and Swanson, 1998). Thin markets generate price, income and consumption risks for semi-commercial farmers. If, in addition, farmers have no market insurance mechanisms to enable them to cope with risk *ex post*, they manage risk *ex ante* through choosing more diverse crop and livestock combinations or producing more than would be optimal in the absence of risk (Roumasset *et al.*, 1979; Sadoulet and de Janvry, 1995; Moschini and Hennessy, 2000).

Fafchamps (1992) demonstrated that a large covariance between price and income exists because food prices are stochastic, especially for smaller farmers. Smaller farmers, who are more risk averse because they have no alternative insurance mechanisms, choose to be self-sufficient in food production in order to insure themselves. They allocate land or labour to produce a range of food crops and varieties rather than specializing in a few cash crops. As markets are integrated and price risk declines, agricultural productivity increases and transaction costs fall. Consequently, the need to self-insure through producing more food on the farm lessens, freeing farm resources for production of cash crops.

Goeschl and Swanson (1998) show theoretically that as markets develop and become integrated, farmers' demand for agricultural biodiversity as either a production input or a provider of consumption goods subsides. In their model, the integration of output and input markets within rural communities and across regions with more heterogeneous natural environments is the fundamental force driving changes in farmer demand. When markets do not function well or are not well integrated, agricultural biodiversity on farms is often the only instrument available for farm households to manage income and consumption risk through production choices. Improved access to markets that function better provides farmers with alternative means for coping with risk.

As market-related risks decline with economic change, any remaining agricultural diversification reflects agroecological heterogeneity and production sources of uncertainty. Uncertainty is inherent in agricultural production, with its time lags, biological and natural processes. In Hungary, though production sources of risk such as rainfall variability are believed to be moderate, there is considerable agroecological heterogeneity in the study sites (Juhász *et al.*, 2000; Gyovai, 2002; Csizmadia, 2004).

A second strand of applied economics literature relates farmer access to market infrastructure with crop biodiversity levels measured on farms (see review of key references in Chapter 1). Brush *et al.* (1992) found that market access, along with insurance and financial capital, were crucial determinants of farmer adoption of modern potato varieties in the Andes. They argued that continued cultivation of potato landraces compensated for market imperfections and satisfied household demand for diversity in diet. With more recent data from Cajamarca in Peru, Winter *et al.* (Chapter 9) also found that more remote households continue to depend more on the diversity of the potato cultivars they grow. Similarly, the level of market integration was a significant determinant of whether Turkish farmers grow wheat landraces in Turkey (Meng, 1997). Distance to markets resulted in higher levels of within and between species diversity on farms in the *milpa* systems of the Sierra Norte de Puebla, Mexico (Van Dusen, 2000; Chapter 4). Smale *et al.* (2001) observed a negative relationship between infrastructure development in a community (including transportation, communication and education) and maize landrace diversity in Guanajuato, Mexico. The diversity of rice varieties cultivated on Nepalese farms increases with the distance of the farm households to the nearest market (Gauchan, 2004; Chapter 10). Other chapters in Part III of this book explore the relationship between aspects or components of market development and crop biodiversity.

The studies described in Chapter 1 and the chapters in Part III investigated the relationship of market development with agricultural biodiversity on farms using revealed preferences observed in survey data from household farms. This chapter applies a choice experiment.

The Choice Experiment Approach

Since most of the outputs, functions and services that home gardens generate are not traded in the markets, non-market valuation methods must be used to determine the value of their benefits. Farmers earn non-market benefits in terms of utility rather than market prices. The preferences of farmers, who are both producers and consumers of home garden outputs, determine the implicit values they attach to home gardens and their attributes.

Of the range of environmental valuation approaches the choice experiment method is most appropriate for valuing home gardens because it enables estimation not only of the value of the environmental asset as a whole, but also of the implicit value of its attributes (Hanley et al., 1998; Bateman et al., 2003; Scarpa et al., 2003a). A relatively new addition to the portfolio of stated preference approaches, the choice experiment method is grounded theoretically in Lancaster's model of consumer choice (Lancaster, 1966) and has an econometric basis in models of random utility (Luce, 1959; McFadden, 1974).

Lancaster proposed that consumers derive satisfaction not from goods themselves but from the attributes they provide. Consider a farmer's choice for a home garden and assume that utility depends on choices made from a set C. A choice set, C includes all possible home garden options. The farmer is assumed to have a utility function of the form:

$$U_{ij} = V(Z_{ij}, F_i, E_i) + e_i \quad (3.1)$$

For any farmer i, a given level of utility will be associated with any alternative home garden j. Utility derived from any of the home garden alternatives depends on the attributes (Z) of the home garden, the social and economic characteristics of the farmer (F) and the farmer's social, economic and agroecological environment (E).

The random utility approach is the theoretical basis for integrating behaviour with economic valuation in the choice experiment. The utility of a choice comprises a systematic component (the first term on the right hand side of Eq. (3.1)) and an error component, e_i. The error component is independent of the systematic component and follows a predetermined distribution. Choices made between alternatives are a function of the probability that the utility associated with a particular option (j) is higher than that for other alternatives. Assuming that the relationship between utility and characteristics is linear in the parameters and variables function, and that the error terms are identically and independently distributed with a Weibull distribution, the probability of any particular alternative j being chosen can be expressed in terms of logistic distribution. Equation (3.1) can then be expressed as a conditional indirect utility function and be estimated with a conditional logit model (McFadden, 1974; Greene, 1997; Maddala, 1999):

$$\begin{aligned}V_{ij} =\ & \beta + \beta_1 Z_1 + \beta_2 Z_2 + \dots \\ & + \beta_n Z_n + \beta_a F_1 + \beta_b F_2 + \dots \\ & + \beta_m F_j + \beta_n E_1 + \beta_o E_2 + \dots \\ & + \beta_z E_p \end{aligned} \quad (3.2)$$

The alternative specific constant (ASC) term, β, captures the effects on utility from a change in any attribute not included in choice-specific attributes. The vectors of coefficients β_1 to β_n, β_a to β_m and β_n to β_z are attached to: (i) the vector of attributes of the home garden (Z); (ii) the vector of interaction terms between the home garden attributes and social and economic characteristics relating to the farmer (F); and (iii) the vector of interaction terms between the home garden attributes and social, economic and agroecological characteristics of the environment in which the farmer is located (E). Social and economic characteristics enter the utility function as interaction terms with the choice attributes since they are constant across choice occasions for any given farmer.

Few choice experiments have analysed the demand for components of agricultural biodiversity.

Other choice experiments have investigated the demand for agricultural products obtained with specific production techniques, such as organic non-GMO eggs (Kontoleon, 2003; Kontoleon and Yabe, 2004), and beef produced with hormones (Lusk *et al.*, 2003). Scarpa *et al.* (2003a,b) analysed the demand for the pig landraces in Mexico and cattle landraces in Kenya. The methodology applied here was developed by Birol (2004) and has been used by Birol *et al.* (2004) to estimate the private value of home gardens to farm families. That study was the first to use a choice experiment approach to explore the demand for a micro-agroecosystem and the components of its agricultural biodiversity. The private value of agricultural biodiversity in home gardens was estimated in terms of farmer willingness to accept (WTA) compensation, conditional on the social and economic characteristics of the farm family.

The demand for agricultural biodiversity in home gardens also depends on the social and economic characteristics of the settlements in which the farm families reside. Settlement characteristics cannot be affected by the decision of any individual family in the community during the short term. The analysis in this chapter holds constant the characteristics of farm families, focusing instead on the effects of economic development factors that vary among settlements and are 'exogenous' to individual farm families at any specific point of time.

Data Collection

Survey design

Data are drawn from home gardens across 22 settlements in three regions of Hungary. The survey design consisted of two stages. In the first stage, three study sites were selected in ESAs identified by the NAEP. Secondary data from the HCSO (2001) and NAEP were used to purposively select areas with contrasting levels of market development and varying agroecologies associated with different farming systems and land-use intensity. In each selected site, agri-environmental programmes are being implemented and the Institute of Agrobotany had identified high levels of agricultural biodiversity (in terms of crop genetic diversity) during preceding collection missions (Már, 2002).

In the second stage of the sample design, all settlements within each site were sorted based on population sizes and an initial sample of 1800 households was sampled randomly from a complete list of all households compiled from telephone books and village maps. Since a minimum final sample of 100 per ESA was thought necessary for data analysis, and the response rate to a mail survey was expected to be low, the team decided to draw 600 households per site. An initial screening survey was sent to the total sample of 1800 households to identify all those who are engaged in home garden management. The initial response rate to the screening survey was only 13%. The final sample was augmented through personal visits to listed sample households with the assistance of key informants in each settlement. All households sampled had home gardens. A total of 323 farm households were personally interviewed in August 2002 with a household survey instrument (Chapter 8). A subset of all respondents (277) was interviewed for the choice experiment. Findings reported in this chapter are statistically representative of the study sites and other sites in rural Hungary to the extent that they share characteristics in common.

The three study sites (Dévaványa, Örség-Vend and Szatmár-Bereg) are depicted in Fig. 3.1. The stratified design enables hypotheses to be tested about the impacts of market integration and economic development on the agricultural biodiversity maintained in home gardens.

Choice sets

A choice experiment is a highly 'structured method of data generation' (Hanley *et al.*, 1998) relying on carefully designed tasks or 'experiments' to reveal the factors that influence choice. Experimental design theory is used to construct profiles of the good in terms of attributes and attribute levels. Profiles are then assembled in choice sets and are presented to respondents who are asked to state their preferences.

Attributes and levels were identified with NAEP experts and agricultural scientists, drawing

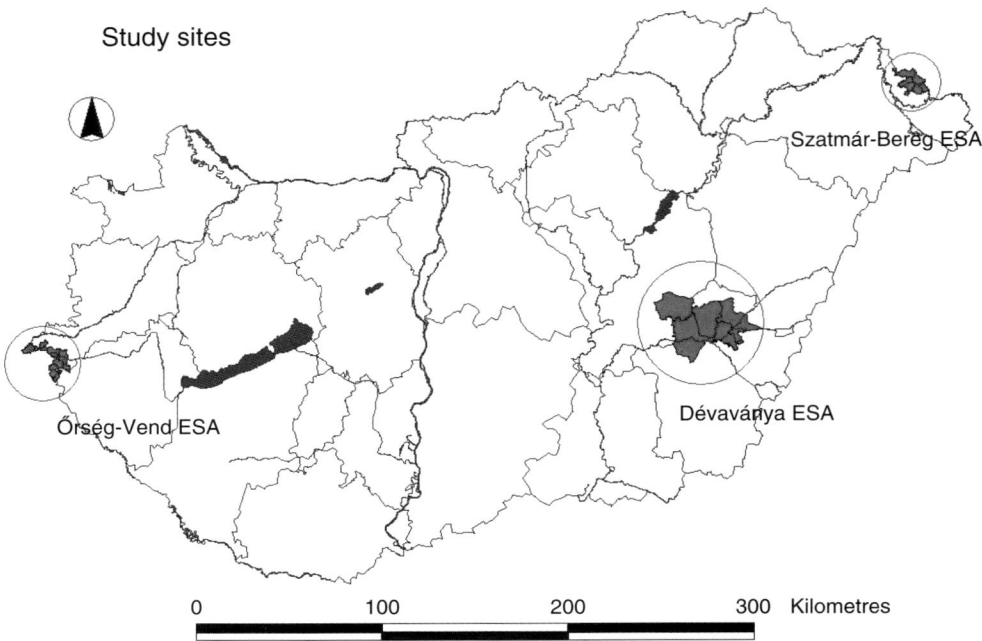

Fig. 3.1. Location of the study sites.

on the results of informal and focus group interviews with farmers in each ESA. Four key components or 'entry points' to agricultural biodiversity were identified: (i) crop variety diversity (richness of crop varieties); (ii) crop genetic diversity (cultivation of landraces as compared to only modern varieties); (iii) agrodiversity (integrated crop and livestock production); and (iv) soil microorganism diversity (use of organic production practices).

The total number of crop varieties grown in a garden of fixed size is an indicator of crop variety richness. Crop variety diversity is one of the most crucial components of agricultural biodiversity. Inter- and intracrop diversity in field crops, trees and vegetables were considered. Presence of a landrace in the home garden expresses crop genetic diversity. Preliminary molecular biological analyses conducted on bean landrace samples collected from the home gardens of the households in the sample reveal that the majority of these landraces are distinct, contain rare and adaptive traits and are genetically heterogeneous (Már, 2002; Már and Juhász, 2002). The traditional method of integrated crop and livestock production represents agrodiversity,

or diversity in agricultural management practices (Brookfield and Stocking, 1999). Organic production takes place if crops are grown without any industrially produced and marketed chemicals, such as pesticides, herbicides, insecticides, fungicides or soil disinfectants. Previous experiments found that use of organic production methods resulted in soil microorganism diversity (e.g. Lupwayi *et al.*, 1997; Mäder *et al.*, 2002). The expected percentage of the annual household food consumption supplied by the home garden represents the family's dependence on its own production in the home garden (Table 3.1).

A large number of unique home garden prototypes can be constructed from this number of attributes and levels using experimental design theory. Main effects, consisting of 32 pair-wise comparisons of home garden prototypes, were recovered with an orthogonalization procedure. Although exclusion of interaction effects in the experimental design may introduce bias into main effects estimations, main effects usually account for more than 80% of the explained variance in a model (Louviere, 1988; Louviere *et al.*, 2000). Moreover, the aim of this choice

Table 3.1. Home garden attributes and attribute levels used in the choice experiment.

Home garden attribute	Definition	Attribute levels
Crop variety diversity	The total number of different crop varieties grown in the garden.	6, 13, 20, 25
Landrace	Whether or not the home garden contains a crop variety that has been passed down from the previous generation and/or has not been purchased from a commercial seed supplier.	Home garden contains a landrace vs home garden does not contain a landrace.
Agrodiversity	Integrated crop and livestock production, representing diversity in agricultural management system.	Integrated crop and livestock production vs specialized crop production.
Organic production	Whether or not industrially produced and marketed chemical inputs are applied in farm production.	Organic production vs non-organic production.
Self-sufficiency	The percentage of annual household food consumption that it is expected the home garden will supply.	15%, 45%, 60%, 75%

experiment was to investigate farmer demand for each home garden attribute independently of the other home garden attributes. An advantage of the choice experiment approach relative to revealed preference approaches is that the effects of each attribute on respondents' demand for the environmental good can be separated, avoiding collinearity between the attributes (Adamowicz et al., 1994, 1997; Adamowicz and Boxall, 2001).

The main effects were randomly blocked to six different versions, two with six choice sets and the remaining four with five choice sets. In face-to-face interviews, each farmer was presented with five or six choice sets. Each set contained two home gardens and an option to select neither garden. The farmers who took part in the choice experiment were those responsible for making decisions in the home garden. Enumerators explained the context in which choices were to be made (a 500 m^2 garden) and that attributes of home gardens had been selected as a result of prior research and were combined artificially. They defined each attribute to ensure uniformity. Overall, a total of 1487 choices were elicited from 277 farmers taking part in the choice experiment. An example of one of the choice sets presented to farmers is shown in Fig. 3.2.

Site Description

Twenty-two settlements (5 in Dévaványa, 11 in Őrség-Vend and 6 in Szatmár-Bereg) were included in the study. Secondary data for settlement characteristics were compiled from the Hungarian Central Statistical Office (HCSO) National Census (2001), Statistical Yearbooks (2001a,b,c,d) and the Hungarian Ministry of Transport and Water (2001). Data are summarized in Table 3.2.

Of the three sites, Dévaványa, located on the Hungarian Great Plain, is closest to the economic centre of the country. Soil and climatic conditions of this region are well suited to intensive agricultural production. Settlements are large in both populations and areas, so that population densities are also relatively high. Labour migration is not a major problem in Dévaványa, although the number of inhabitants is stagnating. The unemployment rate in Dévaványa (12.4%) is slightly higher than the Hungarian average. NAEP aims to conserve the wildlife found in this ESA. Dévaványa is statistically different from the other two ESAs in most indicators of urbanization and market integration, including presence of a train station, distance to the nearest market (both in kilometres and minutes by car), number of primary and secondary schools, food markets and the number of shops and enterprises.

> Assuming that the following home gardens were the ONLY choices you had, which one would you prefer to cultivate?
>
> | Home garden characteristics | Home garden A | Home garden B | |
> |---|---|---|---|
> | Total number of crop varieties grown in the home garden | 25 | 20 | Neither home garden A nor home garden B: I would NOT cultivate a home garden |
> | Home garden has a landrace | No | Yes | |
> | Home garden production is integrated with livestock production | Yes | Yes | |
> | Home garden crops produced entirely with organic methods | No | No | |
> | Expected proportion (in %) of annual household food consumption met through food production in the home garden | 45 | 75 | |
>
> I prefer to cultivate Home garden A..... Home garden B.... Neither home garden ......
> (please check (✓) one option)

Fig. 3.2. Sample choice set.

Table 3.2. Settlement and environmentally sensitive area (ESA) level characteristics. (Compiled from data reported in the Hungarian Central Statistical Office (HCSO) Census, 2001; Statistical Yearbooks for counties of Békés, Jász-Nagykun-Szolnok, Vas and Szabolcs-Szatmár-Bereg, 2001 (Statistical Yearbook, 2001a,b,c,d); Hungarian Ministry of Transport and Water, Road Department Main Data on Roads, 2001.)

	Mean		
Characteristics	Dévaványa ($n = 5$)	Örség-Vend ($n = 11$)	Szatmár-Bereg ($n = 6$)
Presence of train station	0.8	0.18	0
Distance to nearest food market (km)	0	19.85	18.35
Distance to nearest food market (minutes by car)	0	20.36	17.83
Number of primary schools	2.4	0.36	0.83
Number of secondary schools	1	0	0
Number of food markets	1	0	0
Population	9928.6	373.36	659
Area (km^2)	21964.6	1636.18	2407
Population density	0.45	0.20	0.28
Regional unemployment rate (%)	12.4	4.8	19.0
Inactive ratio ((persons on pensions or maternity leave)/population)	0.37	0.40	0.48
Dependency ratio ((inactive, children, housewives, students)/population)	0.28	0.22	0.27
Number of shops	140.8	4.18	9.67
Number of enterprises	491.2	21.55	22.83
Regional road network (km)	6118.6	8678	3593
Regional area of total road network (km^2)	5621.2	5936	3337

Note: Road data are reported at the regional level.

The two isolated ESAs are more similar to each other than either is to Dévaványa. Located in the southwest, Örség-Vend has a heterogeneous agricultural landscape with poor soil conditions that render intensive agricultural production methods impossible. Settlements are very limited in area and most are far from towns. Population sizes are small. Of the three, Örség-Vend is the least urbanized with fewest shops and enterprises. The population is elderly and declining in numbers, though the unemployment rate of this region is lowest in the country at 4.8%. Örség-Vend supports the lowest dependency ratio. NAEP encourages extensive production methods to conserve the picturesque landscape, natural and semi-natural habitats in this ESA, which also serve as a tourist attraction.

Szatmár-Bereg is situated in the northeast, far from the economic centre of the country. Settlements in this ESA are also small. The declining, ageing population reflects a lack of public investments in infrastructure and employment generation. Roads are of poor quality and the regional unemployment rate is the highest in the country (19%). Szatmár-Bereg also has a significantly higher ratio of inactive to total population than either of the other two sites. NAEP seeks to promote nature conservation in Szatmár-Bereg by establishing a national park (Juhász et al., 2000; National Labour Centre, 2000; Gyovai, 2002).

Econometric Analysis

Conditional logit model with interactions of settlement characteristics

Conditional logit models with logarithmic and linear specifications were compared using data from all three ESAs. The highest value of the log-likelihood function was found for the specification with the crop variety count in logarithmic form. For the population represented by the sample, indirect utility from home garden attributes takes the form:

$$V_{ij} = \beta + \beta_1 \ln(Z_{\text{cropdiversity}}) \\ + \beta_2(Z_{\text{landrace}}) + \beta_3(Z_{\text{agrodiversity}}) \quad (3.3) \\ + \beta_4(Z_{\text{organic}}) + \beta_5(Z_{\text{self-sufficiency}})$$

The coefficient β refers to the alternative specific constant and β_{1-5} refers to the vector of coefficients associated with the vector of attributes representing agricultural biodiversity in home gardens.

A random parameter logit model was also estimated to allow for parameters with farmer-specific errors, random taste variation and correlation in unobserved factors (McFadden and Train, 2000). The data did not support the model. A Swait–Louviere log-likelihood ratio test resulted in failure to reject the null hypothesis that the estimated random parameter logit model was equivalent statistically to the conditional logit model (Birol, 2004).

In random utility models the effects of social and economic characteristics are not examined in isolation but in the form of interaction terms with the attributes. The number of terms proliferates with additional vectors of explanatory variables. For each settlement level characteristic, the following conditional logit model with interaction terms was estimated:

$$V_{ij} = \beta + \beta_1 \ln(Z_{\text{cropdiversity}}) \\ + \beta_2(Z_{\text{landrace}}) + \beta_3(Z_{\text{agrodiversity}}) \\ + \beta_4(Z_{\text{organic}}) + \beta_5(Z_{\text{self-sufficiency}}) \\ + \delta_1(Z_{\text{cropdiversity}} \times E_{\text{settlementchr}}) \\ + \delta_2(Z_{\text{landrace}} \times E_{\text{settlementchr}}) \\ + \delta_3(Z_{\text{agrodiversity}} \times E_{\text{settlementchr}}) \\ + \delta_4(Z_{\text{organic}} \times E_{\text{settlementchr}}) \\ + \delta_5(Z_{\text{self-sufficiency}} \times E_{\text{settlementchr}})$$

(3.3′)

Equation (3.3) was estimated separately for each settlement characteristic introduced in Table 3.2. Table 3.3 reports the coefficients of the interaction terms between home garden attributes and each settlement characteristic. As economic theory suggests and previous empirical studies have demonstrated, higher levels of physical market infrastructure are negatively related to farmer demand for agricultural biodiversity on farms. The interaction effects of shops and enterprises on the demand for richness of crop varieties are negative. Other settlement characteristics that are related to economic development, such as the number of schools and the density of population, also have an inverse relationship to farmer demand for richness in crop varieties. Consistent with these findings, demand for crop variety richness increases with the distance of the settlement from the nearest food market.

Table 3.3. Effects of the settlement level characteristics on farmer demand for home garden attributes.

Settlement level characteristics	Crop variety richness	Landrace cultivation	Integrated crop and livestock production	Organic production	Self-sufficiency	ρ^2	Log likelihood
Area	-0.87×10^{-6}**	-0.63×10^{-5}**	0.59×10^{-5}*	0.41×10^{-5}	-0.83×10^{-10}**	0.134	−1407.4
Population	-0.19×10^{-5}**	-0.14×10^{-4}**	0.11×10^{-4}*	0.66×10^{-5}	-0.18×10^{-9}**	0.135	−1406.5
Population density	−0.064**	−0.68*	0.52*	0.84**	-0.88×10^{-5}**	0.135	−1405.9
Primary schools	−0.01***	−0.082**	−0.0047	−0.0087	-0.55×10^{-6}	0.133	−1409.2
Secondary schools	−0.018***	−0.11**	0.042	−0.0041	-0.12×10^{-5}*	0.135	−1407.2
Food markets	−0.01	−0.1*	0.19**	0.18**	-0.18×10^{-5}**	0.134	−1407.9
Enterprises	-0.44×10^{-4}***	-0.23×10^{-3}*	0.69×10^{-4}	-0.36×10^{-4}	-0.29×10^{-8}**	0.136	−1405.0
Shops	-0.15×10^{-3}**	-0.84×10^{-3}*	0.46×10^{-3}	0.56×10^{-4}	-0.10×10^{-7}**	0.135	−1406.7
Train station	−0.0058	−0.1*	0.13*	0.1	−0.18**	0.132	−1412.0
Distance (km)	0.5×10^{-3}*	0.0054*	−0.0046	−0.0044	0.78×10^{-7}*	0.132	−1405.0
Distance (min)	0.47×10^{-3}	0.0061**	−0.004	−0.0047	0.67×10^{-7}*	0.131	−1412.7
Unemployment rate	-0.51×10^{-3}	0.01	0.055***	0.031**	0.24×10^{-6}*	0.134	−1407.9

*Statistically significant with one-tailed test (a priori hypothesis) at 10% level; **at 5% level; ***at 1% level.
Note: N = 1487.

Similarly, the more densely populated the settlement and the greater the number of schools, enterprises and shops, the less farmers demand landraces in their home gardens. Existence of a train station or food market in the settlement is also negatively correlated with demand for landraces in home gardens. As hypothesized, distance from the nearest market is positively related to the demand for landraces.

The unemployment rate in the settlement is positively related to farmer demand for agrodiversity and organic production. Both of these home garden attributes contribute to labour-intensive production methods and would more likely be undertaken where opportunity costs for employment are low. Demand for organic production increases with denser settlement populations and more food markets per settlement, perhaps reflecting its luxury good nature and cost in terms of output forgone. Proximity to markets implies that households have lower fixed transaction costs for either sale or purchase.

The demand for self-sufficiency in food consumption is greater the more distant the settlements from the nearest market town, since higher transaction costs induce farmers to rely on home-produced goods. Conversely, the more urbanized the settlements and the greater the numbers of shops, markets and train stations, the less produce farmers demand from their home gardens. Demand for self-sufficiency in food consumption increases with the unemployment rate of the settlement. With no cash income through employment, farm families depend more on the food they produce.

Settlement development index and factor analysis

Including all interactions of the settlement level characteristics with five home garden attributes in the conditional logit estimation results in multicollinearity problems (Breffle and Morey, 2000). To overcome this constraint, four indices were constructed from 14 settlement level characteristics. The first index is a settlement development index (SDI) that is similar to the human development index (HDI) used by the United Nations (UNDP, 2003). Each settlement was assigned a score for each characteristic. The settlement with the highest value or the characteristic was awarded a score of 100 and others were ranked proportionately in descending order. The SDI was then calculated for each settlement by averaging over the characteristics indices. According to this index, Gyomaendrőd settlement in

Dévaványa ESA is the most developed settlement, while Kerkáskápolna settlement in Örség-Vend region is the least.

The results of the conditional logit regression estimating the demand for home gardens, including all interactions between the SDI and home garden attributes, can be seen in the first column of Table 3.4. Significant interactions are evident between farmer demand for crop biodiversity (crop variety diversity and landraces) and the SDI, and between the level of self-sufficiency attained through home garden production and the SDI. All coefficients have negative signs. The demand for crop biodiversity, as expressed by the richness of crop varieties and presence of landraces, declines as the local market economy develops. Reliance on home gardens for food also declines.

The three indices in the other columns of Table 3.4 include an urbanization index (URI), a food market index (FMI) and a population density index (PDI), each constructed using factor analysis. Factor analysis collapses the number of variables, classifying them according to their correlations and structure. Though common in social statistics, the approach has been used only recently to assess heterogeneity in stated preference methods (Boxall and Adamowicz, 1999; Kontoleon, 2003; Nunes and Schokkaert, 2003). Indices created through factor analysis are used as independent variables and are interacted with farmers' demand for home garden attributes.

The first factor, labelled 'urbanization', consisted of the number of secondary schools, shops and enterprises in the settlement, area and population. The second factor, named 'food market', was composed of the distance to the nearest market and the presence of food markets in the settlements. The final factor, called 'population density', included the number of train stations and population density.

The second column in Table 3.4 reports the coefficients on interaction terms for urbanization (URI). Significant interactions are apparent between settlement urbanization and farmer demand for crop biodiversity (crop variety richness and landrace cultivation) and between urbanization and the level of self-sufficiency demanded from the home garden. These findings reinforce those observed for the SDI.

Estimated coefficients on interactions between the FMI and the demand for home garden attributes are reported in the third column of Table 3.4. The significant interactions between the FMI and home garden attributes are the same as those observed for the SDI and URI. The more food markets a settlement has, the less the households in that settlement depend on their home gardens for food and the fewer the crop varieties and landraces they seek.

The interactions between the demand for home garden attributes and the PDI are presented in the last column of Table 3.4. The results indicate that reliance on home gardens for

Table 3.4. Interactions between settlement development index (SDI), urbanization index (URI), food market index (FMI), population density index (PDI) and demand for home garden attributes.

Variable	Index = SDI	Index = URI	Index = FMI	Index = PDI
Constant	-0.81^{***}	-0.77^{***}	-0.56^{**}	-0.75^{***}
Crop variety richness	0.30^{***}	0.27^{***}	0.103	0.24^{***}
Landrace cultivation	0.23^{***}	0.22^{***}	0.14^{**}	0.24^{***}
Integrated crop and livestock production	6.34^{***}	0.36^{***}	0.45^{***}	0.31^{***}
Organic production	0.15^{**}	0.17^{***}	0.25^{***}	0.12^{**}
Self-sufficiency	$0.81 \times 10^{-5***}$	$0.78 \times 10^{-5***}$	$0.63 \times 10^{-5***}$	$0.83 \times 10^{-5***}$
Crop variety diversity × Index	$-0.32 \times 10^{-3**}$	$-0.26 \times 10^{-5***}$	$-0.99 \times 10^{-3*}$	-0.02
Landrace × Index	$-0.21 \times 10^{-2*}$	$-0.18 \times 10^{-4**}$	-0.01^{*}	-0.21^{**}
Agrodiversity × Index	0.17×10^{-2}	0.14×10^{-4}	0.96×10^{-2}	0.25^{*}
Organic methods × Index	0.14×10^{-2}	0.85×10^{-5}	0.93×10^{-2}	0.23^{*}
Self-sufficiency × Index	$-0.31 \times 10^{-7**}$	$-0.23 \times 10^{-9**}$	$-0.15 \times 10^{-6**}$	$-0.35 \times 10^{-5**}$
ρ^2	0.135	0.135	0.132	0.133
Log likelihood	-1407.088	-1406.43	-1411.95	-1410.40

*Statistically significant with one-tailed test (a priori hypothesis) at 10% level; **at 5% level; *** at 1% level.
Note: $N = 1487$

household food consumption as well as farmers' demand for landraces decreases with the density of the settlement population. However, the interactions of population with agrodiversity (crop and livestock production) and organic production are positive, underscoring the notion that these forms of production are relatively labour-intensive. Organic production also exhibits some luxury good properties.

Conclusions

The application of a stated preference method in rural Hungary confirms the predictions of economic theory and the empirical evidence from analysis of revealed preferences in a number of other countries with much lower national income levels. As the settlements in which farmers reside develop and the physical infrastructure of their markets becomes denser, they rely less on their home-produced goods for food and the agricultural biodiversity they seek to maintain on their farms diminishes.

In the ESAs studied, farmers residing in the most isolated and economically marginalized settlements also value the agricultural biodiversity and food produced in their home gardens most. Combined, the findings presented in this chapter, Chapter 8, and Birol *et al.* (2004) confirm that farmers in these settlements both demand and maintain agricultural biodiversity. As long as this is the case, the opportunity costs to these farmers of sustaining current levels of agricultural biodiversity are nil.

Implications

Presently, these farming communities are clearly the least cost options for any public programmes or incentive mechanisms aimed at sustaining current levels of agricultural biodiversity in Hungary. As Dyer points out in Chapter 2, the opportunity costs and the private values estimated here will change with economic change. Major changes in markets and incomes are expected to occur in Hungary as a consequence of economic transition and EU membership (Fischler, 2003). Market infrastructure in Hungary has expanded rapidly since transition to the market economy began in 1989. Infrastructure development and new employment opportunities proposed in SAPARD (Weingarten *et al.*, 2004) are expected to augment farmers' access to markets, reducing the dependence of farm families on their gardens for household food consumption and diet diversity.

On the other hand, economic development typically progresses unevenly, and transition to market economy has so far resulted in growing income disparities and rising domestic prices (Wyzan, 1996; OECD, 2002). The already marginalized localities studied here may become even more so. Certain goals related to social equity might be suitably addressed through integrating traditional Hungarian home garden management practices into national conservation programmes in selected sites, with selected farmers. One feasible, publicly financed mechanism is the NAEP of Hungary, which is recently integrated into the NRDP. The agri-environmental measures proposed by these policies and programmes are already underway in the ESAs where this research was conducted. Market-based mechanisms and ecotourism options may also be tractable, although these approaches are not necessarily less costly. The willingness of consumers *without* home gardens to pay for home garden attributes must also be assessed before specific policy recommendations can be formulated. The institutional analysis presented in Chapter 15 also underscores the complexity of the issues and range of stakeholders involved.

Acknowledgements

This chapter is developed from the PhD research undertaken by the first author at University College London. This research is part of the Hungarian On-farm Conservation of Agricultural Biodiversity Project, which is led by the Institute for Agrobotany, Tápiószele, in partnership with the Institute of Environmental and Landscape Management, Szent István University, Gödöllő and the International Plant Genetic Resources Institute (IPGRI), Rome, Italy. We gratefully acknowledge the funds from the BIOECON project supported by the European Commission under the Fifth Framework Programme. The authors thank Györgyi Bela, Ágnes Gyovai, László Holly, Phoebe Koundouri, István Már, György Pataki, David Pearce, László Podmaniczky, Timothy Swanson, Eric Van Dusen and Mitsuyasu Yabe for useful comments and suggestions.

References

Adamowicz, W.L. and Boxall, P. (2001) Future directions of stated choice methods for environment valuation. Paper prepared for the Workshop on Choice Experiments: A New Approach to Environmental Valuation, 10th April, 2001, London.

Adamowicz, W.L., Louviere, J. and Williams, M. (1994) Combining stated and revealed preference methods for valuing environmental amenities. *Journal of Environmental Economics and Management* 26, 271–292.

Adamowicz, W.L., Swait, J., Boxall, P., Louviere, J. and Williams, M. (1997) Perceptions versus objective measures of environmental quality in combined revealed and stated preference models of environmental valuation. *Journal of Environmental Economics and Management* 32 (1), 65–84.

Bateman, I.J., Carson, R.T., Day, B., Hanemann, W.M., Hanley, N., Hett, T., Jones-Lee, M., Loomes, G., Mourato, S., Ozdemiroglu, E., Pearce, D.W., Sugden, R. and Swanson, S. (2003) *Guidelines for the Use of Stated Preference Techniques for the Valuation of Preferences for Non-market Goods*. Edward Elgar, Cheltenham, UK.

Bela, G., Pataki, G., Smale, M. and Hajdú, M. (2003) Conserving genetic resources on smallholder farms in Hungary: institutional analysis. Paper presented at the BIOECON International Conference on 'Economic Analysis of Policies for Biodiversity Conservation', 28–29 August 2003, Venice, Italy.

Birol, E. (2004) Valuing agricultural biodiversity on home gardens in Hungary: an application of stated and revealed preference methods. PhD thesis, University College London, University of London, London.

Birol, E., Smale, M. and Gyovai, Á. (2004) Agri-environmental policies in a transitional economy: the value of agricultural biodiversity in Hungarian home gardens. *Environment and Production Technology Division Discussion Paper* No. 117, International Food Policy Research Institute (IFPRI), Washington, DC.

Boxall, P.C. and Adamowicz, W.L. (1999) Understanding heterogeneous preferences in random utility models: the use of latent class analysis. *University of Alberta Staff Paper 99–02*, Department of Rural Economy, University of Alberta, Alberta, Canada.

Breffle, W.S. and Morey, E.R. (2000) Investigating preference heterogeneity in a repeated discrete-choice recreation demand model of Atlantic salmon fishing. *Marine Resource Economics* 15, 1–20.

Brookfield, H. (2001) *Exploring Agrodiversity*. Columbia University Press, New York.

Brookfield, H. and Stocking, M. (1999) Agrodiversity: definition, description and design. *Global Environmental Change* 9, 77–80.

Brookfield, H., Padoch, C., Parsons, H. and Stocking, M. (2002) *Cultivating Biodiversity: Understanding, Analysing and Using Agricultural Diversity*. The United Nations University, ITDG Publishing, London.

Brush, S., Taylor, E. and Bellon, M. (1992) Technology adoption and biological diversity in Andean potato agriculture. *Journal of Development Economics* 39(2), 365–387.

Commission of the European Communities (2002) Enlargement and Agriculture: Successfully Integrating the New Member States into the CAP. *Issues Paper*, Brussels, 30.1.2002 SEC (2002) 95 Final. Available at: http://europa.eu.int/comm/enlargement/docs/ financialpackage/sec2002–95_en.pdf

Csizmadia, G. (2004) Analysis of small farm useful soil nutritive matter from Dévaványa, Örség-Vend and Szatmár-Bereg regions of Hungary. *Institute of Agrobotany Working Paper*, Tápiószele, Hungary.

Fafchamps, M. (1992) Cash crop production, food price volatility and rural market integration in the third world. *American Journal of Agricultural Economics* 74(1), 90–99.

Feick, L.F., Higie, R.A. and Price, L.L. (1993) *Consumer Search and Decision Problems in Transitional Economy: Hungary 1989–1992*. Marketing Science Institute, Report No. 93–113, Cambridge, Massachusetts.

Fischler, F. (2003) Opportunities in an enlarged Europe. Address presented at the 16th European Industry Strategy Symposium in Villach, 10 February 2003.

Gauchan, D. (2004) Conserving crop genetic resources on-farm: the case of rice in Nepal. PhD thesis, University of Birmingham, UK.

Goeschl, T. and Swanson, T. (1998) Analysing the Relationship between Development and Diversity: The Case of Crop Genetic Resources. *Mimeo*, Faculty of Economics, University of Cambridge, Cambridge, UK. Available at: http://www.landecon.cam.ac.uk/cre/timo/devdiv.pdf

Greene, W.H. (1997) *Econometric Analysis*, 3rd edn. Prentice-Hall, Upper Saddle River, New Jersey.

Gyovai, Á. (2002) Site and sample selection for analysis of crop diversity on Hungarian small farms. In: Smale, M., Már, I. and Jarvis, D.I. (eds) *The*

Economics of Conserving Agricultural Biodiversity On-Farm: Research Methods Developed from IPGRI's Global Project 'Strengthening the Scientific Basis of In Situ Conservation of Agricultural Biodiversity'. International Plant Genetic Resources Institute, Rome, Italy.

Hanley, N., Wright, R.E. and Adamowicz, W.L. (1998) Using choice experiments to value the environment. *Environmental and Resource Economics* 11(3–4), 413–428.

Hungarian Central Statistical Office (1996) Microcensus of the households, 1996. Budapest. Available at: http://www.ksh.hu/pls/ksh/docs/index_eng.html

Hungarian Central Statistical Office (2001) T-STAR Database System of Settlement Statistics (In Hungarian: Település Statisztikai Adatbázisrendszer). Budapest. Available at: http://www.ksh.hu/pls/ksh/docs/index_eng.html

Hungarian Central Statistical Office (2002) Statistical Yearbook, 2001. Budapest. Available at: http://www.ksh.hu/pls/ksh/docs/index_eng.html

Hungarian Central Statistical Office (2004) Hungarian Council of Shopping Centers: Shopping centers, hypermarkets, 2003. Budapest. Available at: http://www.ksh.hu/pls/ksh/docs/index_eng.html

Hungarian Ministry of Transport, Communication and Water Management, Department of Public Roads (2004) Main data on public roads from December 2001, Budapest.

IPGRI (2003) Home Gardens and the *In Situ* Conservation of Plant Genetic Resources. Available at: http://www.ipgri.cgiar.org/system/page.asp?frame=publications/indexpub.htm

Juhász, I., Ángyán, J., Fesus, I., Podmaniczky, L., Tar, F. and Madarassy, A. (2000) *National Agri-Environment Programme: For the Support of Environmentally Friendly Agricultural Production Methods Ensuring the Protection of the Nature and the Preservation of the Landscape*. Ministry of Agriculture and Rural Development, Agri-Environmental Studies, Budapest.

Kontoleon, A. (2003) Essays on non-market valuation of environmental resources: policy and technical explorations. PhD thesis, University College London, University of London, London.

Kontoleon, A. and Yabe, M. (2004) Assessing the impacts of alternative 'opt-out' formats in choice experiment studies. *Journal of Agricultural Policy Research* 5, 1–32.

Kovách, I. (1999) Hungary: cooperative farms and household plots. In: Meurs, M. (ed.) *Many Shades of Red: State Policy and Collective Agriculture*. Rowman and Littlefield, Boulder, Colorado.

Lancaster, K. (1966) A new approach to consumer theory. *Journal of Political Economy* 74, 132–157.

Lankoski, J. (ed.) (2000) *Multifunctional Character of Agriculture*. Agricultural Economics Research Institute, Research Reports 241, Finland.

Louviere, J.J. (1988) *Analysing Decision Making: Metric Conjoint Analysis*. Sage, Newbury Park, California.

Louviere, J.J., Hensher, D.A., Swait, J.D. and Adamowicz, W.L. (2000) *Stated Choice Methods: Analysis and Applications*. Cambridge University Press, Cambridge, UK.

Luce, D. (1959) *Individual Choice Behaviour*. John Wiley & Sons, New York.

Lupwayi, N., Rick, W. and Clayton, G. (1997) Zillions of Lives Underground. *APGC Newsletter*.

Lusk, J.L., Roosen, J. and Fox, J.A. (2003) Demand for beef from cattle administered growth hormones or fed genetically modified corn: a comparison of consumers in France, Germany, the United Kingdom and the United States. *American Journal of Agricultural Economics* 85(1), 16–29.

Maddala, G.S. (1999) *Limited Dependent and Qualitative Variables in Econometrics*. Cambridge University Press, Cambridge, UK.

Mäder, P., Fliessbach, A., Dubois, D., Gunst, L., Fried, P. and Niggli, U. (2002) Soil fertility and biodiversity in organic farming. *Science* 296, 1694–1697.

Már, I. (2002) Safeguarding agricultural biodiversity on-farms in Hungary. In: Smale, M., Már, I. and Jarvis, D.I. (eds) *The Economics of Conserving Agricultural Biodiversity On-farm: Research Methods Developed from IPGRI's Global Project 'Strengthening the Scientific Basis of In Situ Conservation of Agricultural Biodiversity'*. International Plant Genetic Resources Institute, Rome, Italy.

Már, I. and Juhász, A. (2002) A tájtermesztésben hasznosítható bab (*Phaseolus vulgaris* L.) egyensúlyi populációk agrobotanikai vizsgálata (*Agrobotanical Analysis of Bean – Phaseolus vulgaris L. – Equilibrium Populations Suitable for Regional Land Cultivation*). Acta Agraria Debreceniensis, Debrecen, Hungary.

McFadden, D. (1974) Conditional logit analysis of qualitative choice behaviour. In: Zarembka, P. (ed.) *Frontiers in Econometrics*. Academic Press, New York.

McFadden, D. and Train, K. (2000) Mixed MNL models for discrete response. *Journal of Applied Econometrics* 15(5), 447–470.

Meng, E.C. (1997) Land allocation decisions and *in situ* conservation of crop genetic resources: the case of wheat landraces in Turkey. PhD dissertation, University of California at Davis, California.

Meurs, M. (2001) *The Evolution of Agrarian Institutions: A Comparative Study of Post-socialist Hungary and Bulgaria*. University of Michigan Press, Ann Arbor, Michigan.

Moschini, G. and Hennessy, D.A. (2000) Regional unemployment rates in Hungary. Uncertainty, risk aversion and risk management for agricultural producers. In: Gardner, B. and Rausser, G

(eds) *Handbook for Agricultural Economics*. Elsevier Science Publishers, Amsterdam, New York.
National Labour Centre (2000) Regional unemployment rates in Hungary. Budapest, Hungary. Available at: http://www.ikm.iif.hu/english/economy/labour.htm.
Nunes, P. and Schokkaert, E. (2003) Identifying the warm glow effect in contingent valuation. *Journal of Environmental Economics and Management* 45, 231–245.
OECD (2002) *Economic Surveys: Hungary*. Organization of Economic Cooperation and Development, Paris.
Romstad, E., Vatn, A., Rørstad, P.K. and Søyland, V. (2000) *Multifunctional Agriculture: Implications for Policy Design*. Department of Economics and Social Sciences, Report No. 21, Agricultural University of Norway, Norway.
Roumasset, J.A., Boussard, J.-M. and Singh, I. (1979) *Risk, Uncertainty, and Agricultural Development*. Agricultural Development Council, New York.
Sadoulet, E. and de Janvry, A. (1995) *Quantitative Development Policy Analysis*. The Johns Hopkins University Press, Baltimore, Maryland.
Scarpa, R., Drucker, A., Anderson, S., Ferraes-Ehuan, N., Gomez, V., Risopatron, C.R. and Rubio-Leonel, O. (2003a) Valuing animal genetic resources in peasant economies: the case of the Box. Keken Creole pig in Yucatan. *Ecological Economics* 45(3), 427–443.
Scarpa, R., Kristjanson, P., Drucker, A., Radeny, M., Ruto, E.S.K. and Rege, J.E.O. (2003b) Valuing indigenous cattle breeds in Kenya: an empirical comparison of stated and revealed preference value estimates. *Ecological Economics* 45(3), 409–426.
Seeth, H.T., Chachnov, S., Surinov, A. and von Braun, J. (1998) Russian poverty: muddling through economic transition with garden plots. *World Development* 26(9), 1611–1623.
Smale, M., Bellon, R.M. and Aguirre Gomez, J.A. (2001) Maize diversity, variety attributes, and farmers' choices in southeastern Guanajuato, Mexico. *Economic Development and Cultural Change* 50(1), 201–225.
Statistical Yearbook (2001a) *Békés County, HCSO, Management of Békés County*. Béksécsaba, 2002.
Statistical Yearbook (2001b) *Jász-Nagykun-Szolnok County, HCSO, Management of Jász-Nagykun-Szolnok County*. Jász-Nagykun-Szolnok, 2002.
Statistical Yearbook (2001c) *Vas County, HCSO, Management of Vas County*. Szombathely, 2002.
Statistical Yearbook (2001d) *Szabolcs-Szatmár-Bereg County, HCSO, Management of Szabolcs-Szatmár-Bereg County*. Nyíregyháza, 2002.
Swain, N. (2000) Post-socialist rural economy and society in the CEECs: the socio-economic contest for SAPARD and EU enlargement. Paper presented at the International Conference: European Rural Policy at the Crossroads, 29 June–1 July, The Arkleton Centre for Rural Development Research, King's College, University of Aberdeen, UK.
Szelényi, I. (ed.) (1998) *Privatising the Land*. Routledge, New York.
Szép, K. (2000) The chance of agricultural work in the competition for time: case of household plots in Hungary. *Society and Economy in Central and Eastern Europe* 22(4), 95–106.
UNDP (2003) *Millennium Development Goals: A Compact Among Nations to End Human Poverty*. Human Development Report 2003. Available at: http://hdr.undp.org/reports/global/2003/
Vajda. L. (2003) *The View from Central and Eastern Europe*. Agricultural outlook forum 2003, USDA. Available at: http://www.usda.gov/oce/waob/oc2003/speeches/vajda.pdf
Van Dusen, M.E. (2000) *In situ* conservation of crop genetic resources in the Mexican *milpa* system. PhD dissertation, University of California at Davis, California.
Weingarten, P., Baum, S., Frohberg, K., Hartmann, M. and Matthews, A. (2004) The future of rural areas in the CEE new member states. Report prepared by the Network of Independent Agricultural Experts in the CEE Candidate Countries. Available at: http://europa.eu.int/comm/agriculture/publi/reports/ccrurdev/text_en.pdf
WHO (2000) The Impact of Food and Nutrition on Public Health: The Case for a Food and Nutrition Policy and an Action Plan for the European Region of WHO, 2000–2005 and the Draft Urban Food and Nutrition Action Plan. Available at: http://www.hospitalitywales.demon.co.uk/nyfaweb/fap4fnp/fap_26.htm http://www.hospitalitywales.demon.co.uk/nyfaweb/urban/urb_02.htm
Wyzan, M. (1996) Increased inequality, poverty accompany economic transition. *Transition* 2(20), 24–27.

4 An Attribute-based Index of Coffee Diversity and Implications for On-farm Conservation in Ethiopia

E. Wale* and J. Mburu

Abstract

This chapter develops an attribute-based diversity index for coffee types managed by farmers in Ethiopia, the centre of origin and diversity of *Coffea arabica*. The index counts the attributes farmers state are important when selecting and differentiating among types. Farmers do not use names to describe their coffee types, other than distinguishing between those introduced from outside ('Project') and those maintained locally. The theoretical frameworks of a random utility model and a characteristics model are used to relate the diversity index to determinants of coffee diversity, including household, farm and market-related factors. A Poisson regression model is estimated using the household survey data collected from 266 coffee growing farmers in South-western Ethiopia. Data support the hypotheses that market access, credit experience, labour and land endowments, the importance of coffee in farm production relative to other crops and factors related to household vulnerability significantly influence farmers' demand for multiple coffee attributes. The analysis indicates how attribute preferences are likely to change with development-oriented interventions.

Introduction

Many ethno-botanical studies assert that coffee originated from a particular region in Ethiopia (Keffa) where it was given the name 'Kaffa' and the trees were called '*Kafa*'. Ethiopian coffee contains a great genetic variation and serves as the world's major source of germplasm (Sylvain, 1958). In addition, Ethiopia is the only region where *C. arabica* is found as a wild forest species (Berthaud and Charrier, 1988; Worede, 1988).

Though it is hard to prove, *C. arabica* is said to have originated in Ethiopia and as a consequence, its genetic diversity in the country is one of the highest in the world. Ethiopia holds 6% of the world's coffee *ex situ* collection (FAO, 1998; Hawkes *et al.*, 2000). Ethiopian coffee genetic materials are represented in world collections with numerous samples exchanged among gene banks and breeders (Worede, 1988).

Deforestation of ecosystems has led to the disappearance of coffee genetic resources (Dubale and Teketay, 1999). Causes of declining coffee biodiversity in semi-forest and garden coffee under farmers' management include declining prices on the international market and the vulnerability of local (indigenous) coffee trees to plant diseases and drought (Worede, 1988; Dubale and Teketay, 1999; Wale, 2004). Conservation activities must be initiated if this trend is to be reversed.

---

*Corresponding author: Assistant Professor, Department of Agricultural Economics, Alemaya University, PO Box 138, Dire Dawa, Ethiopia; E-mail: *edilegnaw@yahoo.com*; Tel. 251-5-112374; Mobile 251-9-663598; Fax 251-5-115230

Despite the immense role of *C. arabica* in the Ethiopian economy, however, most conservation efforts have proceeded no further than a proposal stage (Dubale and Teketay, 1999; Gole *et al.*, 2002). Still, coffee is the crop that has been given greatest priority by conservationists in Ethiopia, and assigned global importance (Worede, 1985). Coffee seeds can be conserved *in situ* (either through on-farm conservation for garden coffee or in the natural forest for forest coffee) or *ex situ* (either in the field gene bank or in the cold room using *in vitro* facilities). *Ex situ* conservation of coffee is problematic due to the nature of its seed, which has cost implications (Ellis *et al.*, 1990; Dulloo *et al.*, 1999).

Finding the appropriate index of crop diversity has been a methodological challenge in the studies of factors affecting crop diversity for community-based conservation. A number of diversity indices have been employed in the applied agricultural economics literature (Meng *et al.*, 1998), including those developed by Weitzman (1992), Solow *et al.* (1993) and those adapted from spatial indices in the ecology literature (Magurran, 1988). Meng *et al.* (2003) developed an index based on predicting membership in morphology-based groups defined by clustering plant characteristics obtained from experimental trials. Other chapters in this book use spatial indices based on farmer-managed 'units of diversity', as these are defined by the farming system and crop (see Chapter 1).

Farmers generally have a variety concept. In many cases this concept is reflected in a name and corresponding set of traits or attributes that the farmers use to distinguish one variety from another. In some cases, farmers do not name varieties and instead refer to the crop name, the name of the farmer who manages it or an attribute that it possesses. The last case best describes the practices of the coffee growers surveyed in Ethiopia. To depart from other chapters in this book where farmers named varieties, the word 'type' is used in this chapter rather than variety, and it implies an attribute.

Depicting farmers' own classification of coffee types, the diversity index used in this chapter is a count of attributes farmers consider to be important when they select local coffee plants. Since attributes are expressions of genes in plant types and it is genetic expression that guides farmers' choices, the total number of crop attributes demanded is a diversity index that proxies for crop genetic diversity of coffee as it is managed on farms. The index is linked to microeconomic theory through the application of attribute preference analysis to household survey data. The advantages of the index are that it can be used in an economic model of farmer decision making and reflects farmers' knowledge in contexts where farmers distinguish plants by traits.

This chapter focuses on semi-forest and garden coffee diversity under farmers' management. The following sections describe the study context and explain how Ethiopian coffee farmers consider variation in their crop and classify plant types. Next, the attribute-based index is proposed and related to microeconomic theory and characteristics models. The theoretical framework for analysing farmer choice is then presented, linking farm household needs, demand for coffee attributes and on-farm management of coffee diversity. The econometric model follows with a description of variables used in the econometric analysis and hypotheses. Following that, regression results are discussed. The last sections draw conclusions about the index used in this chapter and policy implications based on the econometric findings.

Coffee in Ethiopia

Coffee is the single most important export commodity in Ethiopia, and Ethiopia is known as the oldest coffee exporting country in the world. According to the Coffee and Tea Authority (CTA), Ethiopia is today the third largest exporter in Africa (next to Ivory Coast and Uganda). During the country's history, invasions and other conflicts have at times had a negative impact on the country's coffee exports (CTA, 1999). Of the major economic importance to the nation, coffee contributes about 10% of the gross domestic product (GDP), 12% of the agricultural output, 70% of the country's foreign exchange earning, 10% of the government revenue and an estimated 25% of the livelihood of Ethiopian population (CTA, 1999).

In Ethiopia, coffee is found from 550 m (hot humid plain of Gambela) to 2550 metres above sea level (masl) (cool climate of Yeju Wollo mountains). The range in elevation of production is believed to reflect wide genetic variability that is compatible with different climatic conditions. The altitude in which the crop is grown for high economic return lies between 1200 and 2000 masl. Major coffee soil types in Ethiopia are (in descending order): Nitosol, Acrisol, Luvisol, Vertisol and Lithosol.

Coffee is mainly grown in five regions: (i) Sidamo; (ii) Keffa; (iii) Wellega; (iv) Illubabor; and (v) Hararghe. The first three account for more than 70% of the coffee grown (Dercon and Ayalew, 1995). Coffee in Ethiopia is grown in three major production systems: (i) forest (about 5–10%); (ii) semi-forest (about 20%); and (iii) garden (about 70%). Most of the highly suitable coffee growing areas in Ethiopia are located in the South West regions. Sidamo, which presently provides a quarter of the total production, has only 0.6% of the highly suitable, 3.5% moderately suitable and 8.2% marginally suitable areas. Hararghe has nil of the highly suitable, 11.2% moderately suitable and 6.4% marginally suitable areas. Coffee production in this region declines from time to time due to moisture stress, susceptibility to coffee berry disease or because it is replaced by the highly remunerative production of Chat (*Cata edulis*). Despite these constraints, Hararghe coffee is grouped as one of the best quality (dry processed) coffees fetching the highest price in the world market (twice that of those grown in other parts of the country).

The diversity of the Ethiopian coffee genetic stock has enabled it to withstand emerging production problems. The outbreak of coffee berry disease in 1971 did not lead to the abandonment of coffee production in Ethiopia (Gebre-Egziabher, 1990). As a result, coffee production has not been as susceptible to hazards like leaf rust, which wiped out plantations in Sri Lanka during the 19th century (Dubale and Teketay, 1999).

Smallholders account for the largest share of production. In Ethiopia, coffee is not exclusively a 'rich man's crop'. Nor do most coffee growing farmers specialize in growing only coffee. The few commercial coffee enterprises were nationalized in the mid-1970s.

Given these potentials and constraints, the objective of coffee breeding in Ethiopia is to create new or improved varieties for smallholder growers. The desirable features sought by breeders include: adaptability to agroecological regions, especially to hot and dry regions facing frequent drought; high yield potential; resistance to pests and diseases (coffee berry disease and coffee leaf rust); bigger seed size; and improved organoleptic qualities (bean size, taste, colour, flavour, etc.). The most important method of coffee breeding (for *C. arabica*) is selection.

Data Design

The data were collected from eight Peasant Associations (PAs) in Mana (Harro and Kella Guda PAs), Goma (Bulbulo, Kilole Kirkir and Yachi Urechi PAs) and Seka Chekorsa (Gibe Boso, Hallo Sebeka and Sebeka Debiye PAs) districts of Jima Zone, South-western Ethiopia. A total of 266 farmers (on average 33 per PA) were sampled using stratified random sampling.

Jima administrative zone, South-western Ethiopia, was selected purposively based on the relative importance of coffee, as were the three districts. Within districts, PAs were also purposively selected based on both the importance of coffee and representation of agroecological conditions.

In consultation with Agricultural Bureaux and development agents working with farmers, the survey sample was structured to cover a representative sample of villages and farmers with respect to a wide range of agroecological and economic variables. Variables included: prevalence of drought; rainfall distribution pattern; distance from markets; distance from extension and input supply services; relative importance of different income sources; and prevalence of poverty. In consultation with the respective development agents, the most important sources of heterogeneity among households were identified as land size, poverty status, sex of the household-head, income source outside agriculture and household size. Households were stratified by groups according to rosters kept in each PA, and individual farm households were sampled randomly in proportion to stratum representation. All farmers in the sample are growing local coffee varieties. The data were collected through personal interviews with a structured questionnaire.

Coffee Choice Decisions

Ethiopian smallholder farmers use coffee attributes when deciding which trees to maintain or replace. They distinguish among coffee types based on their attributes, and refer to them according to the attributes they have observed during their lifelong experience in growing the crop. Among farmers surveyed, most stated that they have maintained their local coffee trees for more than 20 years (average 25.4). Farmers' estimates ranged from a single year to over a century, with trees handed over through generations. Local coffee trees are inheritable, and interviews suggest that they have heritage value.

Coffee trees take over 10 years to reach their maximum production level and during these years the annual yields vary considerably. Farmers' incentive to replace indigenous trees by the more uniform improved seedlings is mainly a function of their capacity to survive until the coffee is ready for harvest. Farmers do not generally replace all their coffee trees at once, and tree replacement takes place over a longer time horizon. Removing old coffee trees and replacing them by new ones is done progressively to minimize the effect on cash income and household well-being. As a consequence, farmers' choices over coffee types and their decisions to replace or change planting material are not as flexible as they are for annual crops. Compared to an annual crop, switching to new coffee seedlings incurs costs in terms of a long time lag until production commences (Wale and Virchow, 2003). *C. arabica*, the most prevalent species of coffee found in most farmers' fields in Ethiopia, is self-pollinated. Smallholder farmers propagate coffee either from seedlings (especially improved coffee types) or by vegetative cuttings.

Farmers grow coffee principally for cash, though they also consume it. The cash generated from growing coffee is allocated for many purposes, such as clothing, school fees, tax payments, health expenditures, *eddir*[1] contributions, construction of shelter, purchasing agricultural inputs or buying other agricultural goods. As a cash crop, coffee is produced mainly for the international market and price decline is a major concern for most coffee farmers. Price collapse matters more for the decision to switch from coffee farming to other enterprises and less for the choice of one coffee type versus another. The market discriminates little among types.

Farmers recognize very well that all coffee types have different capacities to produce the attributes they care about, and they want to grow the types that best fit their household needs and resource endowments. When there is a more diverse choice set of coffee types available to them, it is reasonable to expect that farmers have greater chances of obtaining the types that fit their requirements and situation.

To gain insight into the relative importance of coffee attributes, farmers were asked to rank their importance in use and replacement decisions. Attributes ranked most highly were agronomic or production-related traits, such as yield potential, disease resistance, yield stability and environmental adaptability. This makes sense given that coffee is primarily a cash crop. Marketability was ranked lower, consistent with the observation that the market signals for types are not strong. Fertilizer response is less highly ranked because not many farmers apply fertilizer. The drinking quality of the coffee has the lowest rank, since coffee, though consumed by farm households, is not a food staple (Fig. 4.1).

Farmers were also presented with two scenarios related to yield potential and stability. In the first scenario, a single coffee type has a yield potential of 12 quintals/ha in a good season and 6 quintals/ha in a bad season. In the second, there are three types with an overall yield potential of 10 quintals/ha in a good season and 8 quintals/ha in a bad season. From these two options, 82% chose the second, seeking a lower maximum yield but less damaging harvest in a bad season. This finding suggests that farmers are prepared to sacrifice yield potential in order to be certain that the yields are not too low in a bad season, depending on the relative frequency of good and bad years.

Farmers do use a name for local coffee types (*Begeja*) as a category, as distinct from improved types (*Project*). Of the coffee growers surveyed, 173 have planted both local and improved coffee trees,

---

[1] An indigenous institution established by the local community to discharge different social and economic responsibilities, e.g. labour sharing, rural finance, funeral and other self-help institutional arrangements.

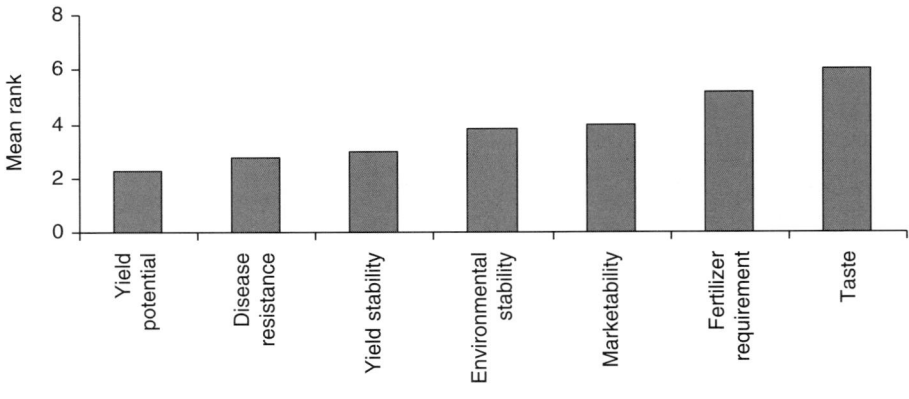

Fig. 4.1. Ranking coffee attributes in farmers' use and replacement decisions. Adapted from 2001/2002 original survey data. Note: 1, highest score; 7, lowest score. Number of respondents ranges from 128 to 257.

86 had only local coffee trees and only 7 grew exclusively improved types. One-fourth of farmers had purchased improved seedlings regularly and 45% had purchased them only once. Thus, 70% of sampled farmers have grown improved coffee trees at least once during their lifetime.

When farmers had to choose between improved and local seedlings given equal chances, 73% opted for improved coffee trees mainly because of disease resistance and higher yields. Slightly under one-fifth (19%) opted for indigenous coffee trees, citing less intensive management and better adaptation to the local environmental conditions as their major reasons. The remainder (8%) of farmers expressed interest in growing both local and improved varieties.

Figure 4.2 shows how farmers ranked indigenous coffee trees compared to improved types. On the whole, the yield potential and disease resistance of the local coffee trees were

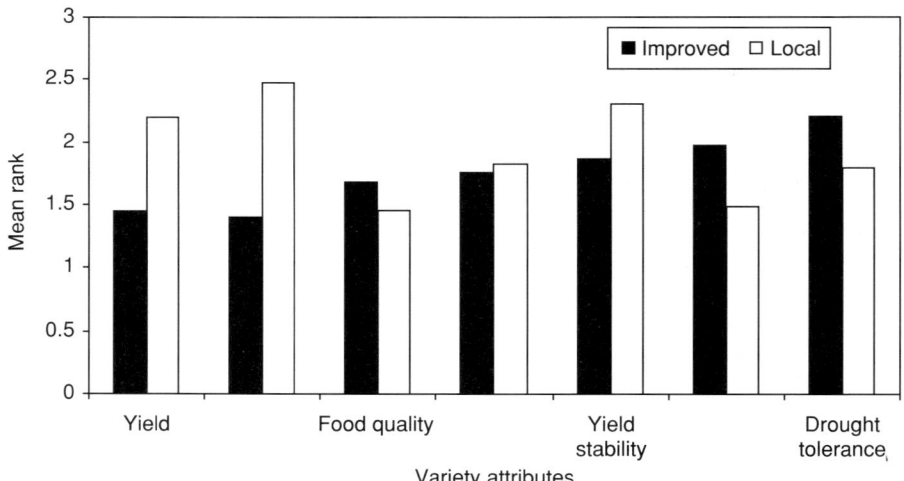

Fig. 4.2. Farmer comparisons of improved varieties and local coffee types. Adapted from 2001/2002 survey data. Note: 1, good; 2, medium; and 3, bad.

ranked lower than for improved types, but their yield potential without fertilizer, tolerance of drought and drinking quality were ranked higher. This suggests that coffee breeders have succeeded in developing better genetic resistance to biotic stress among improved types, but that local plants may be more tolerant of abiotic stresses. The superior rank of improved types in terms of yield stability probably also reflects disease resistance.

Farmers noted anecdotally that unlike indigenous coffee trees the improved coffee trees need more intensive care. The foremost motivation for planting improved trees appears to be expectations of higher yield potential, rapid growth (early maturity) and longevity of production. Most farmers say that they keep growing local coffee because it requires less labour-intensive management (such as weeding and cultivation) and the trees produce something under any weather conditions.

Conceptual Approach

Characteristics models

The conceptual approach in this chapter draws from the Lancaster theory of consumer choice and characteristics models (Lancaster, 1966; Ratchford, 1975; Ladd and Suvannunt, 1976). According to these models, products are differentiated according to their attributes. Consumers demand products because of the utility their attributes provide, rather than the products themselves.

Recently, characteristics models have attracted renewed interest not only in microeconomic theory of diversity (e.g. Nehring and Puppe, 2002) but also in applications concerning farmer preferences for crop varieties and land allocation decisions (e.g. Smale *et al.*, 2001). Adoption models have included technology characteristics (e.g. Adesina and Zinnah, 1993; Adesina and Baidu-Forson, 1995) and attribute-based utility models have been applied to farmers' seed preferences (Baidu-Forson *et al.*, 1997). The models presented in Chapters 3 and 7 draw explicitly from the Lancaster theory of consumer choice and characteristics models. Chapter 14 refers to variety traits or attributes in terms of 'genetic services'.

Farmers have multiple household needs and no single variety is likely to satisfy all of them (Bellon, 1996; Brush, 2000). Ethiopian coffee farmers are consumers of seeds as inputs, and also consume some of the harvested coffee. Due to their heterogeneity in terms of objectives, endowments and constraints, different farmers (like different consumers) derive different levels of utility from the same production attributes (Lancaster, 1966; Hendler, 1975). For instance, a farmer constrained by land shortage may favour a coffee type with less demand for shade trees but another farmer residing in a water stress area may favour a different type that is more productive with shade trees. As a result, different farmers will demand different combinations of attributes based on their household needs.

Coffee attributes and utility are indirectly linked through the production process in characteristics models. Farmers' demand for coffee types can be considered as a derived demand. Subject to the exogenous factors affecting utility, farmers' demand for a variety is determined by the attributes it embodies and the importance of these attributes in addressing the goals of the farm household (Smale and Bellon, 1999). If Z^i designates the vector of attributes, the presence or absence of a certain desirable attribute in the farmers' choice set depends on its importance. In the case of smallholder coffee farmers in Ethiopia, varieties are not named and plant types, other than the categories recognized as local and improved, are distinguished according to attributes.

Let X^1 to X^n denote the vectors of household consumption goods (own produced and purchased) considered in conventional utility functions and $V_{FV_1}^{1}$ to $V_{FV_n}^{n}$ denote the vectors of local coffee attributes consumed by the respective households. Then, the utilities of the 'n' respective households can be written as:

$$U^1(X^1 + V_{FV_1}^1) \text{ to } U^n(X^n + V_{FV_n}^n) \qquad (4.1)$$

Maximizing utility subject to the budget, technology and farm physical constraints yields an 'optimum' levels of attributes and associated levels of production from each coffee type, implying an allocation of land among respective types. The number of local coffee types maintained on-farm and the count of attributes

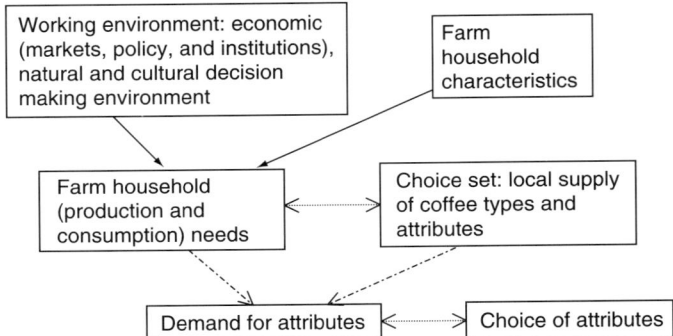

Fig. 4.3. Farmers' working environment, needs and demand for multiple attributes.

embedded in those types are all outcomes of the production process. Household, farm, income and price vectors determine overall demand for local coffee attributes. The count of attributes represents the observable range of variation over which coffee farmers make choices, and these choices in turn determine the level of coffee genetic diversity on their farms.

As presented in Fig. 4.3, the demand of farm households for coffee types is influenced by contextual features such as the policy environment, the economic environment and the natural environment where they farm. In a subsistence-oriented farm economy, as the number of household needs increases, the criteria used to make crop variety choices increase (Teshome *et al.*, 1999). The supply of variety (type) attributes from which farmers choose is constrained by the crop genetic diversity that is available to them in the locality (Smale *et al.*, 2001). The number of attributes in the choice set of any individual farmer represents a subset of the total supply in the region. Farmers' demand for attributes, variety (type) choice and management decisions in turn determines the level of local crop diversity. The wider the range of attributes demanded, the better will be the chance of survival of a large number of varieties (types) of the crop.

An attribute count index of diversity

There are many shortcomings in using variety count as a diversity index (Meng *et al.*, 1998). First, farmers often give different names to the same variety (Wale, 2004) so that the count is overstated. Second, any count index treats each unit as contributing equally to crop diversity. With a variety count a single variety with three desirable attributes would count for less than two varieties, each with only one desirable attribute. Depending on the structure of genetic diversity in a crop, a farmer who plants a single heterogeneous variety with many desirable attributes might be maintaining more functional diversity than a farmer who plants several named, but similar, varieties.

Most importantly, the variety concept, and the diversity concept, must be understood in the context of a specific farming system and crop reproductive system. There is overriding reason why an attribute count makes more sense as an index than a variety count in this chapter. The survey data confirm that most of the coffee trees managed by farmers have been inherited from their ancestors. Due to infrequent turnover and longevity of coffee trees, farmers do not name coffee varieties. They group those in their possession into local types (*Begeja*), improved types (*Project*) and within. This contrasts with the case for annual crops such as sorghum, where farmers named as many as five local varieties (Wale, 2004). Coffee farmers describe their trees in terms of attributes.

Given this empirical context, a count of the coffee attributes that farmers state are important to them is used as a diversity index. This index represents farmers' demand for attributes conditioned on all those that are locally available and known to them. The premise of the index is that the greater the number of coffee attributes desired by the farmer, the greater will be the coffee diversity maintained on farms. Based on this premise demand for multiple attributes implies a high level of coffee diversity on farms.

Several features of this index are worthy of note. Of course, actual utilization depends not only on stated preferences but on the actual supply of genetic materials and farmers' access to them. Whether or not the attribute count is related to diversity as measured by other tools, such as plant descriptors or molecular markers, is not known. None the less, the attribute count measures the diversity that matters to farmers and that is related to the performance of the crop. Like any count or richness index, however, all units in the attribute index are weighted equally.

Explaining farmers' choices

The approach that explains farmers' choice of attributes draws not only from Lancaster's theory of consumer choice (Lancaster, 1966), but also from Roy's (1952) safety first model and random utility theory (Thurstone, 1927; McFadden, 1974).

The demand for attribute i (T_i) leads to a choice of coffee types, which, in turn, leads to production combinations. The utility function of the farmer can be represented as a function of the aggregate sum of coffee attributes preferred (T_i), i.e.

$$V(T_i \mid \omega_i) \tag{4.2}$$

where V is conditional on a set of exogenous factors (ω_i). When households both produce and consume their coffee, their preferences over attributes affecting consumption and production will affect utility levels they attain from growing coffee types.

If all attributes are mutually independent, i.e. the consumption of one attribute does not reduce (increase) the utility from the other, Keeney and Raiffa (1977) have shown that the multi-attribute utility function can be treated as additive, i.e.

$$u(y_1,\ldots,y_n) = \sum_{i=1}^{n} k_i u_i(y_i) \tag{4.3}$$

where u and u_i are utility functions scaled from zero to one, the k_i values are scaling constants with $0 < k_i < 1$. Additivity may not hold true for ordinal utility from attributes (e.g. more utility from expected yield potential could mean less utility from yield stability).

According to random utility theory, a farmer plants a given set of coffee types if and only if the utility he/she derives from the attributes contained in that particular set is greater than the utility that he/she would have from other combinations. Having a total of S desirable variety attributes, the probability that choice T is selected from the available choice set S is given by the maximum of the utility from all the possible combinations ($U_{c1}, U_{c2}, \ldots, U_{cn}$) or symbolically:

$$P_S^i(T) = P\left(U_{cT}^i = \max(U_{c1}, U_{c2},\ldots, U_{cn})\right) \tag{4.4}$$

The choices depend on observable and unobservable characteristics of the farmer and the farmers' production environment. The outcome of the choice is random as it is not possible to predict with full certainty the choice that a randomly selected farmer will make.

Farmers demand both production and consumption attributes, although, as reported above, the demand for production attributes is expected to be greater based on the limited role of coffee in the diet and lack of market differentiation by type. The demand for production attributes (like yield potential, disease resistance, yield stability) emanates from the need of farmers to cope with production constraints. The demand for consumption attributes (like food taste, colour and aroma) is derived from the utility that the crop gives farmers as consumers.

Sources of risk and attitudes towards risk also affect demand for coffee attributes. Safety first (Roy, 1952) or survival algorithms are one way of representing farmers' behaviour in the presence of risk. In these approaches, farmers seek to minimize the probability of disastrous outcomes or shocks, or maximize the chances of meeting some minimum consumption requirements. The reaction of farmers to risk associated with crop production or farm income depends on their sensitivity to negative consequences, which, in turn, is highly conditioned by their capacity to meet their survival needs from their internal endowments of wealth or sources of expected non-farm income that do not co-vary.

One of the major farm income sources for the sampled farmers is income from coffee farming. As agriculture is a risky business, the level of future farm income is not known for sure at each period. Each year, demanding multiple coffee attributes is one of the means of stabilizing farm income from coffee sales. Previous research results show that households with a lower value of assets are less able to bear risk and allocate more resources to higher-risk activities (Dercon, 1996). Other factors held constant, to the extent that maintaining coffee trees with multiple attributes reduces risk related to farm income, less wealthy farmers will choose to grow more. The greater the minimum cash requirements, or the higher the number of dependants, the more reliant the farmer will be on planting trees with multiple desirable attributes.

Variable Definitions and Hypotheses

Dependent variable

Farmers' demand for coffee diversity is examined using an attribute count index as a response variable. Coffee diversity for each farm is given by

$$\left(\sum_{i=1}^{m} T_i\right) \quad (4.5)$$

where $T_i = 1$ if the attribute is important to the farmer and zero otherwise and m stands for the maximum number of attributes the village is endowed with. Coffee attributes, which farmers considered to be important, were listed first in group interviews. Based on this list household-heads were asked: 'which of the following attributes are important for your decision to maintain local coffee trees on the farm?' The list included: yield potential; yield stability; very good harvest during a good season; at least some harvest during a bad season; fertilizer responsiveness; tolerance to environmental stress – drought, frost, wind, soil fertility; disease resistance; marketability; early maturity, taste; other (please specify).

As with any count or richness index, no weights were assigned to attributes to reflect their relative importance. All attributes are assumed to be equally important to farmers and to coffee diversity. It is noteworthy that attributes as perceived by farmers are not mutually exclusive in terms of genetic traits. On average, about four desirable attributes for coffee trees were mentioned by respondents, and the count per household ranged from one to seven. Table 4.1 shows the absolute frequency of responses.

Explanatory variables

Explanatory variables and hypotheses are shown in Table 4.2. Farmers' demand for coffee attributes is expected to differ based on market access, human capital and physical resource endowments, risk-related factors and other fixed, village-level effects.

Different environmental stresses prevail among localities, including pests, diseases and drought. The supply of coffee attributes as well as other institutional and market-related factors varies across villages. Since there are numerous interrelated factors, the sign of village-level effects is difficult to predict.

Market access and lower transaction costs simplify farmers' lives and counteract the need for them to be self-sufficient in goods and their attributes (de Janvry et al., 1991). When farmers have access to markets, purchases can substitute for on-farm production of attributes so that farmers demand only the consumption attributes that the market fails to supply – such as preferred taste, colour or aroma. Production specialization

Table 4.1. Frequency of farmers' aggregate demand for coffee attributes. (From 2001/2002 survey data.)

Attribute count	Frequency	Per cent	Cumulative per cent
1	37	13.91	13.91
2	48	18.05	31.95
3	56	21.05	53.01
4	37	13.91	66.92
5	23	8.65	75.56
6	37	13.91	89.47
7	28	10.53	100.00

Table 4.2. Definitions of explanatory variables and expected signs. (From 2001/2002 original survey data.)

Variable	Definition	Mean	SD	Expected sign
Market access				
Time to market	Walking time needed to reach the nearest market (minutes)	43.73	32.48	+
Household and farm resource endowments				
Education	Education level of the household-head (grades)	2.98	3.20	−
Credit experience	Years using credit	3.18	5.06	−
Experience	Experience in farming (years)	22.33	11.9	+
Plots	Number of plots on farm	2.44	1.30	+
Land size	Land size operated by household (Fechasas = approximately 1/4 ha)	6.39	5.57	+
Labour	Number of full time equivalent workers	4.25	1.97	+
Land in coffee	Proportion of land allocated for coffee	0.498	0.28	+
Vulnerability to risk				
Assets	Value of livestock per consumption requirement (Birr per consumption equivalent)	410.24	612.9	−
Cash income	Off-farm plus non-farm income of households (Birr) earned during year preceding survey	491.20	1774.5	−
Survival	Amount of money required to survive (Birr)	4249.5	3728.7	+
Dependent	Number of dependents (non-workers) per household	3.07	2.33	+
Village fixed effects				
Harro	Village dummy (for Harro)	0.14	0.35	+, −
Kelaguda	Village dummy (for Kela Guda)	0.13	0.33	+, −
Kelokiri	Village dummy (for Kilole Kirkir)	0.13	0.33	+, −
Gibeboso	Village dummy (for Gibe Boso)	0.11	0.31	+, −
Halosebe	Village dummy (for Halo Sebeka)	0.15	0.35	+, −
Sebekdeb	Village dummy (for Sebeka Debiye)	0.14	0.34	−

Note: Villages Bulbulo and Yachi Urechi are omitted category. Consumption requirements are based on the equivalence scale for different group members (NRS, 1989).

may be encouraged if markets favour one coffee type or attribute over another.

The importance of coffee in farm production, expressed by the proportion of land allocated to coffee, and the total farm size are hypothesized to increase the demand for coffee attributes. The number of plots on a farm is often associated with the number of microenvironments. If coffee types respond differently to production environments, and these responses are considered as attributes, having a larger number of plots could induce a demand for more attributes.

The labour and human capital endowments of the household affect the range of coffee attributes that can be maintained on the farm. Human capital is measured in terms of the educational level, farming experience and credit experience of the household-head. These factors are thought to lower the demand for coffee attributes since they lead to other employment or income-generating opportunities. The endowment of labour was computed by subtracting the number of children below nine and inactive household members because of age or permanent sickness from the total number of household members. Household members aged 16–59 were assigned a value of 1. Children aged 9–15 were assigned a value of 0.40, and household members above the age of 60 were given a value of 0.60. That is, children attending school are assumed to spend 40% of their time on household production. Labour conversions were made based on the discussions held with key informants during the survey. Labour endowments are expected to enable the cultivation of a broader range of coffee types and their related attributes.

Here, the amount of money required for household survival, the value of livestock per household consumption requirement, off-farm plus non-farm income per household and the total number of dependents (non-workers) per household are the variables used as proxies for the ability to cope with vulnerability. Based on the conceptual approach and theoretical principles, it is hypothesized that the attribute-based index: (i) decreases with the value of livestock per consumption requirement; (ii) decreases with the increase in off-farm and non-farm income; (iii) increases with the money required for survival; and (iv) increases with the total number of dependents per household.

Econometric Method

The attribute count is an integer index representing the diversity of coffee trees farmers choose to plant from those available to them. Employing the relationship between the repeated binomial and Poisson distributions (Pudney, 1989) the count can be expressed as a Poisson process.

The Poisson regression model is the most commonly used model for count data (Cameron and Trivedi, 1998). In this study, the Poisson regression model is specified as:

$$y_i = e^{x_i'\beta} + \varepsilon_i = e^{(\beta_0 + \beta_1 x_{1i} + \beta_2 x_{2i} + \ldots + \beta_k x_{ki})} + \varepsilon_i \quad (4.6)$$

where y_i refers to the attribute count index and the x_is are the explanatory variables.

One of the basic assumptions of the Poisson regression model is that the variance of dependent variable equals its expected value. Regression-based tests (Cameron and Trivedi, 1990) revealed that the response variable was underdispersed, which can result in spurious rejection of the null hypothesis that a regression coefficient is equal to zero. The easiest way to solve this problem, employed here, is to estimate the Poisson model with the robust (sandwich) covariance matrix (StatCorp, 2001) that remains consistent under the violation of the equi-dispersion assumption (Winkelman, 1995).

Findings

Estimation results are reported in Table 4.3. Predicted attribute diversity at the mean of the regressors is 3.5, which is very close to the average attribute count (3.69). Diagnostic statistics for goodness of fit reveal that the data support a Poisson regression (goodness-of-fit $\chi^2 = 178.4$ and Probability $P > \chi^2 (242) = 0.99$).

A simple way to interpret the regression results is by using the concept of factor change (Long, 1997). For instance, if the walking distance of the household residence from the market increases by 1 min, the expected value of the attribute index increases by 0.31%. If the number of plots on the farm increases by one, the expected number of coffee attributes demanded increases by 5%. Location in Kelaguda reduces the expected number of coffee attributes demanded by a factor of 0.78 (= exp [−0.246]) or decreases it by 22% (0.78−1), holding all other variables constant.

As expected and as shown in other chapters of this book, distance from the market increases the count of attributes demanded by farmers, and by implication, coffee diversity on farms. Households located farther away from the market must satisfy their consumption preferences from their own production.

Human capital variables, including education, experience and credit experience, bear a weak influence on the attribute index. Only credit experience is statistically significant. As farmers' experience with credit increases, their demand for numerous attributes declines, suggesting that they are able to specialize in fewer coffee types or can substitute income earned through other means for income earned from coffee. The importance of coffee in farm production has a large and positive marginal effect on the attribute count, though farm size has none. As hypothesized, the number of plots on the farm bears a positive and significant influence on the number of coffee attributes demanded, perhaps as plots relate to production attributes. Farmers who are capable of meeting the labour requirements for cultivating multiple coffee types also express a demand for more attributes.

Among the vulnerability-related factors, all variables have the expected signs and each is statistically significant except for income (off-farm

Table 4.3. Factors predicting farmers' demand for coffee attributes. (From 2001/2002 original survey data.)

Variable	Coefficient	Marginal effects (Dy/Dx)
Market access		
Time to market	0.0031***	0.0108
Household and farm resource endowments		
Credit experience	−0.0098*	−0.034
Education	−0.004	−0.014
Farming experience	0.00061	0.0021
Plots	0.050**	0.175
Land size	0.0067	0.024
Labour	0.031**	0.110
Land in coffee	0.314***	1.098
Vulnerability to risk		
Assets	−0.00017**	−0.00058
Income	−0.00003	−0.00009
Survival	0.000016**	0.00005
Dependants	0.0346***	0.121
Village fixed effects		
Harro+	−0.376***	−1.155
Kelaguda+	−0.246**	−0.786
Kelokiri+	−0.0003	−0.0011
Gibeboso+	−0.347***	−1.065
Halosebe+	0.0018	0.0061
Sebekdeb+	−0.168	−0.555
Constant	0.745	—
Number of observations = 261		
Wald χ^2 (20) = 163.09 $P > \chi^2$ = 0.00		
Log likelihood = −485.85 Pseudo R^2 = 0.0886		

Note: ***Significant at 1%; **significant at 5%; *significant at 10%.
Note: Dependent variable is attribute count index. Poisson regression is estimated with Huber/White standard errors and covariance. (+)Dy/Dx is for discrete change of dummy variable from 0 to 1.

and non-farm) variable. The number of dependants, in particular, has a marginal effect of large magnitude. Farmers failing to satisfy their household needs due to higher cash requirements for survival or lower value of livestock assets per consumption requirement have more of an incentive to diversify their demand for coffee attributes.

Conclusions

Ethiopia is the recorded centre of origin and diversity for coffee, and coffee is of great economic importance to the nation and as a source of cash for many smallholder farmers. This chapter has proposed an attribute count index as a diversity index for local coffee in Ethiopia. The coffee farmers surveyed have often inherited trees and replace them infrequently. Though they distinguish between local and improved coffee by name (improved coffee is called *Project*), they refer to different local types according to attributes, most of which related to yield. The attribute index is linked to farmer decision making by drawing on theoretical concepts advanced in microeconomic theory, including characteristics models, safety-first models of risk behaviour and random utility. The attribute-based index reflects expression of genes, although there is not a unique correspondence between attributes as farmers describe them and traits based on either single or multiple genes, as these recognized by plant breeders.

An econometric model was then specified to explain farmers' demand for coffee attributes and a Poisson regression was estimated with data from 266 farmers of Jima Zone, South-western Ethiopia. Data support the hypotheses that market access, credit experience, labour and land endowments, the importance of coffee in farm production relative to other crops and factors related to household vulnerability to risk significantly influence farmers' demand for multiple coffee attributes.

Because of their effect on farmers' preferences and relative prices, rural development interventions such as infrastructure development, credit and income-generation programmes will alter the role of coffee in farm income and village economies. These changes will, in turn, affect the relative importance farmers ascribe to different coffee attributes and therefore the types they choose to grow. For instance, if irrigation, better markets and alternative income sources are made available to most small-scale farmers, there may be less demand for attributes like drought resistance, yield stability and environmental adaptability. Coffee trees with these attributes may be substituted with others. As some attributes become less essential to them, farmers will also demand new combinations.

Implications

The role of breeders should be to build a portfolio of improved varieties with diverse desirable attributes that are broadly compatible with farmers' preferences. This avoids the dangers of relying heavily on too few improved varieties and increases the chance that farmers will accept new types. When public funds are limited, breeding should be targeted to the farmers whose objectives and constraints lead them to specialize in coffee types with single desirable characteristics, such as resistance to a particular disease. In the context of this research, the target farmers would be labour constrained but relatively wealthy, with access to markets and credit.

The attribute-based index of coffee diversity relates farmers' demand for crop functions that are important to them with economic factors that are influenced by development interventions and policies, such as markets, credit, education and asset accumulation. In this way, prospects for conserving diverse coffee types can be linked with the design of rural development interventions that are neutral or beneficial for maintaining diversity. The attribute-based index can also be employed to predict which farmers in a community, and which communities, are most or least vulnerable to loss of distinct types that are of functional importance to them. The index has potential applicability to situations where farmers distinguish plants based on attributes.

Acknowledgements

Some of the ideas in this chapter were drawn from the doctoral dissertation of the first author at the University of Bonn, Center for Development Research (ZEF). The advice and support of Professor Holm-Müller (University of Bonn) and Professor Manfred Zeller (University of Göttinger) are gratefully acknowledged.

References

Adesina, A.A. and Baidu-Forson, J. (1995) Farmers' perceptions and adoption of new agricultural technology: evidence from analysis in Burkina Faso and Guinea, West Africa. *Agricultural Economics* 13, 1–9.

Adesina, A.A. and Zinnah, M.E. (1993) Technology characteristics, farmers' perceptions and adoption decisions: a tobit model application in Sierra Leone. *Agricultural Economics* 9, 297–311.

Baidu-Forson, J., Waliyar, F. and Ntare, B.R. (1997) Farmer preferences for socioeconomic and technical interventions in groundnut production system in Niger: conjoint and ordered probit analyses. *Agricultural Systems* 54(4), 463–476.

Bellon, M.R. (1996) The dynamics of crop intraspecific diversity: a conceptual framework at the farmer level. *Economic Botany* 50(1), 26–39.

Berthaud, J. and Charrier, A. (1988) Genetic resources of *Coffea arabica*. In: Clarke, R.J. and Macrae, R. (eds) *Coffee: Agronomy*. Elsevier Applied Sciences, London.

Brush, S.B. (2000) The issues of *in situ* conservation of CGRs. In: Brush, S.B. (ed.) *Genes in the Field: On-Farm Conservation of Crop Diversity*. Lewis Publishers, IDRC and IPGRI, Boca Raton, Ottawa and Rome, pp. 3–26.

Cameron, A.C. and Trivedi, P.K. (1990) Regression-based tests for over-dispersion in the Poisson model. *Journal of Econometrics* 46, 347–364.

Cameron, A.C. and Trivedi, P.K. (1998) *Regression Analysis of Count Data*. Cambridge University Press, Cambridge, UK.

CTA (1999) Ethiopia: cradle of the wonder bean – *Coffea arabica*. Coffee and Tea Authority, Addis Ababa, Ethiopia.

de Janvry, A., Fafchamps, M. and Sadoulet, E. (1991) Peasant household behavior with missing markets: some paradoxes explained. *Economic Journal* 101, 1400–1417.

Dercon, S. (1996) Risk, crop choice, and savings: evidence from Tanzania. *Economic Development and Cultural Change* 45, 485–513.

Dercon, S. and Ayalew, L. (1995) Smuggling and supply response: coffee in Ethiopia. *World Development* 23, 1795–1813.

Dubale, P. and Teketay, D. (1999) The need for forest coffee germplasm conservation in Ethiopia and its significance in the control of coffee diseases. Paper presented at the Coffee Berry Disease Workshop 13–15 August 1999, EARO, Addis Ababa, Ethiopia.

Dulloo, M.E., Guarino, L., Engelmann, F., Maxted, N., Newbury, H.J., Attere, F. and Ford-Lloyd, B.V. (1999) Complementary conservation strategies for the genus *Coffea* with special reference to the Mascarene Islands. *Genetic Resources and Crop Evolution* 45, 565–579.

Ellis, R.H., Hong, T.D., Roberts, E.H. and Tao, K.-L. (1990) Low moisture content limits to relations between seed longevity and moisture. *Annals of Botany* 65, 493–504.

FAO (1998) *The State of the World's Genetic Resources for Food and Agriculture*. FAO, Rome, Italy.

Gebre-Egziabher, T. (1990) The importance of Ethiopian forests in the conservation of Arabica Coffee genepool. *Mitteilungen Institut Allgemeine Botanik* 23a, 65–72.

Gole, T.W., Denich, M., Demel, T. and Vlek, P.L.G. (2002) Human impacts on *Coffea arabica* genetic diversity in Ethiopia and the need for its *in situ* conservation. In: Engels, J., Rao, V.R., Brown, A.H.D. and Jackson, M. (eds) Managing plant genetic diversity. *Proceedings of the International Conference on Science and Technology for Managing Plant Genetic Diversity in the 21st Century*, 12–16 June 2000, pp. 237–247.

Hawkes, J.G., Maxted, N. and Ford-Lloyd, B.V. (2000) *The Ex Situ Conservation of Plant Genetic Resources*. Kluwer Academic Publishers, New York.

Hendler, R. (1975) Lancaster's new approach to consumer demand and its limitations. *American Economic Review* 65, 194–200.

Keeney, R.L. and Raiffa, H. (1977) *Decisions with Multiple Objectives: Preferences and Value Trade-offs*. John Wiley & Sons, New York.

Ladd, G.W. and Suvannunt, V. (1976) A model of consumer goods characteristics. *American Journal of Agricultural Economics* 58, 504–510.

Lancaster, K.J. (1966) A new approach to consumer theory. *Journal of Political Economy* 74, 132–157.

Long, J.S. (1997) *Regression Models for Categorical and Limited Dependent Variables. Advanced Quantitative Techniques in the Social Sciences Vol. 7*. Sage, Newbury Park, California.

Magurran, A. (1988) *Ecological Diversity and its Measurement*. Princeton University Press, Princeton, New Jersey.

McFadden, D.L. (1974) The measurement of urban travel demand. *Journal of Public Economics* 3, 303–328.

Meng, E.C.H., Smale, M., Bellon, M.R. and Grimanelli, D. (1998) Definition and measurements of crop diversity for economic analysis. In: Smale, M. (ed.) *Farmers, Gene Banks and Crop Breeding: Economic Analyses of Diversity in Wheat, Maize, and Rice*. Kluwer Academic Publishers, Boston, Massachusetts, pp. 19–31.

Meng, E.C.H., Smale, M., Rozelle, S.D., Hu, R. and Huang, J. (2003) Wheat genetic diversity in China: measurement and cost. In: Rozelle, S.D. and Sumner, D.A. (eds) *Agricultural Trade and Policy in China: Issues, Analysis and Implications*. Ashgate, Burlington, Vermont, pp. 251–267.

Nehring, K. and Puppe, C. (2002) A theory of diversity. *Econometrica* 70(3), 1155–1198.

NRS (1989) *Recommended Dietary Allowances*. National research service, sub-committee of the 10th edition of the RDAs. National Academy Press, Washington, DC.

Pudney, S. (1989) *Modelling Individual Choice: The Econometrics of Corners, Kinks and Holes*. Basil Blackwell, Oxford, UK.

Ratchford, B.T. (1975) The new economic theory of consumer behaviour: an interpretive essay. *Journal of Consumer Research* 2, 65–75.

Roy, A.D. (1952) Safety first and the holding of assets. *Econometrica* 20, 431–449.

Smale, M. and Bellon, M.R. (1999) A conceptual framework for valuing on-farm genetic resources. In: Wood, D. and Lenne, J.M. (eds) *Agrobiodiversity: Characterization, Utilization and Management*. CAB International, Wallingford, UK, pp. 387–408.

Smale, M., Bellon, M. and Gómez, J.A.A. (2001) Maize diversity, variety attributes and farmers' choices in South-eastern Guanajuato, Mexico. *Economic Development and Cultural Change* 50(1), 201–225.

Solow, A., Polasky S. and Broadus J. (1993) On the measurement of biological diversity. *Journal of Environmental Economics and Management* 24, 60–68.

StatCorp (2001) Stata 7: Users' guide.

Sylvain, P.G. (1958) Ethiopian coffee – its significance to world coffee problems. *Economic Botany* 12, 111–139.

Teshome, A., Fahrig, L., Torrance, J.K., Lambert, J.D.H., Arnason, J.T. and Baum, B. (1999) Maintenance of Sorghum (*Sorghum bicolor*, Poaceae) landrace diversity by farmers' selection in Ethiopia. *Economic Botany* 53(1), 69–78.

Thurstone, L. (1927) A law of comparative judgement. *Psychological Review* 34, 273–286.

Wale, E.Z. (2004) The economics of on-farm conservation of crop diversity in Ethiopia: incentives, attribute preferences and opportunity costs of maintaining local varieties of crops. PhD dissertation, Faculty of Agriculture, University of Bonn, Bonn, Germany.

Wale, E.Z. and Virchow, D. (2003) Crop diversity as the derived outcome of farmers' survival first motives in Ethiopia: what role for on-farm conservation of sorghum genetic resources? Contributed Paper presented at the 25th Conference of the International Agricultural Economists Association held in Durban, South Africa 17–22 August 2003. Available at: http://www.iaae-agecon.org/conferences_paper.asp.

Weitzman, M.L. (1992) On diversity. *Quarterly Journal of Economics* 107(2), 363–405.

Winkelmann, R. (1995) Duration dependence and dispersion in count data models. *Journal of Business and Economic Statistics* 13, 467–474.

Worede, M. (1985) Crop genetic resource activities in Ethiopia. Paper presented at the International Symposium on South East Asian Plant Genetic Resources, 20–24 August, Jakarta.

Worede, M. (1988) Diversity and the genetic resource base. *Ethiopian Journal of Agricultural Sciences* 10(1–2), 39–52.

5 Missing Markets, Migration and Crop Biodiversity in the *Milpa* System of Mexico: A Household-farm Model

M.E. Van Dusen

Abstract

This chapter elaborates a household model to address the way that social and economic factors can affect the process of genetic erosion in the Mexican *milpa* system, counting minor intercropped species along with principal staple crops. The study presented in Chapter 2 was implemented in one of the sites analysed here. A number of models presented in Part III are based on the household model developed in this chapter. The household model joins a farm production function, a household consumption function and market-related constraints on market availability in order to create an inclusive general model. Testable hypotheses are developed from the theoretical model, which are applied in reduced form using a limited dependent variable approach. Migration is one of the most important economic forces in rural Mexico, and the econometric application focuses on how migration can affect diversity through both income and labour market effects. Data indicate that for these farmers, production and consumption decisions cannot be separated. Migration does affect diversity but is differentiated between internal and international migration. Migration within Mexico appears to support crop diversity in the *milpa* system through remittances, while international migration reduces it through displacing household labour. Specifications with household and village-level variables for the migration are compared and offer similar findings.

Introduction

Genetic erosion refers to the loss of crop genetic resources such as the rare genes and gene complexes often found in locally adapted landraces. Researchers have documented genetic erosion in cradle areas of crop domestication where the abandonment by farmers of traditional cultivars accompanied the specialization and intensification of agricultural development (Frankel *et al.*, 1995). However, genetic erosion does not occur solely through direct competition between traditional and improved varieties of the same crop. For example, it can occur among both principal and secondary crops in farming systems based on multiple crops. Secondary crops are of economic as well as biological interest. For example, in the Mexican *milpa* (maize–bean–squash intercrop) system, genetic diversity may be conserved within the principal crop, maize, but also within secondary crops of global importance for genetic resource conservation, including beans, squashes, chillies and tomato. When specialization among, as well as within, species shapes the levels of crop biodiversity maintained on farms, studies focusing on a single species are likely to produce econometrically biased estimates and misleading policy prescriptions.

Rural labour markets can have important effects on the cropping decisions made by farm households, affecting crop biodiversity on farms. Combining farm work with off-farm income, including national and international migration, is an increasingly important feature of rural

© CAB International 2006. *Valuing Crop Biodiversity: On-farm Genetic Resources and Economic Change* (ed. M. Smale)

economies like that of Mexico. Farm labour is lost or displaced when family members take jobs off the farm or in other locations. At the same time, income earned off the farm can be invested in farm production and used to finance home production of goods consumed by the family. In areas of rural Mexico with high levels of crop genetic diversity, the income from regional, national and international migration can have significant impacts on local rural economies. Studying the dynamics of rural environmental and resource conservation issues without taking into account the impacts of migration omits a critical aspect of economic change.

This chapter develops a theoretical model to explain the crop choices of rural households (Van Dusen, 2000; Van Dusen and Taylor, 2005). An econometric method is applied to test the effects of market integration on crop biodiversity in the *milpa* system of Mexico. The specific hypotheses developed in this paper are related to the impacts of migration and off-farm labour on the diversity of the *milpa* system, within and among crops. A Poisson regression is used in order to investigate the determinants of the total number of varieties planted across crops. Original household-farm data from the Sierra Norte de Puebla, Mexico is used to model farmer behaviour regarding *in situ* conservation in a context of multi-dimensional diversity and heterogeneous ecological and market environments.

Conceptual Model

For the purpose of this chapter, crop biodiversity conservation is defined as the cultivation of multiple crops within the *milpa* as well as multiple varieties of each. Diverse and complex multi-cropping systems are part of the ecosystem that generates biodiversity (farmer- or breeder-recognized) in individual crops. The key questions addressed are (i) what variables explain cultivation of the *milpa* as an intercropping system versus the alternative of specializing in single crops? and (ii) which farmers continue to plant minor varieties of individual crops? Minor crop varieties are those planted by few farmers on relatively small areas, and are those assumed most likely to be lost in a process of genetic erosion. The goal of the analytical framework is to identify economic, behavioural or ecological determinants of crop-biodiversity management by farmers on a multi-dimensional level, distinguishing it from previous studies in applied economics, which focused on genetic diversity within individual crops (Brush *et al.*, 1992; Meng *et al.*, 1998; Smale *et al.*, 2001). Previous studies employed models that were to a large extent adaptations of variety adoption models.

The policy focus is the impact of rural labour markets, off-farm labour and migration on resource use. Rural households rely on combining farm activities with off-farm labour and temporary and permanent migration. In rural Mexico, both domestic and international migration to the USA constitute a large percentage of rural incomes. There have been a range of predictions concerning the effects of migration on smallholder agriculture in Mexico: from subsidizing inefficient production practices to allowing households to invest in more productive technologies, or displacing agricultural production through disruptions to local labour markets (*inter alia*, Durand and Massey, 1992; US Commission on Immigration Reform, 1997; Taylor and Yunez, 1999). While there is an active literature to explore the impacts of migration on rural development, this chapter contributes to a new research direction – the impacts of migration on the environment and natural resource management. The natural resource considered here is the biodiversity of cultivated crop plants in the Mexican *milpa* system.

The unit of analysis is the farmer who decides whether or not to plant an additional crop given several potential objectives and constraints. The conceptual analysis yields an empirical model that nests alternative objectives and constraints in order to test competing hypotheses. The model is estimated using limited dependent variable techniques and a set of model specifications. A similar nested modelling approach was utilized by Smale *et al.* (1994) to explore decisions between local and modern varieties of maize.

Model motivation

One rationale for growing multiple varieties is decreasing returns to scale of production in a given crop. An example in the case of the *milpa* is decreasing marginal productivity of land in maize production on a given farm. Decreasing returns to scale imply the existence of some other fixed factor of production, besides land, that must be allocated between crops. Some examples are farmer time, land quality and distance from markets (resulting in increased transport costs). In the context of imperfect labour markets, the household may be limited to using only family labour, alternatively, family and hired labour may be imperfect substitutes.

A fixed endowment of land, land quality or other inputs results in a decreasing marginal value product of labour. If the household can allocate labour to a second crop, e.g. beans, it will do so until the marginal value products of family labour are equated between the two activities at an endogenous 'shadow' family wage. Another example of decreasing returns to scale involves soil heterogeneity and the matching of varieties to soil conditions (e.g. Bellon and Taylor, 1993).

Market imperfections are endemic in rural areas of less developed countries (LDCs). Missing or incomplete markets result from high transaction costs in factor or output markets (Stiglitz, 1989; de Janvry et al., 1991). Figure 5.1 illustrates the effect of a missing market for an output. A production possibility frontier (PPF) represents the technologically efficient production mixes available to a household that allocates scarce resources between crops a and b. If there are markets for both a and b, the household is guided by the (exogenous) market price line WX, the slope of which equals the negative of the ratio of the price of crop a to the price of crop b. In the case illustrated here, optimality with perfect markets implies a corner solution $(0, Q_b^*)$.

If there is a missing market for crop a (including a missing market for a specific quality or trait), all household consumption demand for crop a must be satisfied entirely from own production. The household's subjective valuation of good a is reflected in a shadow price, ρ_a, which is shaped by the household's marginal utility of the good as well as indirectly by the household's production and consumption constraints. The household shifts from using the exogenous prices (P_a, P_b) and producing only crop b to producing at the constrained level Q_b^c corresponding to the point of tangency between

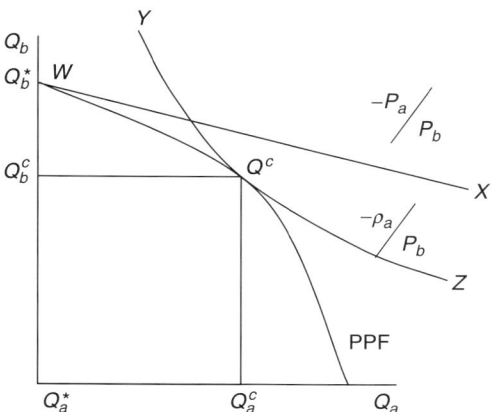

Fig. 5.1. Production possibility frontier (PPF) for two goods with outputs Q_a and Q_b, with a missing market for good a.

the price line YZ (now determined by the exogenous price of good b, P_b and the endogenous price of good a, ρ_a) and the PPF. At this constrained optimum the household produces both the crops (Q_a^c, Q_b^c). The curvature of this price line reflects diminishing marginal utility of household consumption of the non-market crop.

Although it may not appear to be profitable based on regional output prices, the market-constrained household cultivates a certain amount of the crop to satisfy its consumption demand. A simple example for the case of *milpa* is if the household desires to consume a certain amount of home produced maize. While the household may consume mostly purchased maize, it may produce a small quantity of local maize at a level of factor input that reflects an endogenous valuation of maize above the market price.

Missing markets for factor inputs may affect production choices. If hired labour is not available or an imperfect substitute with family labour, the household is limited by its endowment of family labour available for crop production. Households also may be constrained by a lack of liquidity or access to credit to invest in market inputs like land rent, fertilizer or harvest labour. Any of these conditions limit the household's ability to substitute own factors for purchased inputs. In Fig. 5.1 the imperfect substitutability of factor inputs can contribute to the curvature of the PPF. Where a given factor limits the ability of a household to specialize in a single product, this factor will increase the curvature of the possibility frontier and the probability of producing multiple products. A household with a given endowment of land with different qualities will be more likely to plant multiple crop varieties than a household with land of a single quality that can be specialized in a single variety. A household using family labour may be able to efficiently produce maize, beans and squash, while a household using only hired labour may specialize in a single crop.

In the absence of a perfect insurance market, risk or uncertainty may lead the household to plant a portfolio of varieties instead of specializing. In the absence of risk and assuming perfect product markets, the household would specialize completely in crop b. If crop b is characterized by high yield risk, however, the household's valuation of crop b, ρ_b will be endogenous rather than market determined and will reflect the variance of yield and risk preferences. Graphically, as production of crop b increased, the household's subjective valuation of good b would diminish, and the relevant price line would resemble the curved line in Fig. 5.1. In the case of migration and remittances, a household with large components of off-farm income or remittances has less need for crop and variety diversification.

The unobserved household shadow price, ρ_a, transmits information from the household's consumption preferences and market constraints into its decisions about in which activities, and how much, it should participate. A missing market for an output brings the production of a good directly into the household's utility function, so that factors affecting the utility function also affect crop allocations. Factors affecting household risk aversion and exposure to risk, including access to formal or informal insurance, thus may influence land allocation among crops and varieties. The functioning and access to factor markets, and the substitutability of factors of production, also may determine the set of crops that a household grows.

Household-farm model

The household farm is the basic unit of management where decisions and actions are taken that affect crop diversity. It is the consumer who consumes both household production and goods purchased with income from production or wage labour. It is also the producer, combining its labour, land and other capital with purchased inputs to produce agricultural commodities either for consumption or sale to markets subject to resource and market constraints (Singh et al., 1986; de Janvry et al., 1991; Taylor and Adelman, 2002). In the next section, an agricultural household model is presented to identify the effects of key variables on crop and variety choices.

In the basic model the household obtains utility from consuming crops $i = 1, 2, \ldots$ I, any or all of which it may also produce. Let X_i denote consumption of good i and let consumption of all other market goods be denoted by Z. Household utility is affected by exogenous socio-economic, cultural or other characteristics denoted by a vector Φ_{HH}. Households maximize utility subject to a full income constraint, with income composed of farm income, exogenous income $\overline{Y}$ and an endowment of family time T valued at the market wage, w.[1] Households choose which of j crops, $j = 1 \ldots J$ to produce and the output of each crop, Q_j. Farm income is the value of production (at market prices) net of market input costs. Household production is carried out subject to technological constraints embedded in a cost function, $C(\mathbf{Q}; \Phi_{Prod})$, where Φ_{Prod} is a vector of exogenous farm characteristics. Market constraints on production and/or consumption are functions of exogenous characteristics Φ_{Market}.

This model can be represented mathematically as:

$$\max_{X, Q} U(\mathbf{X}, Z; \Phi_{HH}) \qquad (5.1)$$

$$Z = p(\mathbf{Q} - \mathbf{X}) - C(\mathbf{Q}; \Phi_{Prod}) + \overline{Y} + wT \qquad (5.2)$$

$$H_i(\mathbf{Q}, \mathbf{X}; \Phi_{Market}) = 0 \qquad (5.3)$$

---

[1] In a model focusing on migration the household could choose time allocation to different off-farm activities.

Market constraints, represented by the functions H(·), could take many forms. Under certain market conditions reflected in Φ_{Market}, such as high transaction costs, consumption demands must be met from own production; i.e. if the constraint is binding, $Q_i - X_i = 0$. In this case the market constraint would take the form, for $i = 1 \ldots I$,

$$H_i(Q_i, X_i; \Phi_{\text{Market}}) = Q_i - X_i \qquad (5.3')$$

if market i is missing,

$$H_i(Q_i, X_i; \Phi_{\text{Market}}) = 0 \quad \text{otherwise.}$$

As motivated in Fig. 5.1 above, these constraints represent costs of transacting in markets for consumption goods. When the household can supply itself with good i in the market with no transaction costs, constraint H(·) drops out. When transactions costs, characterized by Φ_{Market}, force the household to satisfy consumption from own production the function H(·) binds. Market characteristics Φ_{Market} determine whether a household faces transactions costs for each crop i the household consumes. In the case of migration this means that a required input (family labour) is required (in fixed quantities in a Leontief sense) for the production of certain products (Q_j). When the market restricts the availability of this input, it restricts corresponding outputs and thus the choice of crop activities.

The household chooses a vector of consumption levels, **X**, and output levels, **Q**. Letting λ denote the shadow value of income and γ, a $(1 \times I)$ vector of shadow values γ_i on the market constraints for goods $i = 1, \ldots I$, the Lagrangian corresponding to this general model is:

$$L = U(\mathbf{X}, z; \Phi_{HH}) \\ + \lambda \left[\begin{pmatrix} p(\mathbf{Q} - \mathbf{X}) - C(\mathbf{Q}; \Phi_{\text{Prod}}) \\ + \overline{Y} + wT \end{pmatrix} - z \right] \\ + \gamma(\mathbf{X} - \mathbf{Q}; \Phi_{\text{Market}})$$

(5.4)

The first-order conditions are:

For all consumed goods $X_i \quad U'_{X_i} = \lambda p_i - \gamma_i$ (5.5)

For all produced goods $Q_j \quad C'_{Q_j} = p_j - \dfrac{\gamma_i}{\lambda}$ (5.6)

For all tradable goods $\quad p_i = \overline{p}_i$ and $\gamma_i = 0$ (5.7)

(where $\overline{p}_i$ is an exogenous market price)

For nontradable goods $\quad p_i = \rho_i$ and $X_i - Q_i = 0$ (5.8)

where ρ_i, the unobserved shadow price for good i, is determined by the internal equilibrium of supply and demand for good i:

$$\rho_i = \frac{U'_{X_i}}{U'_z} = \frac{\gamma_i}{\lambda}, \quad \lambda, \gamma_i > 0 \qquad (5.9)$$

As motivated in Fig. 5.1, above, constraint (3) represents transactions costs in obtaining consumption good X_i. When the household can transact for good i in the market without transaction costs, constraint H(·) drops out (i.e. the shadow value on the constraint, γ_i, is 0; note that this model collapses to the standard agricultural household model presented in Singh *et al.* (1986) when this constraint is not binding for all i. However, when transactions costs force the household to satisfy consumption from own production, the constraint is binding. The market characteristics, Φ_{Market}, determine whether a household faces transactions costs for each crop i.

The general solution to the household maximization problem when the constraints bind yields a set of constrained optimum-production levels, Q^c, and consumption levels, X^c:

$$\mathbf{Q} = Q_j^c(p, \Phi_{HH}, \Phi_{\text{Prod}}, \Phi_{\text{Market}}) \qquad (5.10)$$

$$\mathbf{X} = X_i^c(p, Y^c, \Phi_{HH}, \Phi_{\text{Prod}}, \Phi_{\text{Market}}) \qquad (5.11)$$

where Y^c denotes full income associated with the constrained optimal production levels Q^c. For some crops the optimal production level may be 0; therefore, the outcome on Q^c will determine which of the j crops the household chooses to produce.

The crop biodiversity within a given household farm is the result of the choice of which crops and varieties to produce, subject to constraints. This 'diversity outcome' in the constrained case takes the form of a derived demand for number of varieties, $D^c = D(Q_j^c(p, \Phi_{HH}, \Phi_{\text{Prod}}, \Phi_{\text{Market}}))$, resulting from the farmers' utility maximization subject to income, production and market constraints. In the special (and, for most empirical contexts in this book, unrealistic) case of perfect markets the diversity outcome simplifies to the unconstrained $D^* = D(Q_j^*(p, \Phi_{\text{Prod}}))$. The perfect market case is nested within the general agricultural household model of both production and diversity. Note that $D^c < I$ ($D^* < I$) when $Q_j^c = 0$ ($Q_j^* = 0$) for one or more crops.

Testable hypotheses

The starting point for determining diversity outcomes is to test for the separability of effects of consumption and production decisions on diversity. In a separable or recursive model only market prices and production constraints affect production decisions; the effect of other variables on diversity outcomes is insignificant (Singh et al., 1986). The null hypotheses (H1 and H2) to test for market separability are:

$$\frac{\partial D}{\partial \Phi_{HH}} = 0, \quad \frac{\partial D}{\partial \Phi_{Market}} = 0 \quad \text{(H1, H2)}$$

where Φ_{HH} and Φ_{Market} are vectors of exogenous variables other than market prices and production constraints (Benjamin, 1992; Skoufias, 1994).

If we reject the null hypothesis of non-separability, we then consider how the development of local markets affects household cultivation of traditional *milpa* crops and varieties. A missing market for one variety (or for some trait embodied in that variety) can induce a household to produce the variety if the household's subjective valuation of the variety (Eq. (5.9)) is sufficiently high. Given other resource (e.g. land and human capital) constraints, a change in production of one such variety may alter the household's entire set of production choices inside and outside the *milpa*.

Different markets can affect the household's diversity behaviour in different ways. Two key markets of interest are commodity markets for outputs of the *milpa* or consumption substitutes and the local labour market. Commodity market integration may decrease the household's level of diversity by providing the household with access to diversity for consumption or it could increase diversity by enabling the household to supply diverse varieties to the market and receive a price premium. One explanation for the maintenance of crop diversity is the lack of a market for quality of locally produced items. High quality maize may not be marketed, and low quality maize is marketed because its quality is hidden. Personal interviews in the market at Cuetzalan, Puebla, revealed that some merchants reported a local village as a source for maize that actually was imported. Inexpensive, low quality maize was imported both by the government store and by private traders. While farmers are able to market some of their maize through local channels, they are unable to sell large quantities and obtain a price premium for quality. One extreme example of a missing market is the case where households derive utility from growing their own food.

Labour market effects will be tested by extending the approach proposed by Benjamin and Skoufias to crop biodiversity outcomes. Under the null hypothesis of perfect labour markets, family demographic variables, including household size, should not affect crop biodiversity levels on farms. Imperfect labour substitutability has potentially important ramifications for the *milpa*, an intercropping system that is more labour intensive (and less intensive in purchased inputs) than mono-cropping. It also may have ramifications within the *milpa* because of variations in labour intensity among *milpa* crops. Finally, influences of imperfect credit and risk markets will be tested by estimating the effect on crop biodiversity of the level of household wealth, a proxy for risk aversion, exposure to risk and probably also access to liquidity. When a household has off-farm opportunities with a high wage (and possible low income variance), an inability to hire labour may induce a switch into a less labour-intensive cropping system.

The market for migrant labour is differentiated, possibly translating into differential impacts among crops and varieties. Regional and national migration typically involves absence of weeks to months, but international migration is generally undertaken for longer periods of 1 year or longer in this region of Mexico. This provides an opportunity to test specific hypotheses:

$$\frac{\partial D}{\partial Labor_{HH}} = 0,$$
$$\frac{\partial D}{\partial Migration_{National}} = 0, \quad \text{(H3)}$$
$$\frac{\partial D}{\partial Migration_{International}} = 0$$

The relevance for crop biodiversity is that while temporary migrants can return to the village, invest remittances in their own *milpa* production and enjoy consumption of household *milpa* products, international migrants are removed from local production altogether. International migrants are absent from local labour markets and household-farm production.

Survey and Field Site

The data are from a household survey conducted in 1999 in the Sierra Norte de Puebla, a mountainous region of Mexico delimited, and isolated, by two major river valleys. Commerce in this region is dominated by two major regional markets at Cuetzalan and Zacapoaxtla, Puebla, which are connected by a federal highway. The location of the Sierra Norte de Puebla is shown in Chapter 2, Fig. 2.1.

The region is characterized by dependence on two major market towns serving as commercial poles, and is served by one major highway with two branch roads. There is also a basic correspondence to an administrative region in the provision of different levels of government services from the commercial poles. Within the sample area the principal regions are dictated by the topography and climate of the region. *Tierra Caliente* (hot lands) are roughly lower than 1200 metres above sea level (masl), and are characterized by subtropical vegetation and include a lowland transition zone to the coastal plain of Veracruz state. *Tierra Fria* (cold lands) are those above 1200 masl and are characterized by cool, subtropical and mid-altitude climate and a transition towards the higher altitude zone of the high plateau of Puebla state. The principal agronomic and economic differences are that the *Tierra Caliente* lands can grow: (i) coffee, the most important cash crop in the region and (ii) two cycles of maize in a year. The distribution in climate zones is nine villages with a total of 118 households in *Tierra Fria* and 15 villages with 163 households in *Tierra Caliente*.

Maize is the most commercial crop in the Sierra Norte de Puebla. Although most households ultimately are net buyers of maize, many sell small quantities throughout the year, and maize makes up a large share of consumption expenditures. Nevertheless, there is a high volume of low quality substitutes for maize, including imported grain available at government Distribuidora e Impulsora Comercial Conasupo (DICONSA) stores. Beans are semi-commercial, and almost entirely for home consumption, except for a seasonal green market for one variety. Squash is completely non-commercial and produced entirely for home consumption.

Beans and squash are dependent upon the intercropping system for their existence in the *milpa*, and this may make them particularly sensitive to the availability of family labour, the key input to intercropping. The effects of the economic variables can be different for each crop. The reason that crop biodiversity in the *milpa* system may be dependent upon family labour is that these intercropped plants need to be tended, or avoided, while the work to cultivate the principal maize crop is undertaken. When hired labour is used, the goal is usually to work as fast as possible, with a small team of men cultivating and weeding to finish a parcel in a set number of days. The weed pressure is intense in this high rainfall area, and several weeding/cultivations are applied per cycle. Where intercrops are planted the weeding must be done selectively, around the desired intercrops of beans and squash. There are additional delays in the work of *doblar*, where the maize plants are turned down for drying and the bean plants must be handled in a way that allows them to keep growing up the stalks.

The survey was carried out as a part of the McKnight Foundation Collaborative Crop Research *Milpa* Project, composed of a joint Mexico–USA research team of botanists, biologists, crop breeders and social scientists. The author surveyed 281 households in 24 villages in the Sierra Norte de Puebla. The survey sample was structured to cover a representative sample of villages in the study area. The villages were chosen to incorporate a wide range of geographic, agroecological, agronomic, market and cultural diversity. Within each village households were selected at random. Further information on the survey is detailed in Van Dusen (2000). Chapter 2 provides an example of a stated preference analysis applied in one of the study sites analysed here, Zoatecpan.

Data

The unit of analysis for diversity is the farmer designation of how many separate bean, squash or maize types are planted. Defining varieties is particularly problematic in the case of maize. Only two households in the entire survey sample had more than one of the principal white maize types grown in the region, and all households referred to differentiation by colour as their principal distinguishing characteristic. As part of the in-depth qualitative study of the Sierra Norte de Puebla, local maize racial types and variety histories were recorded (presented in Van Dusen, 2000). Mexican maize experts in the research team were able to identify local races contributing to local populations but could not identify unique racial types or differentiate between farmers. Within the sample survey morphological characteristics such as number of rows, cob length and shape of grain were recorded based on respondent assessment. However, these did not provide sufficient variation or correspond to unique varieties. It became clear that populations managed by farmers were combinations of different local racial complexes that could not be differentiated by names or sets of traits.

While originally it was planned to related farmer names for varieties to genetic analyses with seed samples drawn from each household, the cost of genetic analysis was prohibitive and it was unclear what would be used as a metric. At the time of the fieldwork the literature on racial categorization was based on principal components analysis of morphological characteristics, and not on neutral molecular markers. Using neutral molecular markers to assess genetic diversity, Pressoir and Berthaud (2004) found high levels of diversity but low levels of variation between farmers and between villages in the State of Oaxaca. They conclude that the variation within a single farmer's population is bigger than that between farmers (making it difficult to differentiate between farmers) and assert that 'a maize landrace should not be considered as a separate entity, but rather as an open genetic system'. Similarly findings have also been reported by Rice (2004) in the State of Nayarit.

For this study it became more useful to use the definition of farmer population, following Louette's idea of a 'seed lot', and using information from the sample survey on what the farmer considers distinct. This allows us to maintain a linkage between the theoretical model about farmers' behaviour and an observed level of diversity, as compared to a latent level of diversity that is not observed by the farmer.

Richness, or counts of the number of crops and varieties the household plants in the *milpa*, is the diversity concept used at the household level (defined in Chapter 1). Summary statistics for richness indices are shown in Table 5.1. The first variable, total crop varieties, is constructed by summing together the number of varieties of maize, beans and squash. While the maximum number of crop varieties grown per village is ten, the most grown by any household is nine. The average number is far lower (2.4), and the median is lower still, at only two crop varieties per household.

Table 5.1. Summary statistics for dependent variables.

Dependent variables	Mean	SD	Minimum	Maximum
Total varieties	2.41	1.82	0	9
Total maize varieties	1.01	0.73	0	4
Total bean varieties	0.66	0.73	0	3
Total squash varieties	0.74	0.88	0	3

Explanatory variables are classified into three groups following the model specification: (i) household characteristics, Φ_{HH}; (ii) farm production characteristics, Φ_{Prod}; and (iii) market characteristics, Φ_{Market}. Variables are defined and hypotheses summarized in Table 5.2. Descriptive statistics for explanatory variables are shown in Table 5.3.

The set of variables Φ_{HH} describes household characteristics. The variable 'age' is included to test whether older farmers have a higher propensity to maintain crop biodiversity because of traditional practices or taste preferences. A quadratic age term is also included to test whether the very oldest farmers are conserving less. Education is included to reflect possible differences in technology as well as the degree of cultural integration.

Family size, defined as the number of adults living in the household, represents the pool of family labour available to the household for cultivating the *milpa* and other activities. The sign of the effect of family size on diversity is expected to be positive if minor varieties of crops and the intercropped *milpa* system are intensive in family labour and perfect hired substitutes are not available (Benjamin, 1992). The indigenous language variable is between 0 and 1 and is an index of the degree to which family members communicate in

Table 5.2. Definition of explanatory variables and hypotheses.

	Definition of variable	Expected sign
Household characteristics		
Age of household-head	Age of household-head	+
Age squared	Age squared	−
Years at school of household-head	Number of years of education of the household-head	+, −
Family size	Number of family members over 15 years old	+
Indigenous language	Percentage of household members primarily speaking indigenous language	+
Wealth	Index of durable goods and house construction	+, −
Farm characteristics		
Maize (hectares)	Area of land planted to maize	+
Plots	Number of different parcels planted to maize	+
High altitude dummy	Indicator variable for households in villages above 1200 masl	+
Household migration		
Family labour intensity	Percentage of *milpa* labour days performed by family members	+
Remittances from permanent migration	Total cash value of remittances to the household from permanent migrants	−
Remittances from temporary migration	Total cash value of remittances to the household from temporary migrants	+, −
Regional income	Total cash value of income from regional wage work	+, −
Village migration		
Family labour – village level	Family labour intensity averaged at the village level	+
Household-head temporary migrants	Percentage of households in village with temporary migrants	+, −
Household-head permanent migrants	Percentage of households in village with permanent migrants	−
Village USA migration networks	Percentage of households in village with USA migration networks	−
Villlage Mexico migration networks	Percentage of households in village with domestic Mexican migration networks	−

the indigenous language. This is a proxy for cultural values; where the cultivation of a diverse *milpa* is an element of cultural identity more than economic calculation, this variable is expected to be positive.

Farm characteristics, Φ_{Prod}, include variables that are hypothesized to affect production decisions (and thus crop biodiversity) in a separable agricultural household model. The number of plots that the household cultivates is a proxy for uniformity of land holdings and incentives to match varieties to different agroecological conditions (assuming these are different on different plots). A dummy variable for high altitude separates the study region into two major climate zones. Finally the area of *milpa* is included because a greater land area may be expected to lead to greater diversity. The average *milpa* is very small – less than 1 ha – below the Mexican average and what would be sufficient for subsistence.

The set of variables that represents market constraints, Φ_{Market}, is used to test the hypothesis that market integration influences household diversity. Both variables constructed at the household level and those calculated at the village level were used. Although household-specific variables exert a more direct influence on household decisions, village-level variables reflect better the impacts of local labour markets. The village-level variables were calculated

Table 5.3. Summary statistics for explanatory variables.

	Mean	SD	Minimum	Maximum
Household characteristics				
Age of household-head	51	13.68	20	96
Age squared	2821	1465.45	400	9216
Years at school of household-head	3.33	2.84	0	15
Family size	5.14	2.18	1	10
Indigenous language	0.40	0.47	0	1
Wealth	6.86	3.93	0	28
Farm characteristics				
Maize (hectares)	0.76	0.88	0	7
Plots	1.15	0.90	0	4
High altitude dummy	42%	—	0	1
Household migration				
Family labour intensity	0.44	0.362	0	1
Remittances from permanent migration	1201	4198	0	49300
Remittances from temporary migration	536	1550	0	16000
Regional income	3756	10244	0	79200
Village migration				
Family labour – village level	0.44	0.172	0.092	0.788
Household-head temporary migrants	0.09	0.082	0	0.33
Household-head permanent migrants	0.05	0.066	0	0.22
Village USA migration networks	0.08	0.077	0	0.27
Villlage Mexico migration networks	0.58	0.166	0.273	1

as village-level averages from the data in the household survey.[2]

Family labour intensity was used in the construction of both household and village variables, calculated as the percentage of total days of labour in *milpa* production that is supplied by family members. The expected sign on this variable is positive because growing intercrops is labour intensive, as explained above. Family income from migration is divided into remittances from temporary and permanent migration. The effects of either variable could be positive if the household uses the income to invest in *milpa* production; the effects could be negative if the household uses the income to substitute hired labour and/or purchased consumption goods for *milpa* production. A third type of income is earned at salaried jobs in the region while living at home. Similar to remittances, the effects of this variable are difficult to predict a priori. Beginning-of-period wealth is proxied by an index constructed from the number of rooms in the home, the construction materials of the floor, walls and roof of the home and the ownership of major durable goods. The wealth index is considered to be independent of productive assets as well as current year income, and therefore exogenous to crop and variety choices.

The village-wide average of family labour intensity, expressing local labour market conditions, is hypothesized to have a positive effect on crop biodiversity in the *milpa* system. Two other variables are the percentage of household-heads in the village that were permanent or temporary migrants at the time of the survey. Averaged over villages surveyed, these percentages (5%, 10%) are lower than those observed in some other regions of Central Mexico. Two variables representing labour networks are the percentage of households in the village with a close family member living outside of the region elsewhere in Mexico and in the USA. All four of these variables are expected to decrease crop biodiversity by drawing family labour out of *milpa* production.

---

[2] For each household the average is calculated using all of the rest of the households in a village except that one – to avoid the endogeneity of influential data points driving the village results.

Econometric Methods

A Poisson regression was applied because of the discrete, count nature of the dependent variable. This econometric approach can be linked to the theoretical model through a random-utility framework involving a series of discrete decisions to grow or not to grow individual crops and varieties in the *milpa*.

Hellerstein and Mendelsohn (1993) proposed two theoretical linkages between utility theory and a Poisson specification. The first is a demand model for an indivisible good where choice is restricted to be a non-negative integer, which is relevant to a wide range of real consumer choices. The second follows the statistical theory outlined above by modelling a series of discrete consumer decisions that sum across an aggregation of choices to a Poisson distribution. Thus, the Poisson specification is used to model the increase in utility from one additional unit consumed. In this context, the Poisson specification lends the model flexibility to explain the diversity of the *milpa* system aggregated across crops as well as within crops.

The Poisson regression model is non-linear and estimates the effect of independent variables x_i on a scalar dependent variable y_i. The density function for the Poisson regression is:

$f(y_i|x_i) = \dfrac{e^{-\mu_i} \mu_i^{y_i}}{y_i!}$ where the mean parameter is a function of the regressors **x**, and a parameter vector, β

$E(y_i|x_i) = \mu_i = \exp(x'_i \beta)$ and
$y = 0, 1, 2... \quad V(y_i|x_i) = \mu_i(x_i, \beta) = \exp(x'_i \beta)$

Following Cameron and Trivedi (1998), the model was chosen based on appropriate selection criteria and the Negative Binomial regression was fitted in order to test for overdispersion. Blocks of independent variables, corresponding to characteristics Φ in the theoretical model, were run separately in order to test joint significance using likelihood ratio tests.

Findings

Three sets of regression results are presented in Table 5.4. The first regression (Model 1) included only household and agroecological variables, without labour market variables. The second (Model 2) adds household-migration variables and the third (Model 3) replaces these with village-migration variables. This approach does not provide a definitive hypothesis test between models with migration variables measured at the two scales of analysis, but it does allow for comparisons. Due to problems with multicollinearity it was not possible to nest both household and village variables.

The econometric findings confirm that all three groups of variables – household, production and market – shape crop biodiversity in the *milpa* system of the Sierra Norte de Puebla. To test for separability of the model, the joint test for the significance of each group of variables is presented. In Model 2, each group of variables – household, agroecological and market – was found to be jointly significant at below the 0.01% level (chi-squared statistics of 11.1 (5 d.f.), 60.9 (3 d.f.) and 63.4 (5 d.f.), respectively. Similar results were found for Model 3. The separability of the model is rejected, supporting the notion that crop biodiversity in the *milpa* system is influenced by market imperfections.

Two individual tests of hypotheses about household characteristics are of particular interest in the full regressions that include migration variables. First, the coefficient for education of the household-head is positive and significant, contrary to expectations. More years in school does not imply a change of production practices. Second, the coefficient on the indigenous language variable is also positive, revealing that the more a family speaks local dialects – Nahuatl, Otomi – and the less they speak Spanish, the more diverse is their *milpa*. Taken together, these results suggest that crop biodiversity in the *milpa* system may be linked to cultural preferences, but not necessarily to cultural (production) practices.

The larger the maize area cultivated by the household, the greater the level of crop biodiversity maintained in the *milpa*. High altitude also influences crop biodiversity positively. The higher, colder region is more heavily dominated by maize and the *milpa* system. The lower, hotter region has more cattle and coffee and more severe pressures from pests.

In both Models, 2 and 3, family labour intensity has a positive and statistically significant coefficient. In the specification with variables

Table 5.4. Poisson regression explaining total *milpa* (maize, beans and squash) varieties.

	No migration		Migration household		Migration village	
	Coefficient	t-ratio	Coefficient	t-ratio	Coefficient	t-ratio
Constant	0.307435	0.75	0.013	0.03	−0.073	−0.16
Household characteristics						
Age of household-head	0.02	1.53	0.016	1.07	0.022	1.44
Age squared	−0.00024	−1.72*	0.000	−0.85	0.000	−1.34
Years at school of household-head	0.00724	0.53	0.043	2.83***	0.031	2.09**
Family size	0.00	0.17	−0.001	−0.10	0.010	0.69
Indigenous language	0.169	2.44**	0.120	1.68*	0.111	1.51
Wealth	—	—	−0.005	−0.58	−0.014	−1.50
Farm characteristics						
Maize (hectares)	0.199	6.33***	0.207	6.35***	0.212	6.14***
Plots	0.06	1.64	0.047	1.24	0.054	1.42
High altitude dummy	0.43	6.54***	0.338	5.08***	0.402	5.33***
Household migration						
Family labour intensity	—	—	0.696	7.65***	—	—
Remittances permanent migration	—	—	−0.00001	−1.30	—	—
Remittances temporary migration	—	—	0.00003	2.03**	—	—
Income from regional employment	—	—	−0.00001	−2.22**	—	—
Village migration						
Family labour – village	—	—	—	—	1.085	4.61***
Household-head temporary migrants	—	—	—	—	0.643	1.64
Household-head permanent migrants	—	—	—	—	−1.352	−2.41**
Village USA migration networks	—	—	—	—	−1.286	−2.85***
Village Mexico migration networks	—	—	—	—	−0.150	−0.74
Deviance R^2	0.17		0.31		0.25	
Hypothesis tests			LRT	P	LRT	P
Household variables = 0 (d.o.f. = 5)			11.10	0.048**	12.02	0.036**
Agroecological variables = 0 (d.o.f = 3)			60.90	0.000***	56.5	0.000***
Market variables = 0 (d.o.f. = 5)			63.40	0.000***	44.5	0.000***

***Significant at 1%; **Significant at 5%; *Significant at 10%.

measured at the household level, this finding suggests that each household decides on the crops to plant within their maize field based on how much of the labour they can do themselves. In the specification with variables constructed at the village level, the finding indicates that crop biodiversity has decreased in villages where more labour is hired for *milpa* production.

Measured at the household level, remittances from temporary migration positively and significantly influence crop biodiversity levels in the *milpa*. Apparently, the labour lost through temporary migration is more than offset by the positive effect of remittance income that can be invested in *milpa* production. Off-farm income from employment elsewhere in the region has a negative and significant sign associated with investing less time and money in *milpa* crop biodiversity.

Expressed at the scale of the village, the extent of temporary migration has no effect on crop biodiversity in the *milpa*. Higher frequencies of permanent migration in a village, and more extensive membership in USA migrant networks, reduce the biodiversity levels observed in individual *milpa*s within the village. More extensive village membership in networks of USA migrants also has a statistically significant negative impact. As the men of the village leave, crop biodiversity in the *milpa* decreases. Presumably, the importance of maintaining minor crops and varieties, and the labour intensity required to do so, declines.

Conclusions

This chapter contributes a proposed methodology and findings of policy relevance for crop biodiversity in the *milpa* system of Central Mexico. While previous studies have focused on simultaneous cultivation of several varieties of the same staple crop, the empirical context of the Mexican *milpa* required an approach that combines major and minor varieties of the principal crop (maize) with minor crops planted in the maize field. Migration is a major policy issue in Mexico, and the results of this study are mixed. Data from the Sierra Norte de Puebla predict that villages with more active networks of USA migrants and more extensive permanent migration will have lower levels of biodiversity in the *milpa* fields of individual households. Within individual households, however, temporary migration contributes positively to *milpa* biodiversity.

The household-farm model presented here can be adapted to a range of empirical settings where *in situ* conservation of biodiversity in cultivated plants is of social concern. In the separable case of the model, farm physical features and market prices alone determine the pattern of crops and varieties grown. In the non-separable case, viewed in this chapter, household and labour market variables are also statistically significant determinants. The formulation is adapted for activity choice within a single crop and summing across crop varieties – each activity participation choice resulting from the same household processes. Where, as in past studies, the issue motivating the empirical analysis is competition between modern and traditional varieties of the same crop, the model in this chapter can be applied with simplifying assumptions.

Previous studies in applied economics have employed a range of diversity indices borrowed from crop science, genetics and ecology (see Chapter 1; Meng *et al.*, 1998). Although these serve well to characterize the diversity found within crop breeding programmes, *ex situ* collections or wildlife reserves, it is difficult to relate them to the crop and variety units that farmers manage and recognize. The advantage of the Poisson specification is that it maps more directly from the underlying behavioural model into the regression equation, through a series of discrete choices. The disadvantage is that it represents diversity as a count, and like other simple counts, treats all units equally.

Another methodological contribution is the importance of incorporating explanatory variables that vary at different scales of analysis. Here, variables measured at two scales describe local labour markets (village) and labour market participation (households). Crop and variety choices occur at the household level, but household incentives and constraints are circumscribed by the economic, social and environmental contexts in which farmers live. Data collected in a sample survey of households can be used to construct a number of village variables. Alternatively, researchers and conservation agencies may be able to rely on secondary information available at the village level without the costs of collecting additional data from households. However, incorporating village effects into empirical analyses does require a sampling scheme that distributes the sample more widely across villages than within villages, with statistical consequences.

Implications

Rural–urban and cross-border migration are crucial components of economic change throughout the world, specifically in Mexico. In this study, variation in market access across the 24 villages included in the survey made it possible to identify the negative impacts of migration and migration networks on the biodiversity of the *milpa* system in the Sierra Norte de Puebla. Remittances from migrants constitute one of the largest sources of income in some areas, and rural labour markets are depressed by the opportunity costs of migration wages. Here, households with migrants leaving and returning in the same year have higher predicted levels of *milpa* biodiversity, while those located in villages with a high propensity for permanent migration to the USA have lower levels. Diverging signs present a policy challenge. The labour-reducing effect of migration may be offset if remittances are invested in the *milpa*. Income from salaried employment within the region is also associated with lower levels of crop biodiversity in the *milpa*. The relationship between migration and crop biodiversity on farms is indirect. While most of the literature about migration has emphasized the development and acculturation impacts of labour displacement (reviewed in Taylor and Martin, 2000), the analysis in this chapter provides a strong starting point for exploring the linkages between migration and local management of environmental goods.

At present, one overarching question remains in analysing the crop biodiversity on farms in Mexico. How and why do small-scale household farms in Mexico continue to grow maize despite the evidence that it is unprofitable to do so? In Chapter 2, Dyer contends that the non-market benefits of maize are non-trivial. Although cultural aspects are not easily built into econometric models, the findings reported here do predict that indigenous households will maintain more diverse crops and varieties in their *milpas*. The findings from this study indicate that rural Mexican households subsidize non-economic maize and *milpa* production with migrant remittances. However, the impacts of this subsidy depend on the nature of the migration. Results reported in this chapter cannot be generalized beyond the Sierra Norte de Puebla, although the methods and means can easily be applied to other parts of rural Mexico.

Acknowledgements

The material in this chapter is based on a doctoral dissertation (Van Dusen, 2000) but this chapter emphasizes the relationship of diversity in the *milpa* system and a critical policy issue, migration. I would like to thank J. Edward Taylor and Steven Brush for their mentorship and Antonio Yunez, George Dyer and Hugo Perales for their collaboration during the research. Financial support for this research was provided by the McKnight Foundation, UC Mexus and the PRECESAM collaboration between UC Davis and the Colegio de Mexico. Institutional support for the fieldwork was provided by the Genetic Resources Conservation Programme at UC Davis, the Jardin Botanico at UNAM and the economics programme, International Center for Maize and Wheat Improvement (CIMMYT).

References

Bellon, M.R. and Taylor, J.E. (1993) 'Folk' soil taxonomy and the partial adoption of new seed varieties. *Economic Development and Cultural Change* 41(4), 763–786.

Benjamin, D. (1992) Household composition, labour markets, and labour demand: testing for separation in agricultural household models. *Econometrica* 60(2), 287–322.

Brush, S.B., Taylor, J.E. and Bellon, M.R. (1992) Technology adoption and biological diversity in Andean potato agriculture. *Journal of Development Economics* 39(2), 365–387.

Cameron, A.C. and Trivedi, P.K. (1998) *Regression Analysis of Count Data*. Cambridge University Press, Cambridge, UK.

de Janvry, A., Fafchamps, M. and Sadoulet, E. (1991) Peasant household behaviour with missing markets – some paradoxes explained. *Economic Journal* 101(409), 1400–1417.

Durand, J. and Massey, D. (1992) Mexican migration to the US. *Latin American Research Review* 27(2), 3–42.

Frankel, O.H., Brown, A.H.D. and Burdon, J.J. (1995) *The Conservation of Plant Biodiversity*. Cambridge University Press, Cambridge, UK.

Hellerstein, D. and Mendelsohn, R. (1993) A theoretical foundation for count data models. *American Journal of Agricultural Economics* 75(3), 604–611.

Meng, E., Taylor, J.E. and Brush, S.B. (1998) Implications for the conservation of wheat landraces in Turkey from a household model of varietal choice. In: Smale, M. (ed.) *Farmers, Gene Banks and Crop Breeding: Economic Analyses of Diversity in Wheat, Maize, and Rice*. Natural Resource Management and Policy series. Kluwer Academic, Boston, Massachusetts; Dordrecht, The Netherlands; and London, pp. 127–142.

Pressoir, G. and Berthaud, J. (2004) Population structure and strong divergent selection shape phenotypic diversification in maize landraces. *Heredity* 92, 95–101.

Rice, E.B. (2004) Conservation and change: a comparison of *in-situ* and *ex-situ* conservation of Jala maize germplasm in Mexico. PhD thesis, Cornell University, Ithaca, New York.

Singh, I., Squire, L. and Strauss, J. (1986) *Agricultural Household Models: Extensions, Applications, and Policy*. World Bank Research Publication, Johns Hopkins University Press, Baltimore, Maryland.

Skoufias, E. (1994) Using shadow wages to estimate labour supply of agricultural households. *American Journal of Agricultural Economics* 76(2), 215–227.

Smale, M., Just, R.E. and Leathers, H.D. (1994) Land allocation in HYV adoption models – an investigation of alternative explanations. *American Journal of Agricultural Economics* 76(3), 535–546.

Smale, M., Bellon, M.R. and Aguirre Gómez, J.-A. (2001) Maize diversity, variety attributes and farmers' choices in southeastern Guanajuato, Mexico. *Economic Development and Cultural Change* 50, 201–225.

Stiglitz, J.E. (1989) Markets, market failures, and development. *American Economic Review* 79(2), 197–203.

Taylor, J.E. and Adelman, I. (2002) Agricultural household models: genesis, evolution and extensions. *Review of Economics of the Household* 1(1), 33–58.

Taylor, J.E. and Martin, P.L. (2000) Human capital: migration and rural population change. In: Rausser, G. and Gardner, B. (eds) *Handbook of Agricultural Economics*. Elsevier, New York.

Taylor, J.E. and Yunez, A. (1999) *Education, Migration and Productivity: An Analytic Approach and Evidence from Rural Mexico*. Organisation for Economic Co-operation and Development (OECD), Paris.

United States Commission on Immigration Reform (1997) *Mexico–US Binational Migration Study Report*. LB Johnson School of Public Affairs, Austin, Texas.

Van Dusen, M.E. (2000) *In situ* conservation of crop genetic resources in the Mexican milpa system. PhD dissertation. Department of Agricultural and Resource Economics, University of California at Davis, California.

Van Dusen, M.E. and Taylor, J.E. (2005) Missing markets and crop diversity: evidence from Mexico. *Environment and Development Economics* (in press).

6 Explaining the Diversity of Cereal Crops and Varieties Grown on Household Farms in the Highlands of Northern Ethiopia*

S. Benin, M. Smale and J. Pender

Abstract

In this chapter, survey data are used to compare the determinants of intercrop (interspecific) and intracrop (intraspecific) cereal diversity on household farms in the highlands of northern Ethiopia using an approach based on the theoretic framework elaborated in Chapter 5. Physical features of the farm, and household characteristics such as livestock assets and the proportion of adults that are men or women, have large and significant effects on both the diversity among and within cereal crops grown, varying among crops. Demographic aspects such as age of household-head and adult education levels affect only intracrop diversity of cereals. Though there are no apparent trade-offs between policies that would enhance one type of diversity (richness) versus another (evenness), those designed to encourage intracrop diversity in one cereal crop might have the opposite effect on another crop. Trade-offs between factors associated with economic development and crop diversity are not evident in this resource-poor system. For example, market-related variables and population density have ambiguous effects on cereal crop diversity. Education positively influences cereal crop diversity. Growing modern varieties of maize or wheat does not detract from the richness or evenness of these cereals when measured at the level of household farms. The village-level analysis presented in Chapter 11 employs data collected in the same sites.

Introduction

Maintaining genetic variation *in situ* as a complementary strategy to *ex situ* conservation is only one social motivation for policy interest in farmer management of crop genetic resources. The potential to secure harvests in difficult growing environments is of more immediate concern to many social planners. In the less-favoured areas of the world where crop production is risky and opportunities are limited for insuring against it through working off-farm, many farm families still depend directly on the diversity of their crops and varieties for food and fodder. The highlands of northern Ethiopia are an example.

Understanding the trade-offs between promoting economic development and assuring that farmers enjoy the benefits of a range of seed production inputs are fundamental for countries like Ethiopia. Ethiopia is a centre of diversity for cereals such as barley, teff, durum wheat, sorghum and finger millet (Harlan, 1992). Often referred to as one of the eight Vavilovian gene centres of the world, Ethiopia is one of the most environmentally troubled countries: national soil loss rate averages about

---

*This chapter has been developed from a paper presented at the International Association of Agricultural Economists conference in Durban, South Africa in 2003. The original paper was accepted for publication in a special issue of *Agricultural Economics* (2004) that is based on the conference proceedings.

21 t/ha/year (Kebede, 1996); most of the soils show negative nutrient balances (Stoorvogel *et al.*, 1993; Elias *et al.*, 1998); and up to 2% of total crop production is lost annually due to soil erosion alone (Kappel, 1996). Related to the environmental problems are low agricultural productivity, poverty and food insecurity. Crop productivity is low, with cereal yields averaging less than 1 t/ha in many parts of the country. An estimated 31% of the population lives on less than US$1/day while 76% live on less than US$2/day (World Bank, 2001). Data from the household income and consumption expenditure survey and welfare monitoring survey conducted in 1995/96 and 1999/2000 show that the incidence, gap and depth of poverty has declined slightly on average, although the situation is still severe (Woldehanna and Alemu, 2002). The highland areas are particularly affected in terms of both growing environment and market infrastructure, two of the generic factors hypothesized to positively affect the diversity of crop genetic resources. For example, soil erosion averages more than 42 t/ha/year (FAO, 1986; Hurni, 1988). On average, farm households in the highlands of Amhara and Tigray regions lived about a 5-h walk from the nearest all-weather road in 1998/99 (Pender *et al.*, 2001).

Since 1991, the Ethiopian government, within the strategy of conservation-based agricultural development-led industrialization, has undertaken many initiatives and a massive programme of investment towards setting the economy on a path towards sustainable long-term growth (Degefe and Nega, 2000; Degefe *et al.*, 2001). More recently, the government has embarked on a sustainable poverty reduction and rural development strategy to ensure food security, but still recognizing agriculture as the potential source to generate primary surplus to fuel the growth of the entire economy (FDRE, 2001, 2002). Ethiopia has also made a national commitment to conserve genetic resources on farms and in gene banks over the past two decades (Worede *et al.*, 2000). Despite these initiatives and a resultant good progress at the national level (Degefe and Nega, 2000; Degefe *et al.*, 2001) there still exist policy and institutional constraints to sustained development in the Ethiopian highlands. Of primary concern are issues related to land policy and tenure, constrained access to rural credit, limited market development (for input and outputs), limited infrastructure development and access (roads, irrigation, services), the approach and capacity limitations in the research and extension system and lack of coordination between programmes promoting agricultural production and resource conservation (Pender *et al.*, 2001, 2002; Benin *et al.*, 2003).

In Chapter 5, Van Dusen explored both intercrop (interspecific) and intracrop (intraspecific) diversity in the microecosystem of the Mexican *milpa*. Comparing the determinants of inter- and intracrop diversity among the cereals commonly grown on household farms in the highlands of northern Ethiopia, this chapter highlights three types of trade-offs that may occur in the design of instruments or policies to support the development and sustainable management of crop biodiversity. First, the same policies may enhance the numbers or 'richness' of cereals and varieties grown but detract from the 'evenness' of their representation on farms. This is a trade-off in conservation goals. Second, to the extent that the determinants of diversity differ among crops, economic instruments or policies designed to enhance the diversity in one crop may have adverse consequences for the diversity of another crop. These trade-offs are related to conservation goals or criteria. A third trade-off reflects the long-standing preoccupation with the relationship between modern varieties and genetic erosion (Frankel, 1970; Harlan, 1972). A negative relationship between cultivation of modern varieties and crop genetic diversity is typically assumed, although some empirical examples suggest that the relationship is more complex and depends on the production environment (Brush *et al.*, 1992; Zimmerer, 1996).

The conceptual framework for the analysis is presented next. The econometric approach follows, including the data, description of variables and related hypotheses. Findings are then presented followed by conclusions, implications and suggestions for further research.

Conceptual Framework

The conceptual framework is based on Chapter 5, the model of the agricultural household presented by Singh *et al.* (1986), and the literature on partial adoption of agricultural innovations (Feder et al., 1985; Feder and Umali, 1993; Smale et al., 1994).

The model in Chapter 5 treats explicitly the richness or count of crops and varieties but can also be applied to decisions about the extent of cultivation. The household farm maximizes utility over a set of consumption items (C_f) generated on the farm, a set of purchased consumption goods (C_{nf}) and leisure (l). The utility a household derives from various consumption combinations and levels depends on the preferences of its members (Ω_{HH}) shaped by the characteristics of the household, such as the age and education of its members and wealth.

$$\underset{C_f, C_{nf}}{\text{Max}}\ U(C_f, C_{nf}, l; \Omega_{HH})$$

subject to

$$Q = F(\alpha, X, L \mid A, \Omega_F)$$

$$T = H + l$$

$$p_f Q_f - p_x X - wL + Y^0 = p_{nf} C_{nf} + wH$$

Levels of farm produce to be consumed on farm (C_f) or sold ($Q-C_f$) are chosen from a vector Q of farm outputs. Decisions are constrained by a fixed production technology, $F(\cdot)$, that combines purchased inputs (X), labour (L), and an allocation of a fixed land area ($A = A^o$) among m crops and n varieties, given the physical conditions of the farm (Ω_F). Each set of area shares (α_{ij}) among m crops and n varieties sums to 1, $\sum_i^m \sum_j^n \alpha_{ij} = 1$, $i = 1,2, \ldots m$, $j = 1,2,\ldots n$, mapping into the vector Q through physical input–output relationships. The choice of area shares implies a level of farm outputs, and vice versa.[1] The function can then be re-expressed as:

$$\underset{\alpha_{11}, \alpha_{ij},\ldots \alpha_{mn} \geq 0; C_f, C_{nf}, X, L}{\text{Max}}\ V(C_f, C_{nf}, l; \Omega_{HH})$$

Interior solutions may be found for each crop and variety. Choice about the allocation of labour (household or purchased) is constrained by total time (T) available for farm production (H) and leisure. Expenditures of time and money cannot exceed full income. Full income in a single decision making period is composed of the net farm earnings (profits) from crop production and income that is 'exogenous' to the season's crop and variety choices (Y^0), such as stocks carried over, remittances, pensions and other transfers from the previous season. The coefficients p_f, p_{nf} and p_x are the prices associated with farm products, purchased consumption goods and inputs, respectively, and w is the wage rate.

As explained in Chapter 5, a special case of the model is profit maximization. When all relevant markets function perfectly, farm production decisions are made separately from consumption decisions. The household maximizes net farm earnings subject to the technology and expenditure constraints. Farm production decisions or optimum allocations (i.e. crop and variety choices (α^*_{ij}), farm labour (L^*) and other farm inputs (X^*) are driven by net returns, which are determined only by market wage rate, input and output prices (w, p_x and p_f) and farm physical characteristics (Ω_F). A diverse set of crops and varieties is still possible when land quality is heterogeneous and yields depend on land quality (Bellon and Taylor, 1993).

The production and consumption decisions of the household cannot be separated when markets for labour, other inputs or products are imperfect. Then, prices are endogenous to the farm household and affected by the costs of transacting in the markets. For a good that is not traded, no surplus is sold ($Q-C_f = 0$) and the shadow price ρ that governs the choices of the household is determined by the internal equation of supply and demand for the good, expressing the household's valuation of the good. Market constraints on production and/or consumption can be expressed as functions of exogenous market characteristics (Ω_M). The specific characteristics of farm households and markets influence the magnitude of transaction costs involved in market exchanges and, through ρ, the household's choices.

When consumption and production decisions are not separable, the household's optimum

---

[1] Since the focus of this analysis is cereal crop production, livestock production has not been treated explicitly. The size of the livestock herd is assumed fixed for the cropping season, although there is a derived demand for crops and varieties through feed and fodder requirements.

choice $\tilde{h}$ (a vector of $\tilde{\alpha}_{ij}, \tilde{C}_f, \tilde{C}_{nf}, \tilde{X}, \tilde{L}$) can be expressed as a reduced form function of farm size, exogenous income, and household, farm and market characteristics:

$$\tilde{h} = \tilde{h}(A^0, \Upsilon^0, \Omega_{HH}, \Omega_F, \Omega_M) \quad (6.1)$$

Equation (6.1) defines the basis for the econometric estimation to examine the factors affecting diversity of cereal crops on household farms. The biodiversity of cereal crops is an outcome of choices made in a constrained optimization problem rather than an explicit choice. Biodiversity (D) is expressed in the following conceptual form, similar to Chapter 5:

$$D = D(\tilde{\alpha}_{ij}(A^0, \Upsilon^0, \Omega_{HH}, \Omega_F, \Omega_M)) \quad (6.2)$$

These factors are the hypothesized conceptual determinants of cereal crop diversity on household farms. In the next section, the data source and cultivation of various cereal crops and varieties on household farms are described. The dependent and independent variables are also described and hypotheses, as these relate to the literature, are discussed. Finally, the regression structure is summarized.

Data

The detailed data-set used in the analysis is ideal for analysing differences in crop biodiversity on household farms because of the relatively large number of communities sampled and range of conditions represented (as noted in Chapter 5). The variables used in this analysis were constructed from data collected in a sample survey conducted among 934 households in Tigray and Amhara regions of northern Ethiopia between 1998 and 2001 (see Fig. 6.1 for map of the

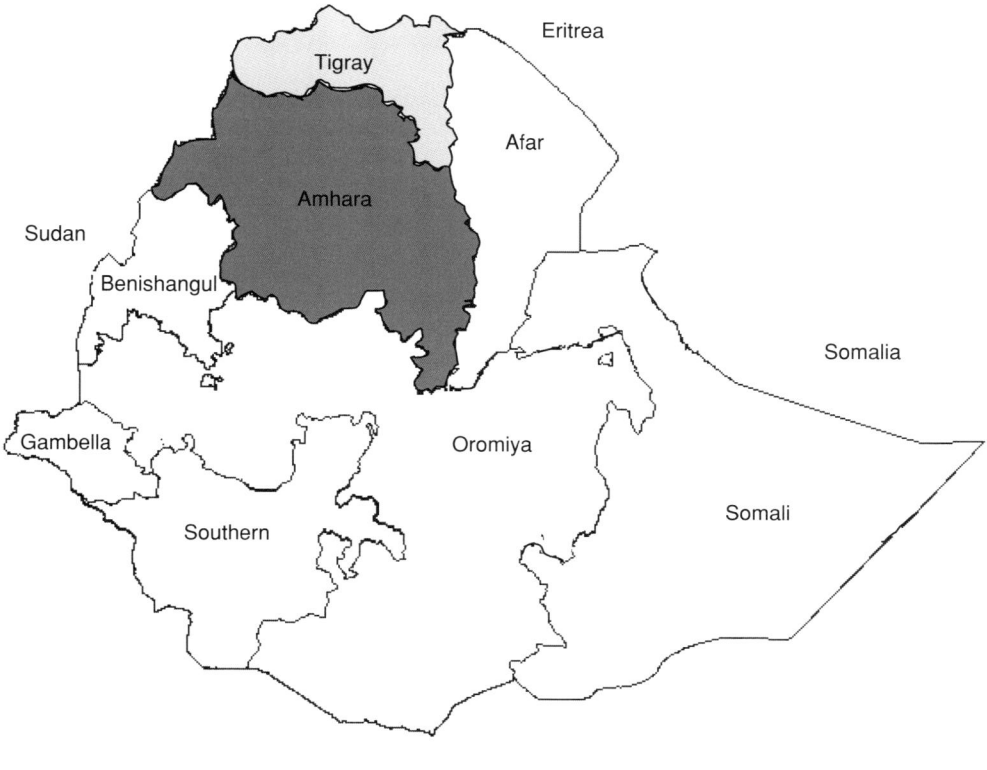

Fig. 6.1. Map of Ethiopia showing survey regions.

regions). First, a stratified random sample of 99 peasant associations (PAs) usually consisting of four or five villages (communities) was selected from highland areas (above 1500 metres above sea level (masl)) of the two regions. The PA is the lowest administrative unit in rural Ethiopia.

Strata were defined according to variables associated with agricultural potential, market access and population density. The stratification scheme was guided by the notion that the strategies for sustainable agricultural development in any given situation depend largely on the comparative advantage of alternative livelihood strategies. In the Ethiopian highlands, and the East African highlands in general, agricultural potential, market access and population pressure are among the primary factors that define comparative advantage (Pender et al., 1999). 'Agricultural potential' represents many factors that influence the absolute advantage of producing agricultural commodities in a particular place, while access to markets is critical for determining the comparative advantage of a particular location, given its agricultural potential. Population pressure affects the land–labour ratio, and may induce innovations in technology, markets and institutions, or investments in infrastructure. Stratification and sampling weights are used in all statistical analyses.

In Amhara Region, secondary data were used to classify the districts (woredas) according to whether the area is drought-prone (following the definition of the Ethiopian Disaster Prevention and Preparedness Commission), access or no access to an all-weather road and the 1994 rural population density greater or less than 100 persons/km^2. The drought-prone districts (48 out of the total 105) are generally located to the eastern part of the region, while the non-drought-prone districts are located to the west and southern tip. Two additional strata were defined for PAs where irrigation projects are found in drought-prone and non-drought-prone areas resulting in a total of ten strata. In each of the ten strata, five PAs were randomly selected (except the irrigated, drought-prone stratum, in which there were only four PAs) for a total of 49 PAs. Then, two villages were randomly selected from each PA for a total of 98 villages. In each village 4–5 households were randomly selected for a total of 434 households. All the plots operated by the selected households were also surveyed.

In Tigray Region, PAs were stratified by whether an irrigation project was present or not, and for those without irrigation, by distance to the woreda town (greater or less than 10 km). Three strata were defined in Tigray, with 54 PAs randomly selected per stratum. PAs closer to towns and in irrigated areas were selected with a higher sampling fraction to assure adequate representation. Four PAs in the northern part of Tigray could not be studied due to the war with Eritrea. From each of the remaining PAs two villages were randomly selected and from each village five households were randomly selected. A total of 50 PAs, 100 villages and 500 households were then surveyed.

In this chapter, only the data from the household (and accompanying plot) surveys have been analysed. After removing observations with missing observations for relevant variables 739 remained. The household data include household composition and assets, access to markets and infrastructure and aspects of crop production during the 1999 season. The survey data were supplemented by secondary geographic and climatic information.

Cultivation of Cereal Crops and Varieties in Survey Sites

In Ethiopia, barley, teff, sorghum and millets are considered as 'old crops', while maize and (bread) wheat are relatively new. Ethiopia is considered the site of origin of teff (*Eragrostis tef*), domesticated before the birth of Christ between 4000 and 1000 BC (Stallknecht, 1997). Teff is a major staple food in Ethiopia and traditionally grown as a cereal crop. The small grains of the teff plant are ground to flour and mainly used for making popular bread resembling a thin pancake or crêpe (*enjera*). It is adapted to diverse agroclimatic conditions. In a few other parts of the world, Southern Africa, India, Australia and South America, teff is grown on a limited basis for livestock forage. In the USA, however, small acreages of teff are grown for grain production and sold to Ethiopian restaurants or utilized as a late planted livestock forage (Stallknecht, 1997). Ethiopia is considered a secondary centre of origin of one of the three physiological races of barley (Fröst, 1974) because barley (*Hordeum vulgare*), although not native to Africa, has been in use in Ethiopia for at least 5000 years (National Research Council, 1996).

Sorghum (*Sorghum bicolor*), known as great millet or guinea corn in West Africa, originated along the Ethiopia–Sudan border (FAO, 1995). Pearl millet, which originated in tropical western Africa, was introduced about 2000 years ago to Ethiopia (and eastern Africa in general), where it has become established because of its tolerance to drought and drier environments (FAO, 1995). Finger millet (*Eleusine coracana*) is believed to have originated in Uganda or a neighbouring region (FAO, 1995). The grains of finger millet are smaller than those of pearl millet, and the shape of the panicle differs. Sorghum and millets are grown in harsh environments where other crops grow or yield poorly.

Ethiopia has been identified by Vavilov as one of the main centres of durum wheat diversity. Bread wheat was introduced in Ethiopia during the 18th century (Rempel, 1997). Although maize is an introduced crop to Ethiopia, the area cultivated has increased substantially over years, mainly due to the greater ability of the crop, compared to others, to respond to use of improved inputs. Maize, which also grows under a wide range of agroclimatic conditions in Ethiopia, was introduced in Ethiopia from the 1600s to 1700s (Haffangel, 1961). The total area of maize cultivated in Ethiopia is now second to teff. In the early 1980s it was fourth after teff, sorghum and barley.

Households cultivated between one and five cereals; 24% cultivated one cereal only, while 40%, 27%, 8% and 1% cultivated two, three, four and five cereals, respectively. Table 6.1 shows the average number of varieties of each crop planted per household as well as the number of households planting each cereal crop, more than one variety of each cereal crop and improved variety of each cereal crop. The greatest number of households (469) cultivated teff followed by barley (352), maize (317), wheat (250), sorghum (110), finger millet (101) and pearl millet (22). The maximum number of varieties of any cereal cultivated by any household was three, with the average number of varieties planted per crop being largest for teff followed by maize, barley, wheat, finger millet, sorghum and then pearl millet. Using the count or number of varieties as a measure of the richness (Chapter 1) of the crop, this finding suggests that greatest diversity per household is maintained for teff. This is followed by maize, barley, wheat, finger millet, sorghum and pearl millet in order of declining richness. Only 52 and 46 households planted a modern variety of wheat and maize, respectively, while a mere 12 households planted a modern variety of teff and only a single household reported a modern variety of barley (Table 6.2). The relationship of growing modern varieties to intracrop diversity was tested only for wheat and maize, since the number of observations was insufficient to estimate the first-stage probit regression for the other crops.

Table 6.1. Number of cereal varieties grown on household farms in the highlands Amhara and Tigray regions, Ethiopia.

	Maize	Wheat	Barley	Teff	Sorghum	Finger millet	Pearl millet
Number of varieties planted							
Mean	0.69	0.54	0.61	0.78	0.30	0.39	0.10
Standard error	0.04	0.04	0.04	0.04	0.04	0.04	0.03
Minimum	0	0	0	0	0	0	0
Maximum	3	3	3	3	2	2	1
Number of households planting							
Cereal	303	250	352	469	110	101	22
More than one variety	30	33	36	62	7	5	0
Improved variety	46	52	1	12	0	0	0
Effective sample size	585	524	638	683	279	253	190

Note: The effective sample size refers to the total number of households in respective communities in which the cereal is cultivated. Data on named varieties of finger and pearl millet were not collected in the Amhara region. Means and standard errors are adjusted for stratification, weighting and clustering of sample.

Table 6.2. Modern and improved varieties of cereals grown by farmers in the highlands of Amhara region, Ethiopia.

Barley	Maize	Sorghum	Teff	Wheat
HB 42	Alemaya compositi	Birmash	Cr-37	BT- B
	Awasa 511	Dinkmash	D2-01-44	Enkoy
	BH 540	Gambella 1107	D2-01-99	ET-13
	BH 660	Is 9302	D2-01-196	HAR 604
	Katumani	P-9401	D2-01-354	HAR 710
	Kuleni	P-9403	D2-01-787	HAR 1685
	PHB 3253	P-9404	D2-01-974	HAR 1709
	Mert	Seriedo	DZ-194	K-6290 Bulk
	Global	—	—	K-62954 A
	—	—	—	Pavon 76

Note: As named by farmers.

At first glance, the number of varieties of cereals (especially sorghum, finger millet and pearl millet) reported per household appears to be low, given that they are among the crops in the 'savannah complex' believed to have originated in a belt that spreads across the Sahelian region in West Africa to the Horn of Africa (Harlan, 1992). Yet, while an individual household may grow relatively few varieties, many varieties of each crop may be found among the households in a community (see Chapter 11). The number of varieties grown by any single farmer is likely to be positively associated with the number of different water regimes in which the farmer plants the crop. In Amhara region, e.g. teff, barley, wheat and maize are grown during the main rains (*meher*), small rains (*belg*) and under irrigation. Finger millet is grown only in the main season, while sorghum and pearl millet are normally grown only in the main season or under irrigation. Moreover, for predominantly cross-pollinating crops, the relationship of variety name to intracrop diversity is not as strong as it is for self-pollinating crops, and diversity is expected to be partitioned more within than among varieties. Pearl millet has very high rates of cross-pollinating relative to sorghum and finger millet, but rates for wheat, barley and teff are lower than any of these. Maize is a highly cross-pollinating species, but modern varieties are also available in the study area.

Econometric Estimation

Method

The general structure of the regression equations is expressed in simple form by

$$D_i = a_i + b_i x + c_i z + e_i \qquad (6.7)$$

D represents either the Margalef index of richness or the Shannon index of evenness (Chapter 1), x is a vector of farm size, exogenous income and household, farm and market/regional factors; z represents adoption of a modern variety, e is unobserved factors; and a, b and c are the parameters to be estimated.

Several estimation problems were encountered in estimating the equations about intracrop diversity. First, a sample selection problem occurs because the diversity index for cereal i exists only when the household cultivates the cereal. Second, a large proportion of households that cultivate the cereal grow only one variety so that both richness and evenness indices are censored at zero.[2] Application of ordinary least squares (OLS) or seemingly unrelated regression (SUR) in this situation yields biased and inconsistent estimates.

The most common approach to dealing with selectivity problems is a technique similar to Heckman's two-step estimation procedure. The probability of growing the cereal would be

---

[2] According to Amemiya (1985), censoring is when the dependent variable takes a limiting value.

predicted in the first stage, a predicted value of the inverse Mills ratio (IMR) would be obtained and the ratio included as an explanatory variable in a second-stage regression (Maddala, 1983). However, since the second stage is a censored regression, the predicted IMR introduces heteroskedasticity because its errors depend on values of the explanatory variables. Unlike in the linear model, heteroskedasticity causes the estimator to be inconsistent (Maddala, 1983). Obtaining the correct standard errors is also complicated by use of the predicted rather than the actual IMR. In the second stage, the censored least absolute deviations (CLAD) estimator, which is robust to heteroskedasticity (Deaton, 1997), was applied. With CLAD, standard errors are computed with bootstrapping.

The third problem is that predicting the effect of modern varieties on intracrop diversity involves endogeneity. Similar to selectivity bias or a treatment effect, including an explanatory dummy variable to represent use of a modern variety gives inconsistent estimates (Barnow et al., 1981; Maddala, 1983; Greene, 1993). Thus, in the second stage of the CLAD regression, predicted probabilities from a first-stage probit regression were used (Barnow et al., 1981).

Identification of the CLAD regression is an important issue, as in many two-stage approaches. The appropriateness of the method depends on finding variables that are correlated with the decision to grow a cereal crop or a modern variety but not correlated with the diversity index. Altitude and walking times to the nearest grain mill, input supply shop and bus service were used as instruments in the probit regressions.

Intuitively, these instruments explain more of the decision of whether or not to grow a crop than of the underlying concept of the diversity index based on how much area to allocate to it. While altitude defines suitability of the area for growing the crop, access to input supply shop or bus service influences access to seeds. Walking times to the nearest grain mill or bus service, on the other hand, influence the cost of consumption of the crop by the household and/or disposal of the surplus, respectively.

To check the appropriateness of using these variables as instruments, they were also used directly in the diversity index models and found to have a zero effect (separately and jointly). Note that, even if the explanatory variables in the first- and second-stage regressions are identical, because the predicted IMRs and probabilities from the first-stage regressions are non-linear functions of the explanatory variables, the CLAD regression is identified under the normality assumptions of the probit model.

The diversity regression equations were estimated across common cereals (including barley, teff, maize, wheat, sorghum, finger millet and pearl millet) and within barley, teff, maize and wheat. Diversity regressions for sorghum, pearl millet and finger millet could not be estimated because the values of the diversity indices were either mostly zeros (since households cultivated only one variety each of these cereals) or data on specific varieties grown were not obtained (see Tables 6.1 and 6.4).

Dependent variables

The dependent variables for the econometric analysis are scalar diversity indices constructed from the choice variable in Eq. (6.2), which is a vector of area shares allocated to crops or varieties of crops. Richness of species or varieties is measured by a Margalef index. The Shannon index combines both richness and relative abundance concepts (Chapter 1). Summary statistics for cereal crop and variety diversity indices are shown in Table 6.3. Note that the mean values are very low and close to the lower bound value of zero, especially at the variety level. This is because a large proportion of households that cultivate a particular cereal grow only one variety (see Table 6.1).

Several caveats must be borne in mind when interpreting the value of these indices. The crops are commonly recognized cereals discussed above: teff, barley, sorghum, millets, maize and wheat. Within these cereal crops, 'variety' is simply understood as a crop population recognized by farmers. This definition encompasses landraces that have been grown and selected by farmers for many years, modern varieties that meet the International Union for the Protection of New Varieties of Plants (UPOV) definition of distinct, uniform and stable as well as 'rusticated' or 'creolized' types, which are the product of deliberate or natural mixing of the two (Wood and Lenné, 1997;

Table 6.3. Summary statistics of indices of cereal diversity on household farms in the highlands Amhara and Tigray regions, Ethiopia.

Cereal	Diversity index	Mean	Standard error	Minimum	Maximum
All cereals	Richness (Margalef index)	0.179	0.008	0.0	0.60
	Evenness (Shannon index)	0.597	0.026	0.0	1.56
Maize	Richness (Margalef index)	0.017	0.006	0.0	0.30
	Evenness (Shannon index)	0.047	0.016	0.0	0.83
Wheat	Richness (Margalef index)	0.019	0.004	0.0	0.23
	Evenness (Shannon index)	0.083	0.014	0.0	0.98
Barley	Richness (Margalef index)	0.017	0.005	0.0	0.23
	Evenness (Shannon index)	0.068	0.018	0.0	1.09
Teff	Richness (Margalef index)	0.021	0.005	0.0	0.31
	Evenness (Shannon index)	0.079	0.018	0.0	0.99

Note: Diversity indices for sorghum, finger millet and pearl millet were not calculated as there were mostly only one variety of each of these cereals grown by any one household. Means and standard errors are adjusted for stratification, weighting and clustering of sample.

Bellon and Risopoulos, 2001). Usually 'named' by farmers, varieties have agromorphological characters, which farmers use to distinguish among them are an expression of their genetic diversity.

The relationship between variety names and genetic variation is generally not well defined. In an economic model of farmer behaviour, however, it is important to establish the relationship between the choice variable itself and the hypothesized explanatory variables.[3] Farmers choose varieties or their observable, distinguishable and expressed traits. As mentioned in Chapters 1 and 5, the more sophisticated the diversity index, the more indirect the relationship between the diversity outcome and farmers' choices and, therefore, between the diversity index and factors that explain the choice.

Strictly speaking, different crop reproduction systems and different farm management practices for these crops make the indices based on named varieties incomparable across crops. Bearing that in mind, the data suggest that the greatest 'farmer-recognized' (based on named varieties) diversity on household farms is maintained for teff, which is consistent with data shown in Table 6.1. Here, however, maize is ranked fourth and wheat is ranked second in intracrop diversity. The low rank for maize is probably a consequence of the larger areas planted to the crop, although it also follows in part from the relative 'newness' of the crop and national efforts to promote modern varieties of maize. The high rank for teff makes sense in a centre of origin and diversity.

---

[3] Named varieties can subsequently be related to the underlying structure of genetic diversity in the community that is identified through agromorphological or molecular analysis with seed samples. Additional work of this type was outside the budget and time frame of this research, although it was implemented in several of the other chapters of this book.

Independent variables and hypotheses

The independent variables are operational measurements of the conceptual factors (A^o, Y^0, Ω_{HH}, Ω_F, Ω_M) shown as explanatory variables on the right-hand side of Eq. (6.2). These are defined in Table 6.3 with related hypotheses and summary statistics. Since estimation consists of a diversity metric over choice variables in a reduced form equation, comparative statics are ambiguous and hypotheses are based on a combination of economics principles and empirical evidence.

Land area (A^0), which is measured by the amount of farmland operated by the household, is hypothesized to have a positive effect on diversity, as greater farm areas can be allocated among more crops and varieties. Exogenous income (Y^0), measured as income from remittances, gifts or aid, may have ambiguous effects, serving either to cope with production risk *ex post* or to intensify production and engage in multiple activities.

Household characteristics (Ω_{HH}) include those related to human capital, labour supply and the life-cycle stage of the household and physical assets. In Chapter 5, Van Dusen found that age of household-head had a quadratic relationship. Younger households may be more willing to try out different crops and varieties, while older households may be more set in their production activities and less likely to try new crops and varieties. However, including the square of age as an explanatory variable introduced severe multi-collinearity, and the variable was dropped from the final regressions. The effect of the gender composition of the household is difficult to predict a priori, while household size is expected to have a positive effect on diversity through its effects on preferences and overall labour capacity. Livestock, as a measure of wealth, may have ambiguous effects. On one hand, it may act as insurance against crop production risk, bearing a negative relationship with diversity. On the other hand, it may have a positive effect on diversity through additional income, enabling farmers to intensify production and engage in multiple activities. Oxen ownership is expected to contribute positively to diversity among cereals through ensuring draught power for ploughing when it is needed.[4] With respect to farm characteristics (Ω_F), slope, erosion and fertility condition, irrigation and degree of fragmentation are used. The genetics and ecological literature suggests that greater heterogeneity in farm conditions will tend to increase inter- and intracrop diversity, while more homogeneity will have the opposite effect (e.g. Marshall and Brown, 1975). Here, it is hypothesized that greater heterogeneity of plots in terms of erosion or fertility and more farm fragmentation[5] increase diversity, while greater flatness is expected to reduce diversity. Irrigation is expected to reduce diversity, as irrigation tends to make farm technology more uniform. Greater distances from the house to the farm may reduce the opportunities to grow more cereal crops because of time requirements in walking.

Market and regional characteristics (Ω_M) are expressed by the average distance from the PA in which households are located to the nearest all-weather road or district town, population density in the PA and regional location of the PA. Market infrastructure operates in several ways that may not be dissociable in a given location at one point in time. For example, the theoretic framework of the household farm predicts that the higher the transaction costs faced by individual households within communities as a function of their specific, social and economic characteristics, the more we would expect them to rely on the diversity of their crop and variety choice to provide the goods they consume. In other words, the more removed a household or community is from a major market centre, the higher the costs of buying and selling on the market and the more likely that it relies primarily on its own production for subsistence. This implies that the more

---

[4] A variance inflation factor (VIF) greater than ten indicates collinearity problems (Kennedy, 1985). There is no collinearity problem associated with using both the number of oxen and total livestock units as explanatory variables, as the VIF with respect to these are 3.81 and 3.73, respectively. See the previous footnote.

[5] Farm fragmentation is measured by three factors according to Blarel *et al.* (1992): the Simpson index ($1-\sum_k \delta^2$), where δ is the share of *k*th plot in total farm size, number of plots and average distance to plots.

Table 6.4. Definition of explanatory variables, summary statistics and hypothesized effects on cereal (inter- and intracrop) diversity on household farms in the highlands Amhara and Tigray regions, Ethiopia.

Variable name	Description	Hypothesized effect Intercrop	Hypothesized effect Intracrop	Mean	Standard error	Minimum	Maximum
Farm size (A^0)	Amount of farmland operated by household (ha)	(+, −)	(+, −)	1.176	0.050	0.01	7.9
Exogenous income (Y^0)	Sum of remittances, food aid, gifts and pension (EB)[a]	(+, −)	(+, −)	111.184	15.745	0.00	1750.0
Household characteristics (Ω_{HH})							
Age	Age of household-head (years)	(+, −)	(+, −)	43.405	0.738	16.00	86.0
Male-headed	Sex of household-head (0 = female; 1 = male)	(+, −)	(−)	0.913	0.016	0.00	1.0
Education	Average number of years of formal education of members 15 years and above	(+, −)	(+, −)	1.827	0.119	0.00	19.5
Household size	Number of household members	(+, −)	(+, −)	5.512	0.160	1.00	15.0
Proportion of males	Proportion of household members that are male	(+, −)	(−)	0.432	0.014	0.00	1.0
Tropical livestock units	Number of tropical livestock units owned by household	(+, −)	(+, −)	3.490	0.153	0.00	17.3
Oxen ownership	Number of oxen owned by household	(+, −)	(+, −)	1.431	0.059	0.00	7.5
Farm characteristics (Ω_F)							
Slope of farmland	Proportion of farmland that is flat	(−)	(−)	0.433	0.022	0.00	1.0
Erosion of farm	Shannon index of areas shares in eroded land classes on farm	(+)	(+)	0.453	0.019	0.00	1.0
Fertility of farm	Shannon index of area shares in soil fertility classes on farm	(+)	(+)	0.397	0.021	0.00	1.0
Irrigation	Proportion of farmland that is irrigated	(−)	(−)	0.030	0.006	0.00	1.0
Farm fragmentation							
Composite index	Simpson index (1− the sum of squared plot area shares)	(+, −)	(+, −)	0.563	0.012	0.00	0.9
Number of farm plots	Number of farm plots operated by household	(+, −)	(+, −)	3.790	0.102	1.00	14.0
Distance to farm	Average walking time from house to farm plots (h)	(−)	(−)	0.589	0.028	0.00	9.0
Market and regional characteristics (Ω_M)							
Distance to road	Walking time to nearest all weather road (h)	(+, −)	(+, −)	3.159	0.152	0.00	24.0
Distance to town	Distance from peasant association (PA) to district town (km)	(+, −)	(+, −)	35.315	1.557	1.00	168.0
Population density	Population density of PA (number/km²)	(+)	(+, −)	128.663	4.102	15.00	379.0
Location in Tigray	Administrative region of PA (Amhara region = 0; Tigray region = 1)	(+, −)	(+, −)	0.174	0.006	0.00	1.0

[a]At the time of the survey (December 1999–August 2001) US$ 1= EB (Ethiopian Birr) 8.50.
Means and standard errors are adjusted for stratification, weighting and clustering of sample.

physically isolated a community or household, the less specialized its production activities.

On the other hand, as market infrastructure reaches a village, new trade possibilities may emerge, adding crops and production activities to the portfolio of economic activities undertaken by its members. Varieties differ in the extent to which they provide agronomic (adaptation to soils, maturity, disease resistance, fodder and grain yield) and consumption (taste, appearance) attributes. When farmers cannot rely on the market, which provides them with the seed that meets their demand for attributes, they may grow a more diverse set of varieties to ensure their needs. At the same time, access to seed markets also enables farmers to combine the attributes of purchased seed types with those selected and maintained by farmers in their own community. Modern varieties may possess traits not found in local varieties (Louette *et al.*, 1997) or may have more uniform grain quality, enabling cash to be earned to satisfy other consumption needs of households (Zimmerer, 1996). Hence, while an area's relative isolation from markets would lead us to predict that modern varieties are less likely to be found or are found to a lesser extent, the number of distinct types may be either greater or fewer when these areas have access to modern varieties, especially when the attributes they offer complement but do not substitute for those provided by local materials.

The ratio of labour to land in the community is associated with the hypothesis that rising population densities induce land-saving technical change or higher output per unit of land. Modern varieties are one form of agricultural intensification. Intensification may also occur in terms of larger numbers of farm production activities undertaken, including more cereal crops.

Finally, regional location, which is a dummy variable, captures the cultural and physical environment in which farmers make their decisions. The physical environment in Tigray region is more degraded and is of lower agricultural potential than that found in Amhara region. For example, the average annual rainfall in Amhara is estimated at 1189 mm, compared to only 652 mm in Tigray. The average size of landholding per household is 1.72 ha in Amhara, compared to 1.05 ha in Tigray.

Findings

Intercrop diversity of cereals

Censored regression results of the determinants of intercrop diversity of cereals are given in Table 6.5. Social and demographic characteristics of the household such as the age and sex of the household-head, the education of its members and its size bear no significant relationship to the diversity of cereal crops they grow. However, labour stock and asset variables are highly significant. Households with larger farms, more male labour or more oxen are associated with greater diversity in cereal crops. The positive effect of more male labour on the diversity of cereal grown by households has very large magnitude, underscoring the labour-intensity of production activities in this less-favoured environment. On the other hand, greater total livestock assets are associated with more specialization or less evenness in cereal crops. More fragmented farms with larger numbers of different plots are associated with more diverse cereal crops that are likely to be more evenly distributed. Households living farther from their farms have lower diversity in cereal crops. Location in Tigray is associated with greater diversity.

Intracrop diversity of cereals

Results of the CLAD regressions about the intracrop diversity of maize, wheat, barley and teff are shown in Table 6.6. Although social and demographic variables were of no significance in determining the diversity among cereal crops (intercrop diversity), they matter for the diversity among varieties.

Larger farms are associated with greater diversity within, as well as among, cereal crops. Households with more exogenous income are also more likely to have other non-farm activities, limiting their ability to engage in more labour-intensive activities associated with growing maize. Younger household-heads and more educated household members are associated with greater diversity in maize, wheat and teff, although the opposite is true for barley. Households-headed by women grow more evenly distributed wheat varieties, while households with

Table 6.5. Censored regression results, factors affecting the intercrop diversity of cereals on household farms in the highlands of Amhara and Tigray regions, Ethiopia.

	All cereals	
Explanatory variable	Richness index	Evenness index
Farm size (A^0)	0.0291**	0.1993***
Exogenous income (Y^0)	−0.0000	−0.0001
Household characteristics (Ω_{HH})		
Age	−0.0003	−0.0023
Male-headed	0.0189	0.0526
Education	−0.0051	−0.0201
Household size	−0.0002	0.0020
Proportion of males	0.1322***	0.3682***
Tropical livestock units	−0.0106	−0.0473***
Oxen ownership	0.0396**	0.1639***
Farm characteristics (Ω_F)		
Slope of farmland	0.0128	0.0691
Erosion of farm	−0.0229	−0.0131
Fertility of farm	0.0274	0.0213
Irrigation	−0.0149	−0.0222
Farm fragmentation		
Composite index	0.0792	0.4529***
Number of farm plots	0.0213***	0.0427***
Distance to farm	−0.0378***	−0.0723*
Market and regional characteristics (Ω_M)		
Distance to road	−0.0003	−0.0025
Distance to town	0.0001	−0.0001
Population density	−0.0001	0.0004
Location in Tigray	0.1427***	0.1612***
Constant	−0.0763	−0.3176*
Number of observations	739	739
Uncensored	577	577
Left-censored	162	162
F	8.89***	10.25***
Pseudo R^2	—	—
Mean (standard error) of index	0.179 (0.008)	0.060 (0.026)

*Statistically significant at the 10% level; **statistically significant at the 5% level; ***statistically significant at the 1% level.

proportionately more women grow more varieties per unit area of wheat, barley and maize.

Households with a larger stock of labour have greater maize diversity, probably because of the labour demand associated with growing the crop, applying fertilizer and harvesting. Households with more livestock assets (including oxen) have lower diversity in teff, but greater diversity in barley and wheat; households with more oxen have more diverse teff and less diverse barley and wheat. Perhaps households with more livestock are concerned with biomass (crop residue) to feed their livestock and so prefer to grow barley and wheat varieties that produce more fodder, while those with more oxen can undertake the intensive ploughing practices associated with teff. Households with greater outside sources of income grow more diverse barley varieties, but the same is not true for maize.

Fragmentation and numbers of plots have conflicting effects among crops. Farms with more flat land have greater diversity in maize,

Table 6.6. Regression (censored least absolute deviation (CLAD)) results, factors affecting the intracrop diversity of maize, wheat, barley and teff on household farms in the highlands of Amhara and Tigray regions, Ethiopia.

	Maize		Wheat		Barley		Teff	
Explanatory variable	Richness index	Evenness index	Richness index	Evenness index	Richness index	Evenness index	Richness index	Evenness index
Farm size (A^0)	−0.0198	0.1618*	0.0989***	0.2920*	0.0183	0.1539*	0.0169	0.0926
Exogenous income (Y^0)	−0.0038***	−0.0232***	−0.0035*	−0.0175**	0.0074***	0.0194***	−0.0024**	−0.0113***
Household characteristics (Ω_{HH})								
Age	−0.0004**	0.0004	0.0001	0.0004	0.0001	0.0003*	0.0000	0.0001
Male-headed	−0.0364	−0.1259	−0.0651	−0.4856*	0.0001	−0.0981	0.0337	0.1816
Education	0.0184**	0.0781*	0.0196***	0.1057***	−0.0036	−0.0253	0.0110***	0.0373*
Household size	0.0095**	0.0663*	0.0051	0.0301	0.0031	0.0071	0.0021	0.0181
Proportion of males	−0.1623***	−0.3186	−0.1608**	−0.9071**	−0.1703**	−0.1130	0.0716	0.2240
Tropical livestock units	−0.0070	−0.0743	0.0397***	0.1734***	0.0264***	0.0408	−0.0090	−0.0585*
Oxen ownership	0.0299	0.2023	−0.0829***	−0.3941***	−0.0712***	−0.1707*	0.0308	0.2104***
Farm characteristics (Ω_F)								
Slope of farmland	0.1084***	0.6599***	−0.0253	−0.2221	0.0076	−0.3052***	−0.0913***	−0.4924***
Erosion of farm	0.1101***	0.6663***	0.0662	0.5218	0.0169	−0.0509	0.0583*	0.2335
Fertility of farm	−0.0952***	−0.2766	0.0134	0.2080	0.0044	0.1175	0.0405	0.0240
Irrigation	−0.1813*	−0.4979	0.6104*	2.2710	0.0213	0.0475	0.1069	0.9719**
Farm fragmentation								
Composite index	0.0181	0.4263	−0.3028***	−1.7204**	0.0118	−0.0276	−0.2129*	−0.5731
Number of farm plots	0.0042	−0.0134	0.0065	0.0867	−0.0411***	−0.0879**	0.0173**	0.0541
Distance to farm	0.0001	−0.1082	−0.0629	−0.3681	−0.0277	−0.0549	−0.0072	−0.0431
Market and regional characteristics (Ω_M)								
Distance to road	0.0192	0.2137**	0.0049	0.0213	0.0094*	0.0279	−0.0233***	−0.1548***
Distance to town	−0.0025**	−0.0242**	−0.0018	−0.0064	−0.0008	−0.0032	0.0007	0.0028
Population density	0.0006**	0.0025**	0.0010**	0.0019	−0.0001	0.0006	−0.0007***	−0.0050***
Location in Tigray	−0.0815	−0.3009	−0.0376	−0.1624	−0.0615*	0.0596	0.0179	0.2743**
IMR (growing cereal)[a]	−0.4513***	−2.3201***	−0.1304	−0.5118	−0.2304***	−0.6242***	−0.2723***	−1.0143***
Prob. (modern variety)[b]	−0.0249	−0.4554	−0.1704	−0.0345	—	—	—	—
Constant	0.2862***	0.3581	0.2672*	1.6500***	−0.0094	−0.0229	0.2665***	1.3289***
Number of observations	303	303	243	243	352	352	469	469
Pseudo R^2	0.48	0.46	0.32	0.21	0.31	0.26	0.16	0.17
Mean (standard error) of index	0.017 (0.006)	0.047 (0.017)	0.016 (0.003)	0.072 (0.013)	0.017 (0.005)	0.068 (0.018)	0.021 (0.005)	0.079 (0.018)

[a]Inverse Mills ratio of growing cereal; [b]probability of growing modern variety of cereal; *, **, and *** mean statistically significant at the 1%, 5% and 10% level, respectively.

but lower diversity in barley and teff. Evenness in the extent of soil erosion on the farm is associated with greater diversity in maize and teff. The greater the proportion of the farm that is irrigated, the greater the specialization in maize types, although the opposite is revealed for wheat and teff.

As predicted, the effects of market-related factors depend on both variable measurement and crop. Households farther away from an all-weather road grow more diverse barley and maize, but less diverse teff. Households in communities located farther away from the district town have less diverse maize. More densely populated communities have more diverse wheat and maize, but less diverse teff. This result is consistent with the notion that these communities have higher food and feed demands and so farmers will choose higher yielding varieties that produce more biomass (maize and wheat, as compared to teff). Location in Tigray implies greater diversity in teff, but less in barley and maize, probably because teff is more adaptable to conditions in which many other crops fail to grow (Worede, 1988). Rainfall is lower and more variable in Tigray than in Amhara region.

Adoption of crops and modern varieties

Cultivation of modern varieties of maize and wheat has no statistically significant impact on the diversity in the maize and wheat varieties grown on household farms (Table 6.6). This finding suggests that modern varieties add traits and attributes that augment the set of traditional varieties provided to farmers, complementing rather than replacing them.

It is interesting to also examine some of the key factors affecting the adoption of cereal crops (maize, wheat, barley and teff) and modern varieties of maize and wheat. Probit regression results of whether or not households cultivated any of these cereals or modern varieties of maize or wheat are shown in Table 6.7. Two factors that generally define the production environment are elevation and regional location. Barley and wheat are more likely to be cultivated at relatively higher elevations of the highlands (above 1500 masl). The opposite holds for teff and maize, which are more likely to be cultivated at relatively lower elevations of the highlands. Similarly, barley and wheat are more likely to be cultivated in Tigray region, while teff and maize are in Amhara region.

The relatively newer crops, maize and wheat, are less likely to be cultivated in high population density areas, suggesting that relatively large areas are needed for profitable production, especially if accompanying technologies are to be adopted. The finding that they are more likely to be cultivated on larger farms supports this view. Maize and wheat are also more likely to be planted on more fertile farms, consistent with the argument that these crops respond better to yield-enhancing technologies.

Households with greater education levels have a greater predicted likelihood of adopting improved varieties of maize as education enhances the ability to better utilize information associated with modern technologies. Those with more adult male labour are associated with greater likelihood of cultivating barley and maize or with adopting improved varieties of maize.

Implications

Trade-offs in diversity goals

No trade-offs are apparent between policies that would enhance the richness of cereal crops, as compared to the equitability of the area allocated to them on individual farms. The direction of the effect of statistically significant factors is the same for both indices. Thus, a policy designed to support one conservation goal would not conflict with the other goal. The same appears to be true for intracrop diversity of any given cereal crop. Different factors are significant in explaining the richness and equitability among varieties grown for any single cereal crop, but they are consistent in sign. In this setting, a programme that conserves the variety richness of any single crop is not likely to have a negative impact on the evenness in land allocated among the varieties on representative farms.

Table 6.7. Regression (probit) results, factors affecting the probability that household farms grow cereals and modern varieties in the highlands of Amhara and Tigray regions, Ethiopia.

Explanatory variable	Maize All varieties	Maize Modern variety	Wheat All varieties	Wheat Modern variety	Barley All varieties	Teff All varieties
Farm size (A^0)	0.2423*	0.7104**	0.0718	0.5328***	0.2082	0.1526
Exogenous income (Y^0)	−0.0000	0.0001	−0.0000	0.0015**	0.0002	0.0000
Household characteristics (Ω_{HH})						
Age	0.0129*	−0.0215	0.0019	−0.0247*	−0.0145**	−0.0008
Male-headed	−0.0382	−0.2325	0.3244	0.5807	−0.3298	0.5024
Education	−0.0292	0.2643***	−0.0610	0.0545	0.0126	−0.0079
Household size	−0.0134	0.0063	−0.0579	0.1821***	0.0862**	−0.0639
Proportion of males	0.9240**	2.4827***	0.6004	0.6302	1.0114***	−0.1233
Tropical livestock units	−0.0166	−0.4819***	−0.0511	0.0109	0.1172*	−0.0310
Oxen ownership	0.2376	1.8495***	0.2313	0.1037	−0.0895	0.0199
Farm characteristics (Ω_F)						
Slope of farmland	−0.3487	1.5153*	−0.0334	−0.1374	−0.0615	−0.0160
Erosion of farm	−0.3389	0.9022	0.0132	−1.1044**	−0.0518	−0.1738
Fertility of farm	0.5114*	−0.1364	0.8238***	−0.2381	−0.2134	−0.1315
Irrigation	−0.0502	−4.3956**	−1.1610	5.9645***	−0.7357	−1.2510
Farm fragmentation						
Composite index	−0.6338	0.1439	0.8894	1.0584	−0.4965	1.3205**
Number of farm plots	0.1416*	0.0426	0.0475	−0.2432*	0.2356***	0.1099
Distance from house to farm	−0.1122	−0.8404	−0.1636	0.1963	−0.3215**	−0.2028
Market and regional characteristics (Ω_M)						
Distance to road	−0.0670	1.6646***	0.0177	−0.0019	−0.0488*	0.0326
Distance to town	0.0015	−0.0480	−0.0033	−0.0005	−0.0017	0.0017
Population density	−0.0035***	0.0054	−0.0030**	0.0032	0.0030**	0.0013
Region	−0.8854***	−2.7827***	0.4740**	0.0850	0.8655***	−0.6373***
Instruments						
Distance to grain mill	−0.0031	−0.0018	−0.0045***	0.0038	0.0024	0.0009
Distance to input supply shope	−0.0024*	−0.0054	0.0004	−0.0015	0.0008	−0.0009
Distance to bus service	−0.0006	−0.0203***	−0.0002	0.0004	0.0015**	−0.0008
Altitude	−0.0012***	—	0.0009***	—	0.0014***	−0.0014***
Inverse Mills ratio (IMR), growing cereal	—	2.4158	—	−0.4142	—	—
Constant	3.1158***	5.1368***	−3.1671***	−2.2631	−5.1313***	2.8819***
Number of observations	565	303	515	243	628	552
F	3.73***	−4.40***	−2.55***	−2.04***	−4.16***	−3.15***

*Statistically significant at the 10% level; **statistically significant at the 5% level; ***statistically significant at the 1% level.
Note: Coefficients and standard errors are adjusted for stratification, weighting and clustering of sample.

Trade-offs in diversity among and within crops

The set of factors that determines the pattern of intracrop diversity varies among cereal crops and some factors are clearly more important for one crop than for another. Thus, policies designed to encourage intracrop diversity in one cereal crop might have the opposite effect on that of another crop.

Policies related to livestock and oxen ownership will affect both the intercrop diversity and intracrop diversity of cereals, but in different ways and differentially among cereal crops. Similarly, farm physical characteristics, market access, population pressure and regional location are related in various ways to both inter- and intracrop diversity of cereals. The incidence of related policies, therefore, would be difficult to predict in a multi-crop, multi-variety context.

Trade-offs between development and diversity

Policies that affect household labour supply and its composition are likely to have a major impact on the intracrop diversity of cereals in the highlands of Amhara and Tigray. If non-farm opportunities arise and fixed labour stocks of adult male labour are drawn out of farm production, intercrop diversity in cereals will probably decline. On the other hand, households with higher proportions of females or female household-heads are more likely to grow cereal crops with greater intracrop diversity than others. Education generally has a positive effect on variety diversity. Educational campaigns, and recognizing the possible importance of women in variety choice and seed management, appear to be relevant for conservation programmes.

At this point, there is no evident trade-off between seeking to enhance productivity through the use of modern varieties and the spatial diversity among named varieties of these two cereal crops in Tigray and Amhara regions of the Ethiopian highlands. So far, introduction of modern varieties has not meant that any single variety dominates or that modern varieties have displaced landraces, most likely because they have limited adaptation and farmers face many economic constraints in this environment (see the 'displacement hypothesis' in Chapter 1).

Instead, as hypothesized, it is just as likely that small amounts of seed of modern varieties diversifies the seed set of these farmers by meeting a particular purpose or filling a particular niche, rather than contributing to uniformity. The obvious reason is that neither the physical terrain nor the market infrastructure network is particularly favourable for specialized, commercial agriculture. This is not to say that the modern varieties introduced in such areas are themselves genetically diverse, but that the traits they add to those of the other varieties grown enable farmers to better meet their production and consumption objectives in this difficult and uncertain growing and marketing situation. These findings confirm that opportunities to pursue development while enhancing cereal crop diversity do occur in areas of the world that are less favoured in terms of environmental conditions and economic infrastructure.

Acknowledgements

The Ministry of Foreign Affairs of Norway and The Swiss Agency for Development and Cooperation provided financial support for the initial research project from which the data used here were obtained. The research project, 'Policies for Sustainable Land Management in Highlands of Amhara and Tigray Regions', was implemented jointly by the International Food Policy Research Institute, the International Livestock Research Institute, the Agricultural University of Norway, the Amhara National Regional State Bureau of Agriculture and Mekelle University. The Food and Agriculture Organization of the United Nations (FAO) supported the analysis. Special appreciation goes to many officials, community leaders and farmers who graciously and patiently participated in the research and responded to our numerous questions.

References

Amemiya, T. (1985) *Advanced Econometrics*. Harvard University Press, Cambridge, Massachusetts.

Barnow, B.S., Cain, G.S. and Goldberger, A.S. (1981) Issues in the analysis of selectivity bias. *Evaluation Studies Review Annual* 5, 43–59.

Bellon, M.R. and Risopoulos, J. (2001) Small-scale farmers expand the benefits of maize germplasm: a case study from Chiapas, Mexico. *World Development* 29, 799–812.

Bellon, M.R. and Taylor, J.E. (1993) Folk soil taxonomy and the partial adoption of new seed varieties. *Economic Development and Cultural Change* 41, 763–786.

Benin, S., Pender, J. and Ehui, S. (2003) Policies for sustainable development in the highlands of Amhara region: overview of research findings. In: Amede, T. (ed.) *Proceedings of the Conference on Natural Resource Degradation and Environmental Concerns in the Amhara National Regional State: Impact of Food Security*, Ethiopian Society of Soil Science, Addis Ababa, Ethiopia, pp. 185–207.

Blarel, B., Hazell, P., Place, F. and Quiggin, J. (1992) The economics of farm fragmentation: evidence from Ghana and Rwanda. *World Bank Economic Review* 6, 233–254.

Brush, S., Taylor, J.E. and Bellon, M.R. (1992) Biological diversity and technology adoption in Andean potato agriculture. *Journal of Development Economics* 39, 365–387.

Deaton, A. (1997) *The Analysis of Household Surveys: A Microeconomic Approach to Development Policy*. Johns Hopkins University Press, Baltimore, Maryland.

Degefe, B. and Nega, B. (2000) *Annual Report on the Ethiopian Economy: Volume I, 1999/2000*. Ethiopian Economic Association, Addis Ababa, Ethiopia.

Degefe, B., Nega, B. and Tafesse, G. (2001) *Second Annual Report on the Ethiopian Economy: Volume II, 2000/2001*. Ethiopian Economic Association, Addis Ababa, Ethiopia.

Elias, E., Morse, S. and Belshaw, D.G.R. (1998) Nitrogen and phosphorus balances of Kindo Koisha farms in southern Ethiopia. *Agriculture, Ecosystems and Environment* 71, 93–113.

FAO (Food and Agriculture Organization of the United Nations) (1986) Ethiopian highlands reclamation study: final report. FAO, Rome, Italy.

FAO (Food and Agriculture Organization of the United Nations) (1995) Food and Nutrition Series 27. FAO, Rome, Italy. Available at: http://www.fao.org/DOCREP/T0818e/Contents

FDRE (Federal Democratic Republic of Ethiopia) (2001) Rural development policies, strategies and instruments. FDRE, Ministry of Information, Addis Ababa, Ethiopia.

FDRE (Federal Democratic Republic of Ethiopia) (2002) Ethiopia: sustainable development and poverty reduction program. FDRE, Ministry of Finance and Economic Development, Addis Ababa, Ethiopia.

Feder, G. and Umali, D. (1993) The adoption of agricultural innovations: a review. *Technological Forecasting and Social Change* 43, 215–239.

Feder, G., Just, R. and Zilberman, D. (1985) Adoption of agricultural innovations in developing countries: a survey. *Economic Development and Cultural Change* 30, 59–76.

Frankel, O.H. (1970) Genetic dangers of the Green Revolution. *World Agriculture* 19, 9–14.

Fröst, S. (1974) Three chemical races in barley. *Barley Genetics Newsletter* 4, 25–28.

Greene, W.H. (1993) *Econometric Analysis*. Macmillan, New York.

Haffangel, H.P. (1961) *Agriculture in Ethiopia*. Food and Agriculture Organization of the United Nations, Rome.

Harlan, J.R. (1972) Genetics of disaster. *Journal of Environmental Quality* 1, 212–215.

Harlan, J.R. (1992) *Crops and Man*. American Society of Agronomy, Madison, Wisconsin.

Hurni, H. (1988) Degradation and conservation of the resources in the Ethiopian highlands. *Mountain Research and Development* 8, 123–130.

Kappel, R. (1996) *Economic Analysis of Soil Conservation in Ethiopia: Issues and Research Perspectives*. Ministry of Agriculture, Addis Ababa, Ethiopia.

Kebede, D. (1996) Brief account on soil conservation research program and its activities in Ethiopia. Paper presented at the Workshop on Soil and Water Conservation Research, 29–30 April 1996, Nazareth, Ethiopia.

Kennedy, P. (1985) *A Guide to Econometrics*. Massachusetts Institute of Technology Press, Cambridge, Massachusetts.

Louette, D., Charrier, A. and Berthaud, J. (1997) In situ conservation of maize in Mexico: genetic diversity and maize seed management in a traditional community. *Economic Botany* 51, 20–38.

Maddala, G.S. (1983) *Limited Dependent and Qualitative Variables in Econometrics*. Cambridge University Press, Cambridge, UK.

Marshall, D.R. and Brown, A.H.D. (1975) Optimum sampling strategies in genetic conservation. In: Frankel, O.H. and Hawkes, J.G. (eds) *Crop Genetic Resources for Today and Tomorrow*. Cambridge University Press, Cambridge, UK, pp. 53–80.

National Research Council (1996) *Lost Crops of Africa: Volume I: Grains*. Board on National Academy Press, Washington, DC.

Pender, J., Place, F. and Ehui, S. (1999) Strategies for sustainable agricultural development in the East African highlands. Environment and Production Technology Department Working Paper 41, International Food Policy Research Institute, Washington, DC.

Pender, J., Gebremedhin, B., Benin, S. and Ehui, S. (2001) Strategies for sustainable agricultural development in the Ethiopian highlands. *American Journal of Agricultural Economics* 83, 1231–1240.

Pender, J., Gebremedhin, B. and Haile, M. (2002) Livelihood strategies and land management practices in the highlands of Tigray. Paper presented at the Workshop on Policies for Sustainable Land Management in the Highlands of Tigray, 28–29 March 2002, Axum Hotel, Mekelle, Ethiopia.

Rempel, S. (1997) Wheat classification, biodiversity and issues for conservation. Available at: http://members.show.ca/oldwheat/resumes/PhDwheatpaper.pdf.html

Singh, I., Squire, L. and Strauss, J. (1986) *Agricultural Household Models: Extensions, Policy and Applications*. Johns Hopkins University, Baltimore, Maryland.

Smale, M., Just, R.E. and Leathers, H.D. (1994) Land allocation in HYV adoption models: an investigation of alternative explanations. *American Journal of Agricultural Economics* 76, 535–546.

Stallknecht, G.F. (1997) *New and Specialty Crop Profiles: Teff*. Centre for New Crops and Plant Products, Purdue University, West Lafayette. http://www.hort.purdue.edu/newcrop/cropfactsheets/teff.html

Stoorvogel, J.J., Smaling, E.M. and Janssen, B.H. (1993) Calculating soil nutrient balances in Africa at different scales. *Fertilizer Research* 35, 227–235.

Woldehanna, T. and Alemu, T. (2002) Poverty profile of Ethiopia: 1995/96 and 1999/00. A Report prepared for the Ministry of Finance and Economic Development (MOFED), MOFED, Addis Ababa, Ethiopia.

Wood, D. and Lenné, J. (1997) The conservation of agrobiodiversity on-farm: questioning the emerging paradigm. *Biodiversity and Conservation* 6, 109–129.

Worede, M. (1988) Diversity and the genetic resource base. *Ethiopian Journal of Agricultural Sciences* 10, 39–52.

Worede, M., Tesemma, T. and Feyissa, R. (2000) Keeping diversity alive: an Ethiopian perspective. In: Brush, S.B. (ed.) *Genes in the Fields: On-farm Conservation of Crop Diversity*. International Development Research Centre, International Plant Genetic Resources Institute, and Lewis Publishers, Ottawa, Rome and Boca Raton, pp.143–164.

World Bank (2001) *World Development Indicators*. The World Bank, Washington, DC.

Zimmerer, K.S. (1996) *Changing Fortunes: Biodiversity and Peasant Livelihood in the Peruvian Andes*. University of California Press, Berkeley, California.

7 Demand for Cultivar Attributes and the Biodiversity of Bananas on Farms in Uganda

S. Edmeades, M. Smale and D. Karamura

Abstract

In comparison to the approach presented in Chapter 5, this chapter presents an attribute-based model of cultivar demand that is also derived within the theoretic framework of the agricultural household. Uganda is one of the largest national producers and consumers of bananas in the world, and is recognized as a second centre of diversity of bananas. Numerous distinct clones of the endemic East African highland bananas are managed by Ugandan farmers, in addition to unimproved, exotic types from South-east Asia and a few recently developed hybrids. High levels of banana cultivar diversity are also observed on individual farms. Reflecting the particular features of the banana plant, cultivar demand is expressed in mat counts and mat shares. A full taxonomy of banana clones is used to construct diversity metrics over mat counts and mat shares allocated to cultivars and use groups. Banana diversity is analysed in terms of two of the three taxonomic levels (cultivars and use groups). Findings underscore the importance of cultivar attributes in explaining the decisions of banana growers in Uganda. Although trade-offs across use groups are revealed when cooking quality and beer quality are considered, production traits are generally more important in explaining cultivar diversity than are consumption attributes. This suggests that maintaining cultivar diversity could be a deliberate strategy for managing abiotic and biotic pressures in this relatively labour-intensive production system with low levels of chemical inputs.

Introduction

Uganda is one of the largest producers and consumers of bananas in the world. Bananas occupy the largest cultivated area among staple food crops in Uganda (1.4 million ha or 38% of total planted area) with more than 75% of all farmers growing banana cultivars (NARO, 2001, unpublished report). Per capita annual consumption of bananas in Uganda is the highest in the world, estimated at roughly 0.70 kg/person/day (INIBAP, 2000). Bananas are consumed as fruit, prepared by cooking, roasting or drying and fermented for the production of banana juice and alcoholic beverages (beer, wine and gin). Bananas are primarily grown as a food crop to meet subsistence needs with excess production sold in local markets. Most banana production takes place on small subsistence farms (plots of less than 0.5 ha) with low input farming methods. The lifespan of banana groves depends on agroecological conditions and management practices, ranging from as low as 4 years in central Uganda to over 30 years in western Uganda, depending on biotic pressures (Speijer et al., 1999).

Uganda is also a second centre of diversity for bananas. Endemic cultivars comprise the vast majority of cultivars in Uganda, prevailing in the East African highlands, including also parts of Tanzania and Kenya. A large number of distinct clones of this endemic type are grown in Uganda, as well as a number of unimproved, exotic types from South-east Asia and a few recently developed hybrids. High levels of banana cultivar diversity are also observed at the household level. Banana cultivars are locally named and are differentiated by their observable (to farmers) characteristics.

In Uganda, banana is primarily a food crop grown to meet subsistence needs of farm families. The multiple end uses of bananas to farm families, as well as biotic and abiotic pressures, influence the mixture and number of distinct banana cultivars grown. Differing consumption preferences, genetic variation and genetic interaction with environments mean that no single cultivar equally supplies the attributes demanded by farm families (Bellon, 1996).

This chapter presents an attribute-based model of cultivar demand that is derived within the theoretic framework of the agricultural household (Edmeades, 2003). The reduced form of this model is similar in variables to that presented in Chapter 5, though the model focuses on cultivar attributes and a planting material constraint. The dependent variables, structure of the underlying crop data and estimation method are also distinct. Reflecting the particular features of the banana plant, cultivar demand is expressed in mat counts and mat shares. A mat is composed of a mother plant and plantlets. A mat share is the share of all mats in a particular cultivar. A full taxonomy of banana clones is used to construct diversity metrics over mat counts and mat shares allocated to cultivars and use groups (see Table 7.1).

The first section of this chapter summarizes the taxonomy, biology and genetics that are essential for understanding the biodiversity of East African highland bananas. The conceptual

Table 7.1. Banana taxonomy.

Cultivar name	Synonym	Genomic group	Common use	Household share (%)	Cultivar share (%)
Atwalira nyina	Nasaba	1	1	0.77	0.03
Bogoya	Musiideke Musindije Eringot	22 (AAA)	3	41.01	2.49
Bogoya Omumyufu	Ekijungu (Red Bogoya) Epiakol	22 (AAA)	3	2.71	0.13
Bikumpu	Mukubakonde Nfuunya bikonde	1	1	4.06	0.51
Butobe	Entobe Kafunze Bujonjo	1	1	10.83	1.84
Ekakot		1	1	0.19	0.00
Ekuron	Okuron	1	1	1.16	0.06
Embururu Embiire		1	2	0.58	0.01
Engongo	Rwamugongo	1	1	4.26	0.38
Enkara	Entundu Endunda Ntuundhu Rwasha	1	4	5.42	0.42
Enkyonkyo		1	1	0.77	0.02
Ensenyuka		1	2	0.77	0.05
Enshenyi		1	1	3.87	4.01
Enshenyi Embiire		1	2	2.71	0.45

Table 7.1. (Cont'd)

Cultivar name	Synonym	Genomic group	Common use	Household share (%)	Cultivar share (%)
Ensika	Engumba	1	2	1.55	0.04
Ensowe	Nsowe	1	2	2.90	0.16
Enzirabahima	Kalyankoko Kibalawo Enjuumba	1	4	7.74	0.77
Entukura	Engabani	1	2	3.29	0.20
Enyabotembe		1	2	2.32	0.15
Enyaruyonga	Naluyonga	1	1	4.06	0.17
Enzinga	Rugata Nalwetinga	1	1	0.39	0.01
Gonja	Wette	22 (AAB)	5	14.31	0.66
Kabula	Enyamaizi Namaji	1	2	5.80	0.52
Kaburuka		1	1	2.13	0.05
Kalyankoko		1	1	0.77	0.03
Kamenyaggali	Kahendagari	1	1	0.58	0.15
Kapusente		1	4	0.97	0.02
Kasese		1	1	0.19	0.00
Katalibwambuzi		1	2	0.58	0.01
Katwalo	Entanzinduka Enzoga Kashenga	1	1	2.90	0.23
Kawanda (fhia) Fhia01 Fhia03 Fhia17 Fhia23		23 (AAAB) (AABB) (AAAA) (AAAA)	4	4.45	1.01
Kayinja		22 (ABB)	2	14.31	4.04
Kazirikwe		1	1	0.19	0.00
Keitabunyonyi		1	1	0.19	0.01
Kibuzi	Enshansha	1	1	32.50	6.38
Kidhozi		22 (ABB)	4	9.48	0.79
Kininira		1	1	1.55	0.07
Kisubi	Egero-gero Kanyamwenge	22 (AB)	2	28.43	3.99
Kivuvu	Boki-boki Ruhumbo Ekalimon	22 (ABB)	4	14.31	1.17

(Continued)

Table 7.1. (Cont'd)

Cultivar name	Synonym	Genomic group	Common use	Household share (%)	Cultivar share (%)
Kiyovu	Kisansa Namayovu Nakayovu Kayovu Mayovu	1	1	11.61	2.10
Km5		23 (AAA)	4	0.77	0.03
Lwaddungu	Ntika Ntwiika	1	1	5.03	0.99
Lyewudhika	Nalwewunzika	1	1	0.39	0.02
Majaga		22 (AAB)	5	0.39	0.06
Makunku	Bukumo Bukunko	1	1	2.51	0.05
Malira Nakasabira		1	1	0.77	0.02
Malira Nalugiri	Soolabasezaala	1	1	1.93	0.13
Malira Nalwela		1	1	2.51	0.10
Malira Omwirugavu		1	1	0.19	0.04
Malira Rufuta		1	1	0.39	0.04
Malira Tatakange	Mukale Sitakange	1	1	0.19	0.01
Manjaya		22 (AAB)	5	0.19	0.00
Mbarara	Rwambarara	1	1	0.58	0.22
Mbidde	Embiire Ibide	1	2	20.12	4.79
Mbwazirume		1	1	37.33	4.92
Mudwale	Mpologoma Mbale Batule Basimirayo	1	1	5.80	1.64
Mugeso		1	1	0.39	0.01
Mukadde alikisa		1	1	0.97	0.37
Mukazimugumba		1	1	0.58	0.01
Musa		22 (ABB)	2	15.86	4.07
Musakala	Nsaagala Namasagala Enshakara Luwata Mayogi	1	1	32.88	4.30

Table 7.1. (Cont'd)

Cultivar name	Synonym	Genomic group	Common use	Household share (%)	Cultivar share (%)
Muvubo	Saga saga Muzhuba Muzubwe Mujuba	1	1	22.44	1.70
Nabusa	Enyeru Enyarweru	1	1	22.63	6.04
Nakabinyi	Kakono	1	1	3.09	0.14
Nakabululu	Mbululu Embururu Butende Enyigit	1	1	43.52	6.39
Nakamali	Malira Omutono Malira Nakangu	1	1	6.96	1.48
Nakasabira		1	1	1.55	0.05
Nakawere	Karinga Kasenene Musenene	1	1	2.13	0.15
Nakinyika	Kifuba Kafuba Enzuma	1	1	19.92	2.53
Nakitembe	Entaragaza Enshembashembe Malira Omunene Malira	1	1	57.83	9.18
Nakyetengu	Kitetengwa Kitika	1	1	5.22	0.49
Nakyewogoola		1	1	0.19	0.00
Nalugolima	Nyarugoroma	1	1	0.39	0.02
Nalukira	Enyarukira	1	2	1.35	0.30
Nalwesaanya		1	2	0.19	0.01
Nalyewurula		1	1	1.16	0.04
Namadhi	Nalusi	1	1	1.35	0.01
Namadhugudha	Namunwe Nyeko-ger	1	1	7.54	0.66
Namafura		1	1	0.19	0.00
Namaliga	Namalevu Kyanakyandiga Kiriga Rwakashita	1	1	4.06	0.28
Nambi		1	1	0.97	0.08
Namwezi	Serunjogi Ngalodabha Mbedha	1	1	5.03	0.36

(Continued)

Table 7.1. (Cont'd)

Cultivar name	Synonym	Genomic group	Common use	Household share (%)	Cultivar share (%)
Nandigobe	Enjagata Nyarwanda Ntinti	1	1	18.96	1.55
Ndyabalangira	Enzirabushera Mulangira Muzirankanja Muziranyama Katetema	1	1	25.73	3.89
Nfuuka	Nyakahangazi	1	1	6.00	1.07
Njoya		1	1	2.13	0.83
Nkago	Ikago	1	1	3.09	0.10
Nsabaana		1	1	0.19	0.00
Ntujo		1	1	0.19	0.00
Okiteng		1	1	0.19	0.01
Oringoi		1	1	0.19	0.01
Shombobureku		1	1	0.39	0.01
Siira		1	1	12.38	0.95
Ssalalugazi		1	1	0.39	0.01
Sukali Ndiizi	Kabaragara Osukari Epusit Kapere	22 (AB)	3	60.74	6.71

Note: Genomic group: 1 = endemic (AAA-EA); 2 = non-endemic; 22 = exotic; 23 = hybrid.
Common use: 1 = cooking; 2 = beer; 3 = dessert; 4 = multi-use; 5 = roast.
Household share: Proportion of households that grow this cultivar out of 517 households in the sample.
Cultivar share: Proportion of mats planted to this cultivar out of 52,321 mats in the sample.

framework is then presented followed by a synopsis of data collection methods and some essential descriptive statistics. Econometric methods and results are reported in subsequent sections. Conclusions are drawn in the final section.

Banana Taxonomy, Biology and Genetics

The origin, or primary centre of diversity, of bananas is believed to be South-east Asia, specifically, Malaysia (Simmonds, 1959). There are six genomic groups found in Uganda. Genomic groups are a scientific classification based on the combinations of genomes A and B that are found in cultivars. NARO (2001, unpublished report) estimates that as much as 85% of bananas grown in Uganda are East African highland bananas. The East African highland banana is classified as genomic group *Musa* spp., AAA-EA, a subgroup particular to this region.

Within the AAA-EA genomic group, banana cultivars are classified by morphological (or observable) characteristics into five distinct clone sets, and two types determined by their use for cooking (*matooke*) or beer (*mbidde*) (Karamura

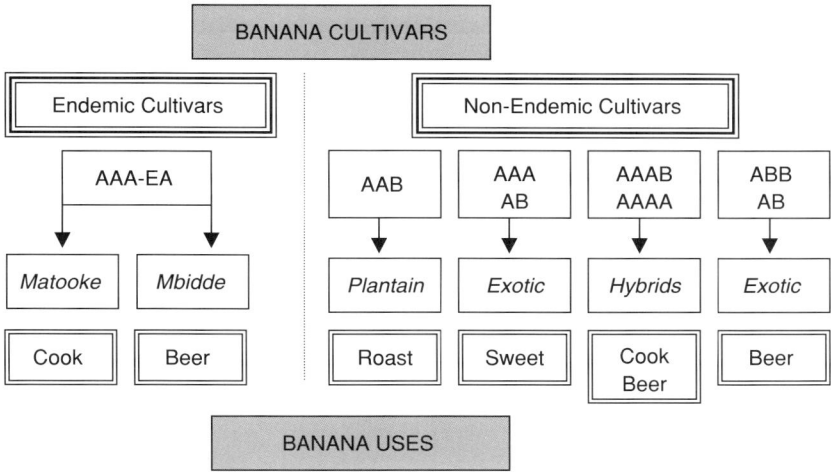

Fig. 7.1. Levels of banana diversity.

and Pickersgill, 1999). Figure 7.1 summarizes the different levels of banana diversity: (i) cultivar (endemic vs non-endemic); (ii) genomic group; and (iii) common use. Figure 7.2 shows the distribution of genomic groups in Uganda, Tanzania and Kenya.

As used in this chapter, the terms 'endemic' and 'non-endemic' are not to be confused with 'traditional' and 'modern'. Differences between endemic and non-endemic cultivars are associated with differences in observable characteristics, genome and common use, and not with improvement status. Although Ugandans clearly consider that endemic cultivars embody important cultural traditions, non-endemic cultivars include both modern types and other long-established banana cultivars introduced from other regions of the world.

The non-endemic bananas grown in Uganda are mostly naturally occurring hybrids of *Musa acuminata* (A genome) and *M. balbisiana* (B genome), which have their origins in Southeast Asia, the first centre of banana diversity. They include exotic beer and sweet bananas (AB, ABB and AAA genomic groups) and roasting bananas or plantains (AAB genomic group). During the last decade, after many years of professional breeding without success, several new improved banana hybrids were introduced in the region (FHIA and IITA hybrids[1]).

Karamura and Karamura (1994) have identified a total of 233 East African highland banana cultivars (genome group AAA-EA), of which 145 are cooking bananas and 88 are beer bananas. Farmers growing 10–15 different banana cultivars in stands of less than 200 banana mats are frequently encountered (Karamura *et al.*, 1999, unpublished manuscript; INIBAP, 2000). Farmers perceive different banana cultivars to be associated with distinct advantages and disadvantages related to both consumption needs and production requirements. Cultivar selection criteria are found to vary among farmers in a given region and across regions according to household production objectives (sale and domestic consumption) (Gold *et al.*, 1998). Insight into the specific traits that motivate farmer selection of a cultivar is limited and primarily derived from on-station research trials rather than from on-farm research. The relationship between morphological or trait diversity and the utility of these traits

---

[1] FHIA stands for Fundacion Hondurena de Investigacion Agricola (Honduras Foundation for Agricultural Research). IITA is the International Institute for Tropical Agriculture located in Ibadan, Nigeria.

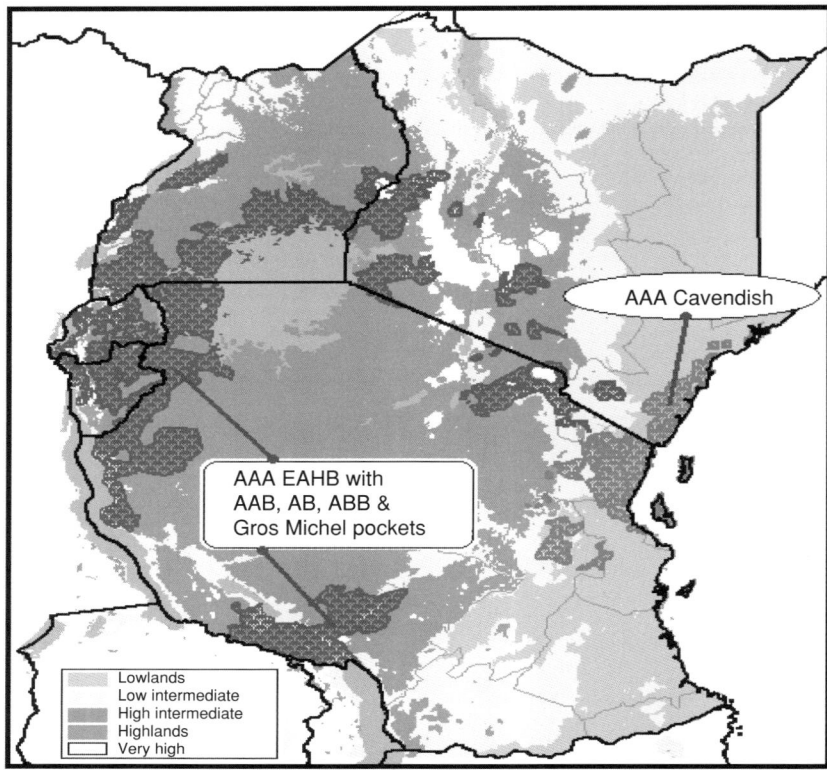

Fig. 7.2. Principal banana growing areas of East Africa showing the terrain and genome differentiation. Source: International Network for Improvement of Banana and Plantain.

to farmers is also poorly understood (Gold *et al.*, 1998; Karamura *et al.*, 1999, unpublished manuscript).

Most banana cultivars and all plantain landraces are triploid genotypes that are almost or fully sterile. Sexual reproduction normally occurs in diploid, tetraploid or hexaploid species. An uneven number of chromosomes (e.g. triploidy) often induces sterility. Bananas are reproduced through vegetative propagation, which constrains banana improvement efforts by means of conventional plant breeding techniques (based on seed production by either self-pollination, cross-pollination or by hybridization) (Persley and George, 1999; Johanson and Ives, 2001). Banana breeding can be achieved in a number of ways: (i) naturally, through mutations and selection by farmers; (ii) traditionally, through vegetative propagation from stems or roots of existing plants (i.e. planting material is multiplied by uprooting a sucker, e.g. root, shoot from an existing banana plant and replanting it); (iii) scientifically, through tissue culture (i.e. the clonal propagation of 'clean' – pest- and disease-free – planting material); or, more recently (iv) with cross-breeding techniques used to produce tetraploid banana hybrids among exotic cultivars.

Large areas with these new hybrids are planted in Tanzania, where substantial yield increases have been achieved (from 5–20 kg to 40–110 kg a bunch) (INIBAP, 2000), though their introduction in Uganda is still in its initial stages. Recent estimates generated with the survey data used in this chapter suggest that only 2% of farmers in the major banana-producing regions of Uganda grow them. Acceptability among farmers is yet to be assessed. Although the production of big bunches is a characteristic of hybrids, they have poor cooking quality, which renders their consumption attributes undesirable to rural households and also to consumers who purchase them in urban markets of the country (NARO, 2001, unpublished report).

A number of pests and diseases affect banana production and lead to significant food and income losses. Among them are weevils, Black Sigatoka disease and Panama disease (or Fusarium wilt). The incidence of pests and diseases has intensified, eliminating susceptible cultivars altogether in some parts of the country (Karamura et al., 1999, unpublished manuscript). Weevils are insects that attack banana cultivars and can cause yield reductions up to 60%. Different levels of susceptibility among cultivars have been observed and the intensity of weevil damage has been found to decrease with elevation (the most severe being between 1000 and 1100 metres above sea level (masl)) (Gold et al., 1994).

Black Sigatoka is an airborne fungal disease that can cause yield losses of around 50% (due to the reduction in the number of fruit per bunch and lower fruit weight) and reduce the longevity of banana farms from 30 years to as little as 2 years (Craenen, 1998). Although it is believed that the potential damage of Black Sigatoka may be limited by altitude, the virulence of this pathogen in highland situations remains unknown. East African highland bananas are highly susceptible, while exotic beer cultivars are found to exhibit some resistance to the disease (Gold et al., 1993; Stover, 2000).

Panama disease (Fusarium wilt) is another fungal disease, which attacks the roots of banana plants. The development of the disease in a single plant is rapid (2 months) and the damage it causes is extensive, with the pathogen persisting in the soil for a long period of time. The spread of the disease is further facilitated by the use of infected planting material by farmers. The exotic brewing cultivars are particularly susceptible to the disease, with the extent of wilt incidence reported to be as high as 67% on some farms. Endemic cultivars also exhibit less susceptibility to Fusarium wilt (Gold et al., 1993).

The planting material of a banana is not a seed but a 'sucker' (shoot or plantlet) that grows from and is a clone of the mother plant. The shoot must be uprooted from an existing mat to reproduce the cultivar, and because of its bulk, transports poorly. We know comparatively little about the mechanisms by which banana planting material circulates among farmers and communities. The data reveals that 60% of banana farmers in the sample supply planting material within or outside their village, while only 18% receive banana shoots from farmers outside the village. The planting material replacement ratio (the number of shoots received to total years of growing the banana cultivar) varies from as high as a mean of six in the low elevation areas of the country to one in high elevation areas. Planting material is typically exchanged between neighbours or family members free of charge or in the form of a gift. At present, there is scant evidence of markets for banana planting material. This is presumably due to the related costs of transactions, the bulkiness of the shoots and the vegetative nature of planting material, which allows farmers to reproduce identical genetic material on their own farms (except for mutations).

The unique characteristics of banana as a species, the diversity of use groups and range of biotic constraints have ramifications for the conceptual framework presented next.

Conceptual Framework

The theoretical model developed by Edmeades (2003) is summarized here. The model borrows from frameworks that consider the role of goods attributes in the utility function (Lancaster, 1966; Ladd and Suvannunt, 1976) or input attributes in the production function (Ladd and Martin, 1976), placing cultivar choice within the decision making framework of the agricultural household (Singh et al., 1986; Sadoulet and de Janvry, 1995).

Household utility is defined over the set of 'intrinsic' attributes of the goods it consumes, rather than from the goods themselves, $\mathbf{Z^C}(\mathbf{X}; \mathbf{x})$, the consumption of other (purchased) goods ($\mathbf{X^G}$) and leisure (or home time, H), given a vector of exogenous household characteristics ($\mathbf{\Omega}_{HH}$). The vector of consumption attributes $\mathbf{Z^C}$ comprises banana bunches ($\mathbf{X}$) from different banana cultivars consumed by a given household and input–output coefficients ($\mathbf{x}$) that map banana bunches consumed (inputs) to levels of consumption attributes (outputs) they possess. While the household can vary the type and amount of banana bunches it consumes, it has no control over the input–output coefficients embodied in the different banana bunches consumed (Ladd and Suvannunt, 1976).

The agricultural household also engages in the production of banana bunches on its farm. Variable inputs (labour, L and cultivar-specific planting material, $\mathbf{V}$) and land allocated to banana production (A^B) are used for the production of banana bunches ($\mathbf{Q}$), given a vector of exogenous farm characteristics ($\mathbf{\Omega}_F$). The choice of planting material is associated with the farmers' perceptions of the 'intrinsic' agronomic traits it provides. The characteristics vector $\mathbf{Z^P}(\mathbf{V}; \mathbf{v})$ defines the association between bunches from different banana cultivars ($\mathbf{V}$) and the relative (fixed) proportions of production attributes ($\mathbf{v}$) they yield (Ladd and Martin, 1976). A one-to-one physical correspondence exists between land area allocated to banana cultivars (A^B) and the count of banana mats from different banana cultivars grown by the household ($\mathbf{V}$).

Household participation in market transactions is conditional on the existence and completeness of markets and the type and magnitude of transaction costs involved (de Janvry et al., 1991). Both input and output markets for bananas are often incomplete or not readily available in rural areas in Uganda. Planting material is either reproduced by farmers or obtained through informal networks and no market price is typically charged for its exchange. Instead, a shadow price for banana cultivars captures their marginal valuation to the household. Family labour is widely used for banana production, implying that leisure is valued by its marginal worth to the household rather than as an opportunity cost derived from a market wage rate.

The cost of investing in banana replanting or replacement is determined by the shadow prices of planting material and labour used to carry out the task. Rotation or replanting of banana cultivars typically occurs with offshoots of established on-farm planting material. Addition of new cultivars occurs in some instances as an exchange and seldom through monetary transactions. The average replacement rate for banana planting material is unknown. However, the perennial nature of banana production and vegetative propagation suggest a slow replacement rate, unless disease pressures are high and farmers require clean shoots.

Production of banana cultivars for household consumption is widespread in Uganda, suggesting that although banana output markets exist, they either fail to capture quality differentials between bunches from different banana cultivars, or other transaction costs prevent households from participating in them. The perishable nature of bananas precludes the possibility of storage, underscoring the need to meet immediate household consumption demand either through market purchase or by production on farms.

The household maximizes utility by consuming non-tradable ($\mathbf{X^{NT}}$) and tradable ($\mathbf{X^T}$) banana bunches it produces on its farm, a number of other goods it purchases at the marketplace and leisure, subject to a full income constraint, a time constraint, a cultivar constraint, a non-tradability constraint and a production technology:

$$\max_{\psi} U\left[\mathbf{Z^c}(\mathbf{X^T}, \mathbf{X^{NT}}; x^T, x^{NT}), \mathbf{X^G}, H | \mathbf{\Omega}_{HH}\right]$$

for $\psi = $

$$(\mathbf{X^T}, \mathbf{X^{NT}}, \mathbf{X^G}, \mathbf{Q^T}, \mathbf{Q^{NT}}, \mathbf{V^T}, \mathbf{V^{NT}}, L^T, L^{NT})$$

s.t.
Full income constraint:

$$\mathbf{P}(\mathbf{Q^T} - \mathbf{X^T}) - \mathbf{P^G}\mathbf{X^G} + I = 0$$

Time constraint:

$$T - (L^T + L^{NT} + H) = 0$$

Planting material constraint:

$$\mathbf{V} - (\mathbf{V^T} + \mathbf{V^{NT}}) = 0 \text{ for } \forall \mathbf{V} \in \tilde{\mathbf{V}}$$

Non-tradability constraint:

$$\mathbf{Q^{NT}} - \mathbf{X^{NT}}(\mathbf{\Omega}_M) = 0$$

Production function:

$$G[\mathbf{Q^T}, \mathbf{Q^{NT}}, \mathbf{Z^P}(\mathbf{V^T}, \mathbf{V^{NT}}; v^T, v^{NT}), L^T, L^{NT} | \mathbf{\Omega}_F] = 0$$

In the non-tradability constraint, market failure is assumed to be on the demand side. That is, household production of banana cultivars with desired attributes is stimulated because household consumption demand for attributes cannot be met through market transactions. Market failure on the supply side can be captured by the inability of the household to sell bananas with specific attributes (because of transaction costs or lack of price differentials for quality). In that case, the non-tradability constraint can be represented as $\mathbf{Q^{NT}}(\mathbf{\Omega}_M) - \mathbf{X^{NT}} = 0$.

The full income constraint captures the budget limitations for the tradable banana bunches. Missing markets for labour are depicted by the explicit lack of wage labour. Thus, the time constraint reflects the total time available for farm production and home activities. The fact that markets are missing for planting material is expressed by the absence of a price for banana plantlets as production inputs. The planting material constraint depicts both the total amount of land available for banana production and the total number of different types of planting material available for planting at the village level ($\tilde{V}$). In other words, the number of banana mats planted is an input to banana production both in terms of planting material used (type of cultivar chosen) and in terms of land used (number of mats of each cultivar planted). Including a planting material constraint (and no land constraint) suffices for the analytical purposes of the model since it encompasses both the physical limitations of available land for banana production as well as household-specific limitations of home-supplied planting material.

Assuming an interior solution (the household consumes both the tradable and the non-tradable banana bunches), an optimal reduced form demand for a specific banana cultivar can be derived from the first-order conditions of the model. When banana markets are incomplete the reduced form equation is

$$\mathbf{V}^* = \mathbf{V}(x, v, \mathbf{P}, \mathbf{P}^G, T, I, \tilde{\mathbf{V}} \mid \Omega_{HH}, \Omega_F, \Omega_M) \quad (7.1)$$

Cultivar demand is defined as the mat count of a given banana cultivar grown by the household. The optimal mat count is determined by household characteristics, production technology and agroecological factors, market-related variables and the revealed (by the household) importance of consumption and production banana attributes.

To relate the reduced form to banana diversity, scalar metrics are constructed over optimal demands for individual cultivars, as in other chapters of Part III:

$$\mathbf{D} = \mathbf{D}\left[\mathbf{V}^*(x, v, \mathbf{P}, \mathbf{P}^G, T, I, \tilde{\mathbf{V}} \mid \Omega_{HH}, \Omega_F, \Omega_M)\right] \quad (7.2)$$

Equation (7.2) is the basis of econometric estimation. Diversity metrics are formulated at the level of banana cultivars as well as for different banana use groups. The data source and descriptive statistics are presented below.

Data

The data for the study, collected in 2003, are drawn from a multi-stage, stratified random sample of rural banana-growing households in Uganda. The sample domain was purposively selected to represent major banana producing areas in eastern, central and south-west Uganda. The sample was stratified according to elevation (below and above 1400 masl, respectively). Prior biophysical information suggests that elevation is correlated with soil fertility and the incidence and severity of pests and diseases, productivity parameters thought to be major sources of variation in observed patterns of cultivar use.

Primary sampling units (PSUs) (27) were defined at the sub-county level – the lowest administrative entity possible to map. PSUs were allocated proportionately with respect to elevation. Secondary sampling units (SSUs) were defined at the village level. One SSU (or community) was selected per PSU using a random number from a list of only those rural villages with over 100 households according to the 1991 Uganda census. A total of 20 households with access to land were selected randomly in each village. A farm household is culturally defined and includes female-headed and child-headed (orphaned) units as well as male-headed households with more than one wife. Survey weights were calculated as the inverse of the product of the sampling fractions at each level, and were used in tabulated descriptive statistics. Among the 540 sampled households, 23 reported no involvement with bananas and were excluded from the sample. The effective sample size for analysis is 517 households.

Site Description

Survey data confirm the high level of banana cultivar diversity both in the aggregate (the country) and at the microlevel (on single farms). A total of 95 banana cultivars are currently grown among sample households. Farmers named the banana cultivars they grow and identified observable levels of major characteristics, including: bunch size, bunch position, finger compactness, size of fingers, maturation period, pseudostem colour and space between hands on the bunch.

Banana cultivars were classified into synonym groups according to banana taxonomy (Karamura and Karamura, 1994). For the purposes of the analysis presented in this chapter, five banana cultivar groups were formulated according to use (1 = cooking; 2 = beer making; 3 = sweet; 4 = roasting; 5 = multi-use).

A full list of cultivars grown by sampled farmers is included in Table 7.1. The majority of the cultivars (86%) are endemic to East Africa (AAA-EA genomic group). The remaining 14% are composed of non-endemic cultivars of the types described in the section above. Most individual cultivars belong to only one use (beer or cooking) group because farmers use them for a single purpose. There is a one-to-one correspondence between use and genomic group for cooking cultivars.

Richness of banana cultivars and use groups is shown in Table 7.2 at the household and village levels, by elevation. Households grow a large number of different banana cultivars simultaneously on their farms, with a maximum of 19 cultivars in the low elevation areas and 27 in the high elevation areas. Both the level of frequency distribution (min, max, mode) and its mean are higher in high elevation areas. Growing three or more different use groups of cultivars appears to be popular. Differences are statistically significant, though not very meaningful between elevation levels. Endemic cooking bananas are the most widely grown use group in the sample – 97% of all households grow at least one cooking cultivar. All households in high elevation areas engage in production of cooking bananas. The diversity within this group is also striking. Farm households most often grow three or more distinct cooking banana cultivars, with an average of four in the low elevation and over six in high elevation areas. The number of distinct cultivars per village ranges from 13 to 38, with an average of 23.

Major cultivars appear to be fairly uniformly distributed across households. That is, the cultivars most frequently grown by farmers (percentage of households) are generally the same as those most widely planted (percentage of mats). Among them, the endemic cooking bananas predominate, highlighting the importance that farmers attribute to banana landraces. However, although popular, even the most dominant banana cultivars occupy less that 10% of all banana mats in the entire sample domain. This is indicative of the tremendous clonal diversity maintained spatially by farmers across the major banana-producing regions of Uganda. The ten most frequently and most widely grown cultivars are listed in Table 7.3.

Table 7.2. Cultivar and use group richness at household and village levels.

	Count of banana cultivars			
Unit of analysis/elevation	Minimum	Maximum	Mode	Mean
Household, cultivars				
Low elevation	1	19	4	6.72**
High elevation	2	27	6	9.07**
Household, use groups				
Low elevation	1	5	4	2.85**
High elevation	1	5	3	3.36**
Household, cooking cultivars				
Low elevation	0	14	4	4.01**
High elevation	1	18	3	6.38**
Village, cultivars				
Low elevation	13	32	20	22.41**
High elevation	17	38	38	28.54**

Note: Using a pairwise t-test the means in the two locations are found to be significantly different from each other at the 1% level (P value < 0.0001). Household subsample sizes are 419 in low elevation and 98 in high elevation. There are 27 communities.

Table 7.3. Most frequently and most widely grown banana cultivars in Uganda.

Cultivar name	Use group	Percentage of households	Cultivar name	Use group	Percentage of mats
Sukali Ndiizi	NES	60.74	Nakitembe	EC	9.18
Nakitembe	EC	57.83	Sukali Ndiizi	NES	6.71
Nakabululu	EC	43.52	Nakabululu	EC	6.39
Bogoya	NES	41.01	Kibuzi	EC	6.38
Mbwazirume	EC	37.33	Nabusa	EC	6.04
Musakala	EC	32.88	Mbwazirume	EC	4.92
Kibuzi	EC	32.50	Mbidde	EB	4.79
Kisubi	NEB	28.43	Musakala	EC	4.30
Ndyabalangira	EC	25.73	Musa	NEB	4.07
Nabusa	EC	22.63	Kayinja	NEB	4.04

Note: The use groups are: EC = endemic cooking; EB = endemic beer; NES = non-endemic sweet; and NEB = non-endemic beer. The household share is the proportion of households in the sample that currently grows the cultivar. The cultivar share is the proportion of all banana mats in the sample that is planted in the cultivar.

Econometric Estimation

Equation 7.2 was estimated econometrically with several definitions of D, corresponding to different diversity concepts and units of analysis. Estimation methods were index-specific and accounted for the limits of the dependent variables. The Shannon index (expressing the concept of evenness) is censored at 0 and the regression was estimated with a tobit model. For the count indices (expressing the concept of richness) the data supported neither the Poisson model nor the Negative Binomial specification. The count is defined over positive, integer values. Truncation at 1 reflects the definition of the dependent variables. Each household grows a positive number of cultivars and a positive number of use groups. The truncated tobit is believed to provide a good continuous approximation for count data when few or no zero outcomes are observed.

Dependent variables

Richness and evenness and indices (**D**) were constructed as indicated in Chapter 1 for banana cultivars as well as for banana use groups. The unique features of banana as a species affect the definition of the dependent variable. Counts of mats and mat population shares, rather than crop area (land) shares, are used as the units over which the indices are calculated. A plant or mat count fits more closely the definition of these indices in the ecological literature, which is based on species counts (Magurran, 1988). Each index represents a vector of diversity outcomes bounded from below at either 0 or 1.

Only the count and Shannon indices were used as dependent variables in the regressions explaining cultivar and use group diversity. Although the Margalef index is well defined in the context of this crop, the mat count is more easily interpreted and preliminary regression results revealed no difference in statistical significance of estimated parameters between the two. In this empirical context, the Berger–Parker index is a weak measure of relative abundance due to the fact that an individual household often grew several cultivars with the same level of dominance. Over two-thirds of households (69% of the sample) allocated roughly the same mat share to more than one banana cultivar. Summary statistics for the two dependent variables of the diversity indices are presented in Table 7.4.

Independent variables

Independent variables are vectors representing farmers' perceptions of the importance of banana attributes, individual and household characteristics, physical features of the farm and banana plantation and market-related characteristics

Table 7.4. Banana cultivar and use group diversity indices.

Index	N	Mean	SD	Minimum	Maximum
Cultivars					
Count	517	7.14	3.61	1	27
Shannon	517	1.56	0.51	0	2.85
Use groups					
Count	517	2.87	1.07	1	5
Shannon	517	0.67	0.38	0	1.48

(Eq. (7.2)). Their inclusion as explanatory variables is based on the theoretical foundations of the conceptual model. The expected effects for most explanatory variables are ambiguous and no a priori theoretical underpinning exists to support the direction of comparative static relationships. Comparative statics do not apply when dependent variables are metrics over choice variables, which is the case of the diversity metric over variety count or variety share. Moreover, the comparative statics of a non-separable agricultural household model are complex. In general, unambiguous signs on the direction of effects cannot be derived for the choice variables. The empirical findings in the published literature on crop biodiversity do provide some maintained hypotheses. Variable definitions and maintained hypotheses are shown in Table 7.5.

The demand for each attribute is defined at the level of the banana production decision maker. Attribute cards with illustrations were used to ensure visual recognition of each attribute, and the respondent was asked to rate each attribute as not important (= 1), indifferent (= 2) or very important (= 3). The relative importance of attributes is believed to affect banana diversity on farms through trade-offs that farmers make when choosing the type and number of banana cultivars they grow. As an alternative hypothesis, the count of all attributes rated as very important was also tested as an explanatory variable, articulating the Bellon (1996) notion that no single cultivar is likely to meet all the concerns of smallholder farmers. In the conceptual framework developed here, we hypothesize that no single cultivar equally supplies the attributes demanded by the agricultural household.

Individual characteristics are summarized for a representative household member who is identified as the person in charge of banana production and management decisions, in contrast to the usual emphasis on the household head. In this sample most household heads were identified as men. However, the gender distribution of banana decision makers appears to be well balanced in the sample, highlighting the role of both women and men in banana production. Women's role in crop biodiversity conservation is often emphasized in the general literature (Howard, 2003). A higher proportion of adults in the household who are women, women household heads or education of women has been associated with higher levels of diversity for wheat, barley and maize in Ethiopia (Chapter 6) and rice in Nepal (Chapter 10). Education of the banana production decision maker, measured in years of schooling, proxies for acquired human capital, with its effect on diversity being ambiguous. Variables related to human capital bore no statistical importance for potato diversity in Peru (Chapter 9).

'Relative experience' is used as an indicator of involvement with bananas while taking into consideration population dynamics in Uganda. Typically, age and experience, both measured in years, tend to be correlated. Older decision makers are typically more experienced. A correlation was not found in the Ugandan sample. Lack of correlation is attributed to mortality among the active adult population from human disease pressure. The expected direction of effect is unclear, however. Although the age of the decision maker has repeatedly been shown to be positively associated with crop and variety diversity in several chapters of this book, this was not the case for cereals in Ethiopia (Chapter 6). The definition of the dependency ratio (number of economically dependent divided by total household size) has been adapted to country-specific population features. A number of households

Table 7.5. Definition of explanatory variables, hypothesized effects and summary statistics.

Variables	Definitions	Expected effect	N	Mean	SD
Attribute importance (for all variables, 1 = not important; 2 = indifferent; 3 = very important)					
Bunch size	Bunch size as a component of banana yield	+/−	517	2.82	0.47
Resistance to BS	Importance of resistance to Black Sigatoka	+/−	517	2.21	0.73
Resistance to FW	Importance of resistance to Fusarium wilt	+/−	517	2.28	0.80
Resistance to WE	Importance of resistance to weevils	+/−	517	2.42	0.80
Cooking quality	Importance of cooking quality	+/−	517	2.56	0.68
Beer quality	Importance of beer quality	+/−	517	2.09	0.89
Utility weight count	Count over all important (= 3) attributes	+	517	3.48	1.45
Individual and household characteristics					
Gender	Sex of the banana production decision maker (1 = female; 0 = male)	+	517	0.62	0.49
Education	Years of schooling of banana production decision maker	+/−	517	5.21	4.02
Relative experience	Ratio of years of experience with banana production to age	+/−	517	0.23	0.21
Dependency ratio	Ratio of dependent household members to household size	+/−	517	0.52	0.24
Assets	Value of livestock owned by the household (in '0000 USh)	+/−	517	42.19	96.18
Exogenous income	Income received by the household in the previous year (in '0000 USh)	+/−	517	90.88	282.60
Extension	Number of contacts of the banana production decision maker with extension agents in previous 6 months	+/−	517	0.54	1.73
Farm characteristics					
Banana area	Total available land for banana production (in acres)	+/−	517	1.02	1.66
Age of plantation	Total number of years the household has grown bananas on their farm	+/−	517	11.91	12.08
Crops	Number of crops grown at the household level	+/−	517	1.94	1.18
Village planting stock	Number of distinct banana cultivars available in village	+/−	517	23.39	5.53
Probability of BS	Perceived (by farmer) frequency of occurrence of Black Sigatoka	+	517	0.18	0.29
Probability of FW	Perceived (by farmer) frequency of occurrence of Fusarium wilt	+	517	0.20	0.28
Probability of WE	Perceived (by farmer) frequency of occurrence of weevils	+	517	0.39	0.33
Rainfall	Average seasonal rainfall (in mm)	+/−	517	90.95	8.12
Market characteristics					
Selling bunches	Market participation as a seller in previous season (1 = sell; 0 = not sell)	+/−	517	0.51	0.50
Buying bunches	Market participation as a buyer in previous season (1 = buy; 0 = not buy)	+/−	517	0.34	0.47
Time to market	Time taken to get to nearest market (in hours)	+/−	517	1.00	0.53

in the sample have no members between the ages of 16 and 64. The boundaries of these age cohorts are those used by sociologists in Uganda. Because of the importance of livestock to household consumption needs and cash requirements the value of animals owned by the household is used as a proxy for wealth. Another indicator of wealth is exogenous income, which is measured as total income received in the previous year. Wealth is often associated positively with crop biodiversity in poorer economic contexts, but not necessarily in higher income countries (see Chapters 1, 5, 6, 8, 9, 10).

The frequency of contact with extension workers controls for acquired knowledge, which differs from experience or observation and formal education. In this context, extension contact provides general information relevant to management of banana plantations and end uses of various banana cultivars, but no particular focus on dissemination of new planting material.

The extent of area planted to bananas captures household-specific scale effects. Farmer objectives determine the direction of this variable's effect. It is established in the biogeographical theory of ecology that, other factors being held constant, larger area is associated with larger number of species encountered (MacArthur and Wilson, 1967, cited in Brush et al., 1992). Economic theory may suggest alternative explanations. If the farmers specialize in the production of a subset of cultivars that sell well in local markets, an inverse relationship between banana area and diversity is plausible. When subsistence needs drive production decisions, larger banana area can be associated with greater diversity if farmers perceive that different cultivars provide distinct advantages or disadvantages related to consumption needs and production requirements. The age of banana plantation is included to control the time dimension of the diversity decision. The direction of the hypothesized effect is ambiguous. Old plantations may be associated with greater diversity due to the longer time span for trying different cultivars. However, plantation age could also be related to lower diversity because farmers select fewer cultivars that they believe are most suitable to their household needs and production environment.

The number of crops grown at the household level is an indicator of farming-system complexity. A more complex cropping system would suggest less specialization in bananas and the possibility of substituting own-produced bananas for those purchased through income from sales of other crops. More crops might also imply more dispersed or fragmented banana mats. Fragmentation, sometimes related to soil type heterogeneity, is hypothesized to relate positively to the biodiversity of crops (Bellon and Taylor 1993; Chapters 5, 6, 9, 10). The total number of banana cultivars in the village proxies for the stock of planting material available to farmers for exchange through formal and informal networks. Generally, a larger number of distinct cultivars in the village indicate a greater local supply that is readily available to households (Smale et al., 2001) and a positive effect on on-farm diversity. However, the extent to which each farmer is willing to diversify could vary across farmers in a village, and some farmers may meet their end-use needs with fewer cultivars than others. The frequencies of occurrence of the airborne disease Black Sigatoka and the soilborne disease Fusarium wilt, as well as weevils, represent biotic pressures to banana production as recognized and experienced by farmers. Since tolerance appears to vary by cultivar, disease pressures are expected to increase demand for a wider set of cultivars. Rainfall is also included as an agroecological characteristic that is important to banana production and varies continuously, as compared to the discrete variable for elevation.

Market participation during the previous year is included to account for market failures that may encourage farmers to grow some cultivars and not others. Although the predicted direction of the effect is ambiguous, some speculations are possible. Semi-subsistent households participating in banana markets as sellers often meet their consumption needs and sales requirements through own production. They may grow a larger number of different cultivars, some allocated to their own consumption, and others to market sales. A derived demand for diversity in cultivars could be motivated by market participation as a seller. By contrast, buyer participation is likely to reduce diversity on farm since it enables households to substitute for on-farm production with market purchases. Households can then fulfil the range of their consumption needs through acquiring bunches at the market place rather than from their own banana plots. The time taken to get to the nearest banana market is used as a transaction cost variable. The farther a household is from a market, the greater the incentive it has to maintain a wider range of distinct banana cultivars in order to satisfy consumption needs. Although most of the empirical literature relates crop biodiversity to relative isolation from markets (review in Chapter 1; Chapters 3 and 5), the analyses presented in Chapters 6 and 11 have demonstrated that effects among indicators of market participation and transaction costs may be offsetting.

Results

Table 7.6 summarizes the marginal effects for banana cultivar diversity defined on household farms.

Inferences about the importance of single banana attributes and the count of all important banana attributes are presented in separate columns, for both richness and evenness regressions. The importance of banana attributes for household-level diversity outcomes appears to be

Table 7.6. Determinants of banana cultivar diversity at the household level.

Explanatory variables	Marginal effects			
	Count (Richness)		Shannon (Evenness)	
Attribute importance				
Bunch size	0.0643	—	0.0938*	—
Resistance to Black Sigatoka	−0.2341	—	−0.0287	—
Resistance to Fusarium wilt	0.3050	—	0.0348	—
Resistance to weevils	0.4403	—	0.1154**	—
Cooking quality	0.2091	—	0.0454	—
Beer quality	0.1522	—	0.0316	—
Utility count – all attributes	—	0.1524*	—	0.0368**
Individual and household characteristics				
Gender (1 = female)	−0.0420	−0.0436	−0.0049	−0.0042
Education	0.1170**	0.1101**	0.0102^	0.0086
Relative experience	−0.3545	−0.1804	0.0873	0.1098
Dependency ratio	−0.1637	−0.1226	−0.0508	−0.0365
Assets	0.0059**	0.0057**	0.0007**	0.0007**
Exogenous income	−0.0008^	−0.0007	−0.0002**	−0.0002**
Extension	0.0594	0.0718	0.0305**	0.0333**
Farm characteristics				
Banana area	−0.0627	−0.0751	−0.0196	−0.0215^
Age of plantation	0.0634**	0.0652**	0.0053*	0.0058*
Crops	−0.1805	−0.2156*	−0.0065	−0.0182
Village planting stock	0.2648**	0.2859**	0.0178**	0.0228**
Probability of Black Sigatoka	0.6276	0.3354	0.1265	0.0650
Probability of Fusarium wilt	0.6640^	0.7162	0.0894	0.0736
Probability of weevils	−0.4345	−0.3136	−0.1164*	−0.0858
Rainfall	−0.0716**	−0.0624**	−0.0110**	−0.0090**
Market characteristics				
Selling bunches (1 = sell)	1.1791**	1.1651**	0.1186**	0.1168**
Buying bunches (1 = buy)	−0.3077	−0.2574	−0.0485	−0.0508
Time to market	0.2407	0.2415	0.0749^	0.0666^
Log-likelihood	−1252.56	−1257.84	−325.36	−337.83

**Denotes statistical significance at the 1% level; *denotes statistical significance at the 5% level; and ^denotes statistical significance at the 10% level.
Note: Levels of statistical significance are according to the hypothesized one-tailed or two-tailed tests.

index-specific. The importance of bunch size has a positive effect on the evenness of the distribution among cultivars and no significant effect on richness of banana cultivars. Much stronger is the effect of resistance to weevils on both the richness and evenness in the banana plantation. Households that attribute great importance to resistance to weevils and Fusarium wilt are likely to grow a larger number of more evenly distributed banana cultivars on their farms. Diversifying cultivars may enhance tolerance to biotic pressures and maximize expected yields on the plantation. More cultivars more evenly distributed may also be associated with introducing new, clean clones.

Neither the importance of consumption attributes (cooking quality and beer quality) nor that of Black Sigatoka is significantly associated with either of the cultivar diversity indices. In the case of Black Sigatoka the result could be interpreted as a lack of recognition of the effects of this new disease. Insignificance of consumption attributes suggests that variation in the diversity of banana cultivars is more likely to be influenced by production side characteristics. Nevertheless, the positive effect of the count over all the important attributes is consistent with Bellon's (1996) hypothesis that no single cultivar offers all attributes demanded by the household. Clearly, combinations of attributes represented by different subsets of cultivars drive banana diversity levels on farms in Uganda.

The gender and experience of the decision maker have no apparent statistical relationship to banana cultivar diversity on household farms. Decision makers with more years of schooling maintain more cultivar-rich banana plantations, perhaps because they have a greater interest in new clones (whether exotic, improved or endemic) or they have more access to information about them. There appears to be no statistical association between banana diversity and the composition of the household in terms of the ratio of dependents. The value of livestock assets is positively associated with both the richness of cultivars and the evenness in the distribution of mat shares within the plantation. Wealth is positively associated with the capacity to bear production risk from biotic pressures, also providing access to new clones, related information and the resources to manage them. Income in cash transfers is negatively associated with banana cultivar diversity, perhaps through the substitution away from growing banana cultivars they can purchase at the marketplace. Extension enhances cultivar evenness significantly. Since extension services in these communities are believed to provide information about cultivars and their management more often than cultivars themselves, this finding reflects the effects of access to banana-related information and general farming support services.

The statistical significance and direction of effects for farm characteristics are similar across indexes, differing in magnitude of effect given the two dependent variables. The extent of banana area has no association with either, except for the weak effects on evenness index. The age of the banana plantation strongly influences diversity under both indexes. Long established plantations appear to be associated with greater richness and relative abundance of cultivars. The significance of the effect underscores the importance of the time component of diversity on household preferences for growing different banana cultivars and the effects of past investments in the plantation stock. Growing a larger number of other crops also tends to reduce the richness of banana cultivars, which is indicative of trade-offs in land allocation. Greater richness in crops other than banana reduces the richness within the banana crop.

A larger pool of distinct types of banana planting material at the village level enhances both the richness and relative abundance of banana cultivars on individual farms. Considering the vegetatively propagating nature of the species, this result is consistent with observations that banana planting material diffuses through strong farmer-to-farmer exchanges, enabling individual farmers to grow a number of different cultivars or grow more of a single cultivar than they would be able to in the absence of a network.

The perceived (by farmers) frequency of occurrence of biotic constraints (Black Sigatoka, Fusarium wilt and weevils) does not appear to have a uniform statistical effect across indexes. Greater probability of occurrence of Fusarium wilt appears to weakly increase richness, while perceived probability of occurrence of weevils reduces the relative abundance of cultivars. This suggests that the likelihood of occurrence plays some role in decisions to plant and how much to plant.

Rainfall has the strongest effect on diversity among the agroecological characteristics. More rain is associated with reduced richness of banana cultivars and the greater relative abundance of particular cultivars. The interpretation of this result is more transparent when rainfall is considered as a proxy for elevation and other stratum-specific physical characteristics. High elevation areas enjoy greater rainfall and are also the major banana production areas in Uganda. The scale of banana production is larger and more commercially driven than in low elevation areas. The finding may reflect regional specialization in banana production as well as on-farm specialization in production of a specific subset of banana cultivars.

Findings confirm that market-related characteristics are important determinants of cultivar diversity on farms. Rather than contribute to specialization, market participation as a seller has a strong positive effect on both evenness and richness and the magnitude of the effect on richness is particularly large. On average, the market demand for banana bunches generates a derived demand for an additional cultivar per farm. Market participation as a buyer does not appear to influence diversity. Buyers are likely to be those that are less able to meet the range of the consumption needs through their on-farm stocks of banana cultivars. As hypothesized in the literature, higher transaction costs (in terms of longer time spent getting to a banana market) are associated with induced demand for diversity on farms to meet consumption needs. This is supported by results for the Shannon index.

Marginal effects for the diversity of banana use groups on farms in Uganda are summarized in Table 7.7. Only inferences for the importance of single banana attributes are reported, since inclusion of the importance of all attributes or subgroups of attributes (e.g. consumption and agronomic attributes) did not yield statistically significant results.

In the use group regressions, the importance of bunch size and resistance to Black Sigatoka are both statistically significant factors. Differences in susceptibility to Black Sigatoka that have been observed by scientists are clearly also perceived by farmers and these perceptions affect the relative composition of the plantation by use group. Cooking and beer quality are also found to be statistically significant in the use group as compared to the cultivar regressions, reflecting similar trade-offs. The magnitudes of the effects are large in explaining the richness of use groups cultivated on farms. When cooking quality is more important as an attribute, households tend to grow fewer use groups, highlighting the importance of cooking cultivars in meeting household subsistence requirements. When beer quality is important, use group diversity increases, suggesting that the beer use group is added to the set of cooking cultivars already grown, rather than substituting for them. The statistical significance of consumption attributes also supports the notion of non-separability in the conceptual model.

No statistically significant relationships among the individual and household characteristics and use group diversity are apparent, with two exceptions. Wealth in livestock assets is again a statistically significant and positive determinant of richness in use groups. Extension seems to also have a weak positive effect on the relative abundance of use groups.

Several of the farm characteristics appear to affect use group diversity. As expected, older plantations have a larger number of use groups. The wider the range of banana cultivars that can be found in the village, the more use groups a farmer plants, though the magnitude of the effect is not so large as in the case of individual banana cultivars. The availability of different types of cultivars in the community allows farmers to diversify use groups grown on the farm, responding to the consumption needs of the household or to market requirements. A higher frequency of occurrence of Fusarium wilt in the plantation leads farmers to grow more use groups. This is understandable, given that Fusarium wilt affects exotic, brewing bananas more than endemic, cooking bananas. Greater likelihood of occurrence of both fungal diseases (Fusarium wilt and Black Sigatoka) appears to increase the evenness among use groups. This result is consistent with the hypothesis that farmers are able to perceive the type and frequency of a biotic constraint and is differential incidence across use groups, diversifying genetic resistance to banana diseases by planting combinations.

Market participation as a seller has a significant effect on both indexes. Selling banana bunches appears to be associated with more banana diversity in terms of use groups. As

Table 7.7. Determinants of banana use-group diversity at the household level.

Explanatory variables	Marginal effects	
	Count (Richness)	Shannon (Evenness)
Importance of single banana attributes		
Bunch size	0.1599^	0.0818**
Resistance to Black Sigatoka	0.0083	−0.0537*
Resistance to Fusarium wilt	0.0347	0.0379
Resistance to weevils	−0.0562	−0.0384
Cooking quality	−0.1204^	−0.0669**
Beer quality	0.1193*	0.0669**
Individual and household characteristics		
Gender (1 = female)	−0.0723	0.0107
Education	0.0150	−0.0005
Relative experience	−0.0003	0.1177
Dependency ratio	0.1380	0.0390
Assets	0.0010*	0.0002
Exogenous income	0.00003	−0.00005
Extension	0.0069	0.0179^
Farm characteristics		
Banana area	−0.0265	−0.0144
Age of plantation	0.0121*	0.0019
Crops	0.0568	0.0233
Village planting stock	0.0322**	0.0026
Probability of Black Sigatoka	0.2083	0.1383*
Probability of Fusarium wilt	0.7693**	0.3228**
Probability of weevils	−0.0022	−0.0164
Rainfall	−0.0074	−0.0013
Market characteristics		
Selling bunches (1 = sell)	0.4488**	0.1295**
Buying bunches (1 = buy)	0.0372	0.0607^
Time to market	0.0291	0.0100
Log-likelihood	−706.86	−280.71

**Denotes statistical significance at the 1% level; *denotes statistical significance at the 5% level; and ^ denotes statistical significance at the 10% level.
Note: Levels of statistical significance are according to the hypothesized one-tailed or two-tailed tests.

hypothesized, households grow some use groups for their own consumption and others, such as dessert bananas, for sale. Buying bunches has a positive, although weak, effect on the relative abundance of use groups. This could be explained by specialization in the production of some use groups (hence increasing their abundance), while preferences for other use groups are met through market purchases. Significance of market participation underscores the role of markets in shaping banana diversity on single farms.

Conclusions

The biodiversity of bananas in Uganda is understood at the taxonomic levels of genomic group, use group and cultivar. This diversity is impressive at all geographical scales of analysis – the household farm, the village and the region. Though banana specialists in East Africa have long made this observation, a recent sample survey establishes this fact statistically for the major banana-growing regions of the country. Findings underscore the importance of cultivar attributes in explaining the decisions of banana growers in Uganda. The differential vulnerability to pests and diseases among cultivars probably explains the effects of these biotic stresses on the diversity of banana cultivars and use groups maintained by farmers. Vulnerability was measured in this study both in terms of farmers' perceptions of the frequency of occurrence in their plantation and the relative importance of the biotic stress. Although trade-offs across use groups are revealed when cooking quality and beer quality are considered, production traits are generally more important in explaining cultivar diversity than are consumption attributes.

Econometric analyses indicate that farmers holding more value in livestock assets are more likely to grow larger numbers of distinct cultivars and use groups, more evenly distributed. In contrast, cash income effects appear to reduce the diversity in both indexes. The availability of large stocks of diverse banana planting material in the community is positively associated with greater richness of cultivars and use groups on individual farms. When the breadth of cultivar attributes demanded by subsistence farmers is limited, dissemination of planting material (either improved or landrace) can have a positive effect on crop biodiversity. The results also suggest that the age of the plantation is positively associated with both cultivar and use group diversity. The older the plantation, the more time for families to accumulate diverse banana types within and over generations of managers. Participating as a seller increases the diversity of both cultivars and use groups on farms. However, reduction in time to market through investments in road infrastructure would offset this effect. Participation as a buyer has a positive impact on use group diversity but detracts from cultivar diversity. As suggested by other chapters in this book, a more comprehensive treatment of the role of markets in empirical analyses of crop biodiversity appears to be needed.

Implications

This chapter contributes to the literature about the East African highland banana by using an economics conceptual framework to explain the patterns of banana diversity found on farms. It contributes to the literature about on-farm conservation by relating banana diversity on farms to the demand for banana attributes. Banana cultivars supply attributes unequally. For semi-subsistent households facing market imperfections, attributes demanded include cooking and beer quality as well as production traits.

In this chapter, as compared with other chapters, analysing the diversity of cereal crops, Shannon evenness indices are constructed from counts of banana mats (mother plant and plantlets). Banana diversity is analysed in terms of two of the three taxonomic levels (cultivars and use groups).

Although many of the results are consistent with empirical findings in other sections of this book, the unique nature of bananas limits the extent of applicability of the findings to other perennial crops (e.g. Chapters 4 and 12). Bananas are also vegetatively propagated and a unique system of reproduction and dissemination of planting material exists among farmers. In Uganda, however, banana production is primarily driven by subsistence needs rather than commercial goals, as is commonly the case for perennial crops such as coffee, cocoa, rubber, tea and other fruits – as well as bananas in other contexts. On one hand, the overwhelming importance of attributes and the extensive biodiversity on farms in Uganda does not lead to the expectation that newly improved banana cultivars will displace local cultivars in the near future. Still, judicious introduction of newly improved banana cultivars will be important if, in addition to relieving productivity and market constraints to banana production, protecting biodiversity in the East African highland banana is of policy concern.

Acknowledgements

This chapter is drawn from PhD research conducted as part of a project with multiple partners, including Uganda, the International Network for the Improvement of Banana and Plantain (INIBAP) and the Banana Research Programme of the National Agricultural Research Organization. The research was funded by the U.S. Agency for International Development. We gratefully acknowledge the contributions of staff in INIBAP and the Banana Research Programme in Uganda as well as those of Professors Mitch Renkow and Dan Phaneuf at North Carolina State University.

References

Bellon, M. (1996) The dynamics of crop intraspecific diversity: a conceptual framework at the farmer level. *Economic Botany* 50, 26–39.

Bellon, M. and Taylor, J. (1993) 'Folk' soil taxonomy and the partial adoption of new seed varieties. *Economic Development and Cultural Change* 41, 763–786.

Brush, S.B., Taylor, J.E. and Bellon, M.R. (1992) Technology adoption and biological diversity in Andean potato agriculture. *Journal of Development Economics* 2, 365–387.

Craenen, K. (1998) *Black Sigatoka Disease of Banana and Plantain: A Reference Manual.* IITA, Ibadan, Nigeria.

de Janvry, A., Fafchamps, M. and Sadoulet, E. (1991) Peasant household behaviour with missing markets: some paradoxes explained. *The Economic Journal* 101, 1400–1417.

Edmeades, S. (2003) Variety choice and attribute trade-offs within the framework of agricultural household models: the case of bananas in Uganda. PhD dissertation, North Carolina State University, Raleigh, North Carolina.

Gold, C., Ogenga-Latigo, M., Tushemereirwe, W., Kashaija, I. and Nankinga, C. (1993) Farmer perceptions of banana pest constraints in Uganda: results from a rapid rural appraisal. In: *Biological and Integrated Control of Highland Banana and Plantain Pests and Diseases.* Proceedings of a Research Coordination Meeting, IITA, 12–14 November 1991, Ibadan, Nigeria.

Gold, C., Speder, P., Karamura, E., Tushemereirwe, W. and Kashaija, I. (1994) Survey methodologies for banana weevil and nematode damage assessment in Uganda. *African Crop Science Journal* 2, 309–321.

Gold, C., Kiggundu, A., Karamura, D. and Abera, A. (1998) Diversity, distribution and selection criteria of *Musa* germplasm in Uganda. In: Picq, C., Foure, E. and Frison, E.A. (eds) *Bananas and Food Security.* International Symposium, November 1998, Cameroon, pp. 163–179.

Howard, P. (2003) *Women and Plants: Gender Relations in Biodiversity Management and Conservation.* Zed Press and St Martin's Press, London and New York.

INIBAP (2000) Networking bananas and plantains. INIBAP Annual Report, Montpellier.

Johanson, A. and Ives, C. (2001) *An Inventory of Agricultural Biotechnology for the Eastern and Central Africa Region.* Michigan State University, Michigan.

Karamura, D. and Karamura, E. (1994) *A Provisional Checklist of Banana Cultivars in Uganda.* National Agricultural Research Organization (NARO) and INIBAP, Uganda.

Karamura, E., Frison, E., Karamura, D. and Sharrock, S. (1998) Banana production systems in eastern and southern Africa. In: Picq, C., Foure, E. and Frison, E.A. (eds) *Bananas and Food Security. International Symposium, Cameroon, 10–14 November 1998*, pp. 401–412.

Karamura, D. and Pickersgill, B. (1999) A classification of the clones of East African highland bananas (*Musa*) found in uganda. *Plant Genetic Resources Newsletter* 119, 1–6

Ladd, G. and Martin, M. (1976) Prices and demands for input characteristics. *American Journal of Agricultural Economics* 58, 21–30.

Ladd, G. and Suvannunt, V. (1976) A model of consumer goods characteristics. *American Journal of Agricultural Economics* 58, 504–510.

Lancaster, K. (1966) A new approach to consumer theory. *The Journal of Political Economy* 74, 132–157.

MacArthur, R. and Wilson, E. (1967) *The Theory of Island Biogeography.* Princeton University Press, Princeton, New Jersey.

Magurran, A. (1988) *Ecological Diversity and its Measurements.* Princeton University Press, Princeton, New Jersey.

Persley, G. and George, P. (1999) Banana, breeding and biotechnology: commodity advances through banana improvement project research, 1994–1998. World Bank Banana Improvement Project Report 2, The World Bank, Washington, DC.

Sadoulet, E. and de Janvry, A. (1995) *Quantitative Development Policy Analysis.* Johns Hopkins University Press, Baltimore, Maryland.

Simmonds, N. (1959) *Bananas.* Longmans Publishers, London.

Singh, I., Squire, L. and Strauss, J. (eds) (1986) *Agricultural Household Models: Extensions, Applications, and Policy.* Johns Hopkins University Press, Baltimore, Maryland.

Smale, M., Bellon, M. and Aguirre Gomez, J. (2001) Maize diversity, variety attributes, and farmers' choices in south-eastern Guanajuato, Mexico. *Economic Development and Cultural Change* 50, 201–225.

Speijer, P., Kajumba, C. and Tushemereirwe, W. (1999) Dissemination and adaptation of a banana clean planting material technology in Uganda. *InfoMusa* 8, 11–13.

Stover, R.H. (2000) Diseases and other banana health problems in tropical Africa. In: Karamura, E. and Vuylsteke, D. (eds) *Proceedings of the First International Conference on Banana and Plantain for Africa, Uganda, October 1996.* ISHS, Leuven, Belgium, pp. 311–317.

8 Explaining Farmer Demand for Agricultural Biodiversity in Hungary's Transition Economy

E. Birol, M. Smale and Á. Gyovai

Abstract

In this chapter, a household farm model is used to predict farmer demand for four components of the agricultural biodiversity found on family farms in Hungary. Family farms in Hungary are known traditionally as 'home gardens'. The analysis is based on survey data from the same sites studied in Chapters 3 and 15, which use stated preference and institutional approaches. The four components analysed are: (i) richness in crops and varieties (crop variety diversity); (ii) cultivation of landraces, as compared to modern varieties (crop genetic diversity); (iii) integration of crop and livestock production (agrodiversity); and (iv) use of organic production methods (soil microorganism diversity). The econometric model is based on the approach presented in Chapter 5, specified in different ways to reflect the definition of the dependent variable. Farm households who are most likely to sustain observed levels of agricultural biodiversity are described statistically. The stratified sample design lends insights into the potential impact of economic transition on the prospects for conserving agricultural biodiversity. Findings can assist those who formulate agri-environmental policy in Hungary to design efficient programmes that incorporate home garden management.

Introduction

Chapter 3 examined the stated preferences of farm families in rural Hungary for four components of agricultural biodiversity found on their homestead fields, traditionally known as 'home gardens'. In Hungary's emerging agricultural system today, these production units may be more appropriately termed small-scale farms. The historical role and significance of home gardens was also sketched in Chapter 3. Chapter 15 elaborates the institutional context of home gardens, and summarizes in detail the policies and stakeholders that will influence the prospects of their survival with Hungary's recent membership to the European Union (EU).

As in the other chapters in Part III of this book, this chapter investigates the factors affecting observed levels of agricultural biodiversity within the theoretical framework of the household farm. Conceptually, the levels observed are understood as reflecting the optimal choices of farm families. Optimal choices reveal the preferences of these families given the multiple constraints they face, including communities with imperfect markets for production inputs and farm produce.

In this chapter, predictions based on the econometric model enable us to profile households that are most likely to sustain current levels of agricultural biodiversity components because they reveal the greatest preference for them. Dynamics of economic change are 'controlled' through sample design in this cross-sectional data set. Methods of this type can assist in designing strategies for on-farm conservation programmes that are cost-effective, efficient and equitable (Meng, 1997; Chapter 1).

The following section describes the structure of family farms in the study sites, including farm fields and home gardens, the characteristics of farm households and the agrobiodiversity components they manage in the study sites. The third section presents the underlying theoretical approach that motivates the econometric models. The econometric models seek to explain variation in levels of four different components of agricultural biodiversity found on Hungarian family farms. Hypotheses and operational variables are then defined. The fourth section presents the econometric findings followed by the profiles of the household farms that are most likely to sustain these components. Conclusions for the design of conservation programmes are drawn in the final section.

Data Source

A detailed description of the sample design is included in Chapter 3 along with a map of study sites. Site regions were purposively selected in environmentally sensitive areas (ESAs) of Hungary to represent contrasting levels of market development and varying agroecologies associated with different farming systems and land-use intensity. In the three sites (Dévaványa, Örség-Vend and Szatmár-Bereg regions), previous collections undertaken by the Institute of Agrobotany disclosed a relatively high frequency of crop landrace cultivation. Households in 22 settlements were randomly selected within sites. A total of 323 farm households were personally interviewed in August 2002 with a household survey instrument. Findings are statistically representative of the three regions and of other ESAs in Hungary bearing similar features.

Site Description

Chapter 3 described the economic and social characteristics of settlements in the three survey sites in detail. Dévaványa region, on the Hungarian Great Plain, is a flat mosaic of cultivated lands and grasslands where soil and climatic conditions are well suited to intensive agricultural production. Dévaványa is the most urbanized and economically developed region with good market infrastructure. Located in the south-west, Örség-Vend region has a heterogeneous agricultural landscape with knolls, valleys, forests, grasslands and arable lands. Poor soil conditions render intensive agricultural production methods impossible (Gyovai, 2002). Szatmár-Bereg region, in the north-east of Hungary, has a landscape consisting of moors, grasslands, forests and arable lands.

Family and farm characteristics of households surveyed are shown in Table 8.1. The average family size is three persons and children are few in all the sites. Örségi households have larger families and more children than those in Dévaványa. Households in Örség-Vend have significantly higher levels of income than those in Dévaványa and Szatmár-Bereg, but the difference between Dévaványa and Szatmár-Bereg is insignificant. The number of family members employed off-farm is higher in Örség-Vend than in Szatmár-Bereg but similar between Örség-Vend and Dévaványa. On average, households in Dévaványa and Örség-Vend spend approximately the same percentage of their income on food but this percentage is statistically higher than in Szatmár-Bereg. Home garden decision makers are elderly, and their average ages do not differ statistically among the three regions. Dévaványa has statistically more experienced and educated home garden decision makers compared to Szatmár-Bereg. Örség-Vend has the smallest percentage of decision makers that have fewer than 8 years of education across the three ESAs. A large proportion of them are retired, though the percentage is statistically lower in Dévaványa. The percentage of home garden decision makers with off-farm employment is higher in Dévaványa than Szatmár-Bereg. A higher percentage of Örségi households own cars compared to the other two regions.

Home gardens are homestead fields adjacent to the family dwellings that were essentially fixed in size from 1958 to 1989. During the period of agricultural collectivization and state ownership (1958–1989), families were allowed to cultivate these fields privately. There are regional variations in sizes and functions of home gardens across ESAs. Redistribution of land and fields since 1989 is also likely to have contributed to heterogeneity in the organization of production. Home gardens in Dévaványa are generally small

Table 8.1. Characteristics of households and home garden decision makers, by region. (From Household Survey, Hungarian On-farm Conservation of Agricultural Biodiversity Project, 2002.)

		Mean (s.e.)		
Variable	Definition	Dévaványa N = 104	Örség-Vend N = 109	Szatmár-Bereg N = 110
Family size**	Number of family members	2.7 (1.2)	3.1 (1.6)	2.8 (1.5)
Participation**	Number of family members that work in home garden	2.1 (1)	2.5 (1.3)	2.4 (1.3)
Children*	Number of family members < 12 years	0.3 (0.7)	0.5 (0.8)	0.4 (0.8)
Off-farm employment**	Number of family members employed off-farm	0.8 (1)	1 (1.1)	(0.7) (1)
Income***	Average monthly income from off-farm employment, pensions, rents, gifts or other benefits	747,778.2 (25,413.2)	92,341.5 (19,986.3)	71,685.6 (40,740.4)
Food expenditure***	Stated percentage of income spent on food consumption	39.2 (15.1)	39.7 (16.8)	32.8 (11.8)
Age	Average age of home garden decision makers	58.5 (13.1)	57.8 (12.4)	56.6 (15)
Experience*	Average years of farming experience of home garden decision makers	42.8 (17.6)	40.7 (17.1)	38.4 (19.6)
Education*	Years of formal education the home garden decision makers have received	10 (2.8)	9.9 (2.7)	9.3 (3.3)
		%		
Off-farm*	Decision makers with off-farm employment	39.4	33.9	30
Retired	Retired decision makers	66.3	72.5	72.7
Less than minimum education**	Decision makers with fewer than 8 years of education	13.5	4.6	21.3
Car***	The household owns a car	41.7	64.2	44.6

*The t-tests and Pearson Chi square tests show significant differences among at least one pair of environmentally sensitive areas (ESAs) at 10% significance level; **at 5% significance level; ***at 1% significance level.

and oriented towards supplying the food needs of farm families. In Örség-Vend they are larger, and mostly include fields, orchards and/or grasslands as a result of the special settlement structure of 'szer' and 'szórvány' in this region. Home gardens are largest in Szatmár-Bereg, where they can contain orchards and/or fields that supply not only the needs of the households but also enable sales to generate cash income.

The likelihood that a farm household cultivates a field in addition to a home garden is greater in Örség-Vend than in either of the other ESAs, though the areas of land owned and cultivated, and cultivated that is also owned are less (Table 8.2). The smallest home gardens and the largest total areas owned and cultivated are in Dévaványa, the most favoured ESA in terms of either soils or infrastructure. Home gardens with the least irrigation and best soil quality are located in Szatmár-Bereg. Örségi home gardens have more irrigation than those in Dévaványa, but they also have the worst soil quality of the three regions surveyed.

As sellers of home garden produce, households in Szatmár-Bereg are more integrated into markets compared to those found in the other two ESAs, with a larger volume of sales. Mean value of sales per garden does not differ

Table 8.2. Home garden, field and market characteristics, by region. (From Household Survey, Hungarian On-farm Conservation of Agricultural Biodiversity Project, 2002.)

		Mean (s.e.)		
Variable	Definition	Dévaványa $N = 104$	Örség-Vend $N = 109$	Szatmár-Bereg $N = 110$
Home garden area**	In square metres	560.9 (683)	16,24.6 (2,872.1)	2,649.2 (3,041.9)
Total fields owned***	"	86,215.7 (319,476.5)	24,561.3 (36,780.2)	40,300.9 (62,608.4)
Total fields cultivated***	"	83,709.1 (321,854)	21,657.7 (43,372)	61,323 (103,984)
Owned land***	Total land cultivated by the household that is also owned by the household (m^2)	78,956.2 (320,233.3)	16,962 (31,441.5)	42,753.7 (64,057.4)
Irrigation**	Percentage of home garden land irrigated	36.1 (45.5)	46 (40.4)	16.6 (28.2)
Sales**	Value of total home garden output sold in market prices in Hungarian Forint per square metre of home garden	5.5 (29.6)	6.6 (49.7)	33 (103.3)
Distance***	Distance of the settlement in which the household is located from the nearest market (km)	0 (0)	19.9 (6.8)	18.4 (3.2)
		%		
Cultivate field**	Household cultivates a field along with the home garden	42.3	59.6	44.5
Good soil**	Home garden soil is of good quality	16.8	9.2	31.2

The *t*-tests and Pearson Chi square tests show significant differences among at least one pair of environmentally sensitive areas (ESAs) **at 5% significance level; ***at 1% significance level.

statistically between the other two regions. Distance to the nearest food market is negligible in Dévaványa since each of the settlements studied in this survey has a market. Households in Örség-Vend and Szatmar-Bereg must travel much greater, and similar, distances (nearly 20 km) to the nearest food markets.

Agricultural biodiversity on home gardens is depicted in terms of four components of the home garden. As explained in Chapter 3, four key components or 'entry points' to agricultural biodiversity in home gardens were identified in key informant interviews: (i) crop variety diversity (richness of crop varieties); (ii) crop genetic diversity (cultivation of landraces as compared to only modern varieties); (iii) agrodiversity (integrated crop and livestock production); and (iv) soil microorganism diversity (use of organic production practices). Crop variety diversity was represented by richness, or a count of all varieties of all crops grown, including field crops, vegetables and trees. Crops included species and underspecies (subspecies, covarieties) of field crops, vegetables and trees. Cultivation of landraces was measured for maize and bean crops, based on previous collections by the Institute of Agrobotany in study regions. Only management of large animals (i.e. pig, cattle, horse and donkey) was taken into consideration since small animals do not require much labour time or land area. Soil microorganism diversity was indicated by the use of organic methods.

Table 8.3 describes the components of agricultural biodiversity found on home gardens in each ESA. The average levels of crop variety diversity maintained by farm families in their

Table 8.3. Agricultural biodiversity found in home gardens, by region. (From Household Survey, Hungarian On-farm Conservation of Agricultural Biodiversity Project, 2002.)

	Mean (s.e.)		
Component of agricultural biodiversity	Dévaványa N = 104	Örség-Vend N = 109	Szatmár-Bereg N = 110
Crop variety diversity**	17 (8.9)	28.1 (12.5)	18.6 (7.5)
		%	
Landrace cultivation**	27	52	52
Agrodiversity	51	62	55
Organic production*	16	17	8

*The t-tests and Pearson Chi square tests show significant differences among at least one pair of environmentally sensitive areas (ESAs) at 10% significance level; ** at 5% significance level.

home gardens are significantly higher for Örség-Vend than in the other two sites. In Dévaványa, the percentage of households growing landraces is half of that found in the other two. Across the three sites, roughly 50–60% of households tend livestock along with crops in their homestead plots with no statistically significant differences.

Use of organic methods is similarly represented in Dévaványa and Örség-Vend, but probably for different reasons. The stated preference study conducted on the same sample of households demonstrated that in Dévaványa households that have access to food markets as well as off-farm employment prefer organic produce, suggesting a good with luxury properties. In Örség-Vend, by contrast, older and poorer households preferred organic production methods since they lack access to chemical input markets (Birol et al., 2004). Only 8% of farmers in Szatmár-Bereg, the region with the largest home gardens and sales of produce, apply organic practices, which is significantly lower than in the other regions.

In Dévaványa region 94 different crops were cultivated in home gardens and fields by households surveyed (Table 8.4). Surprisingly, most of the crops (60) can be found in both gardens and fields. In outlying areas of settlements, fields called 'closed gardens' often serve the same

Table 8.4. List of crops grown in Dévaványa. (From Household Survey, Hungarian On-farm Conservation of Agricultural Biodiversity Project, 2002.)

No.	Hungarian name	Latin name	English name	Home garden	Field
1	Paradicsom	Lycopersicon esculentum	Tomato	x	x
2	Étkezési paprika	Capsicum annuum var. grossum	Sweet pepper	x	x
3	Fűszerpaprika	Capsicum annuum var. longum	Red pepper	x	x
4	Csicsóka	Helianthus tuberosus	Jerusalem artichoke	x	—
5	Uborka	Cucumis sativus	Cucumber	x	x
6	Sárgarépa	Daucus carota subsp. sativus	Carrot	x	x
7	Petrezselyem	Petroselinum crispum	Apiaceous	x	x
8	Zeller	Apium graveolens	Celery	x	x

(Continued)

Table 8.4. (Cont'd)

No.	Hungarian name	Latin name	English name	Home garden	Field
9	Pasztinák	*Pastinacea sativa*	Parsnip	x	—
10	Lestyán	*Levisticum officinale*	Lovage	x	—
11	Cékla	*Beta vulgaris* var. *conditiva*	Red beet	x	x
12	Vöröshagyma	*Allium cepa*	Onion	x	x
13	Lila hagyma	*Allium cepa*	Purple onion	x	x
14	Gyöngyhagyma	*Allium cepa* var. *margaritaceum*	Levant garlic	x	x
15	Póréhagyma	*Allium porum*	Leek	x	—
16	Metélöhagyma	*Allium schoenoprasum*	Chive	x	—
17	Téli sarjadékhagyma	*Allium fistulosum*	Welsh onion/ stone leek	x	—
18	Fokhagyma	*Allim sativum*	Garlic	x	x
19	Patiszon	*Cucurbita pepo* var. *patissonina*	Scallop squash	x	x
20	Padlizsán	*Solanum melongena*	Aubergine	x	x
21	Cukkini	*Cucurbita pepo* var. *giromontiina*	Vegetable marrow	x	x
22	Karfiol	*Brassica oleracea* subsp. *botrytis*	Cauliflower	x	x
23	Brokkoli	*Brassica oleracea* subsp. *botrytis* var. *italica*	Broccoli	x	x
24	Karalábé	*Brassica oleracea* subsp. *caulorapa* var. *gongyloides*	Kohlrabi	x	x
25	Fejeskáposzta	*Brassica oleracea*	Headed cabbage	x	x
26	Kelkáposzta	*Brassica oleracea* var. *sabauda*	Savoy cabbage	x	x
27	Fejessaláta	*Lactuca sativa*	Cabbage head lettuce	x	x
28	Jégsaláta	*Lactuca sativa*	Iceberg lettuce	x	x
29	Hónapos retek	*Raphanus sativus*	Little radish	x	x
30	Fekete retek	*Raphanus sativus* var. *niger*	Black radish	x	—
31	Jégcsapretek	*Raphanus sativus*	Ice radish	x	—
32	Zöldborsó	*Pisum sativum*	Sugar pea	x	x
33	Spenót	*Spinacia oleracea*	Spinach	x	x
34	Spárga	*Asparagus officinalis*	Asparagus	x	—
35	Sóska	*Rumex acetosa*	French sorrel	x	—
36	Rebarbara	*Rheum rhaponticum*	Rheum, pieplant	—	x
37	Kapor	*Anethum graveolens*	Dill	x	x
38	Torma	*Armoracia lapathifolia*	Horse radish	x	—
39	Paradicsompaprika	*Capsicum annuum lycopersiciforme*	Tomato paprika	x	—
40	Almapaprika	*Capsicum annuum*	Apple paprika	x	x
41	Borsmenta	*Mentha piperita*	Peppermint	x	—
42	Tárkony	*Artemisia dracunculus*	Tarragon	x	—
43	Kaliforniai paprika	*Capsicum annuum*	Californian paprika	x	—
44	Díszpaprika	*Capsicum frutescens/ Solanum capsicastrum*	Decoration paprika	x	—
45	Alma	*Malus domestica*	Apple	x	x
46	Körte	*Pyrus communis*	Pear	x	x
47	Birsalma	*Cydonia oblonga*	Quince	x	x

Table 8.4. (Cont'd)

No.	Hungarian name	Latin name	English name	Home garden	Field
48	Birskörte	*Cydonia oblonga*	Pear quince	x	—
49	Szilva	*Prunus domestica*	Plum	x	x
50	Ringlószilva	*Prunus domestica* subsp. *italica* var. *claudiana*	Greengage	x	x
51	Meggy	*Cerasus vulgaris*	Sour cherry	x	x
52	Cseresznye	*Cerasus avium*	Cherry	x	x
53	Öszibarack	*Prunus persica*	Peach	x	x
54	Nektarin	*Prunus persica* var. *nucipersica*	Nectarine	x	x
55	Sárgabarack	*Prunus armeniaca*	Apricot	x	x
56	Földieper	*Fragaria*	Strawberry	x	x
57	Málna	*Rubus idaeus*	Raspberry	x	x
58	Szeder	*Rubus occidentalis*	Blackberry	x	x
59	Ribizli	*Ribes rubrum*	Ribes, currant	x	x
60	Köszméte	*Ribesuva-crispa*	Gooseberry	x	x
61	Josta	*Ribes rubrum* × *Ribes uva-crispa*	Jostabeere	x	—
62	Görögdinnye	*Citrullus lanatus*	Watermelon	x	x
63	Sárgadinnye	*Cucumis melo*	Sugarmelon	x	x
64	Naspolya	*Mespilus germanica*	Medlar	x	—
65	Füge	*Ficus carica*	Fig	x	—
66	Dió	*Juglans regia*	Walnut	x	x
67	Mogyoró	*Corylus avellana*	Hazelnut	x	x
68	Bodza	*Sambucus nigra*	Elder	x	—
69	Szölö	*Vitis vinifera*	Grape	x	x
70	Szedermálna	*Rubus mohacsyanus*	—	x	—
71	Lucerna	*Medicago sativa*	Lucerne	x	x
72	Vöröshere	*Trifolum pratense*	Red clover	x	—
73	Cirok	*Sorghum vulgare technicum*	Broomcorn	x	—
74	Takarmányrépa	*Beta vulgaris* var. *crassa*	Cattle turnip	x	—
75	Öszi búza	*Triticum aestivum*	Winter wheat	—	x
76	Tavaszi búza	*Triticum aestivum*	Spring wheat	—	x
77	Öszi árpa	*Hordeum hexastichon*	Winter barley	—	x
78	Tavaszi árpa	*Hordeum distichon*	Two-rowed barley	—	x
79	Zab	*Avena sativa*	Oat	—	x
80	Napraforgó	*Helianthus annuus*	Sunflower	x	x
81	Dohány	*Nicotiana tabacum*	Tobacco	x	—
82	Cukorrépa, burgundi répa	*Beta vulgaris* var. *altissima*	Sugarbeet	x	—
83	Mák	*Papaver somniferum*	Garden poppy	x	x
84	Sárgaborsó	*Pisum sativum*	Dried peas	x	—
85	Takarmány kukorica	*Zea mays*	Fodder maize	x	x
86	Csemege kukorica	*Zea mays* var. *saccharata*	Sweet corn	x	x
87	Pattogatni való kukorica	*Zea mays* var. *microsperma*	Popcorn	x	x
88	Silókukorica	*Zea mays*	Silage maize	—	x
89	Bab	*Phaseolus vulgaris*/*P. coccineous*	Bean	x	x
90	Főzötök	*Cucurbita pepo* subsp. *pepo*	Pumpkin	x	x
91	Sütötök	*Cucurbita maxima*	Giant pumpkin	x	x
92	Olajtök	*Cucurbita pepo*	Oil pumpkin	x	x
93	Takarmánytök	*Cucurbita pepo* var. *pepo*	Fodder pumpkin	x	x
94	Burgonya	*Solanum tuberosum*	Potato	x	x

function as gardens located adjacent to the dwelling. As a consequence, vegetables are the most frequently grown crops in the fields, although the agroecological conditions are well-suited to intensive production and the areas cultivated by farmers are extensive. Other frequently grown crops include maize, wheat, lucerne, barley, oat and sunflower.

Örség has the highest number of crops grown in home gardens and fields of the three sites (122). Again, a large proportion of the same crops can be found both in the home garden and in the field (Table 8.5). Fields in this region are usually small in size (Table 8.2). Due to special settlement forms, they are located close to households and function similarly to home gardens.

Table 8.5. List of crops in Örség-Vend. (From Household Survey, Hungarian On-farm Conservation of Agricultural Biodiversity Project, 2002.)

No.	Hungarian name	Latin name	English name	Home garden	Field
1	Paradicsom	Lycopersicon esculentum	Tomato	x	x
2	Étkezési paprika	Capsicum annuum var. grossum	Sweet pepper	x	x
3	Fűszerpaprika	Capsicum annuum var. longum	Red pepper	x	x
4	Csicsóka	Helianthus tuberosus	Jerusalem artichoke	x	—
5	Uborka	Cucumis sativus	Cucumber	x	x
6	Sárgarépa	Daucus carota subsp. sativus	Carrot	x	x
7	Petrezselyem	Petroselinum crispum	Apiaceous	x	x
8	Zeller	Apium graveolens	Celery	x	x
9	Pasztinák	Pastinaca sativa	Parsnip	x	—
10	Lestyán	Levisticum officinale	Lovage	x	—
11	Cékla	Beta vulgaris var. conditiva	Red beet	x	x
12	Vöröshagyma	Allium cepa	Onion	x	x
13	Lila hagyma	Allium cepa	Purple onion	x	—
14	Gyöngyhagyma	Allium cepa var. margaritaceum	Levant garlic	x	x
15	Póréhagyma	Allium porum	Leek	x	—
16	Metélőhagyma	Allium schoenoprasum	Chive	x	—
17	Téli sarjadékhagyma	Allium fistulosum	Welsh onion/stone leek	x	—
18	Fokhagyma	Allim sativum	Garlic	x	x
19	Patiszon	Cucurbita pepo var. patissonina	Scallop squash	x	x
20	Padlizsán	Solanum melongena	Aubergine	x	—
21	Cukkini	Cucurbita pepo var. giromontiina	Vegetable marrow	x	x
22	Karfiol	Brassica oleracea subsp. botrytis	Cauliflower	x	—
23	Brokkoli	Brassica oleracea subsp. botrytis var. italica	Broccoli	x	—
24	Karalábé	Brassica oleracea subsp. caulorapa var. gongyloides	Kohlrabi	x	x
25	Fejeskáposzta	Brassica oleracea	Headed cabbage	x	x
26	Lila káposzta	Brassica oleracea subsp. capitata f. rubra	Red cabbage	x	x
27	Kelkáposzta	Brassica oleracea var. sabauda	Savoy cabbage	x	x

Table 8.5. (Cont'd)

No.	Hungarian name	Latin name	English name	Home garden	Field
28	Bimbós kel	*Brassica oleracea* subsp. *acephala* var. *gemmifera*	Brussels sprouts	x	x
29	Kínai kel	*Brassica rapa* subsp. *chinensis*	Chinese cabbage	x	—
30	Fejessaláta	*Lactuca sativa*	Cabbage head lettuce	x	x
31	Tépösaláta	*Lactuca sativa* var. *crispa*	Looseleaf lettuce	x	x
32	Jégsaláta	*Lactuca sativa*	Iceberg lettuce	x	x
33	Endívia	*Cichorium endivia*	Endive	x	—
34	Hónapos retek	*Raphanus sativus*	Little radish	x	x
35	Fekete retek	*Raphanus sativus* var. *niger*	Black radish	x	—
36	Zöldborsó	*Pisum sativum*	Sugar pea	x	x
37	Spenót	*Spinacia oleracea*	Spinach	x	—
38	Spárga	*Asparagus officinalis*	Asparagus	x	—
39	Sóska	*Rumex acetosa*	French sorrel	x	—
40	Mángold	*Beta vulgaris cicla*	Spinach-beet, swiss chard	x	x
41	Rebarbara	*Rheum rhaponticum*	Rheum, pieplant	x	—
42	Articsóka	*Cynara scolymus*	Artichoke	x	—
43	Kapor	*Anethum graveolens*	Dill	x	x
44	Gumós kömény	*Carum carvi*	Caraway	—	—
45	Torma	*Armoracia lapathifolia*	Horse radish	x	x
46	Paradicsompaprika	*Capsicum annuum lycopersiciforme*	Tomato paprika	x	x
47	Almapaprika	*Capsicum annuum*	Apple paprika	x	x
48	Fehér hagyma	*Allium cepa*	White onion	x	—
49	Borsmenta	*Mentha piperita*	Peppermint	x	x
50	Zsálya	*Salvia officinalis*	Sage	x	—
51	Citromfü	*Melissa officinalis*	Lemon balm	x	—
52	Kakukkfü	*Thymus vulgaris*	Thyme	x	—
53	Szurokfü (oregano)	*Origanum vulgare*	Wild marjoram	x	—
54	Borsikafü vagy csombor	*Satureja hortensis*	Summer savory	x	—
55	Tárkony	*Artemisia dracunculus*	Tarragon	x	—
56	Zsidó cseresznye	*Physalis pruinosa*/ *P. peruviana*	Ground cherry	x	—
57	Kaliforniai paprika	*Capsicum annuum*	Californian paprika	x	—
58	Majoranna	*Origanum majorana*	Marjoram	x	—
59	Ánizs	*Pimpinella anisum*	Anise	x	—
60	Metélöpetrezselyem	*Petroselinum crispum* var. *crispum*	Parsley	x	—
61	Díszpaprika	*Capsicum frutescens*/ *Solanum capsicastrum*	Decoration paprika	x	—
62	Rukkola	*Eruca sativa*	Rucola	x	—
63	Alma	*Malus domestica*	Apple	x	x
64	Körte	*Pyrus communis*	Pear	x	x
65	Birsalma	*Cydonia oblonga*	Quince	x	—
66	Szilva	*Prunus domestica*	Plum	x	x
67	Ringlószilva	*Prunus domestica* subsp. *italica* var. *claudiana*	Greengage	x	—

(Continued)

Table 8.5. (Cont'd)

No.	Hungarian name	Latin name	English name	Home garden	Field
68	Meggy	*Cerasus vulgaris*	Sour cherry	x	x
69	Cseresznye	*Cerasus avium*	Cherry	x	x
70	Öszibarack	*Prunus persica*	Peach	x	—
71	Sárgabarack	*Prunus armeniaca*	Apricot	x	x
72	Eper (fa)	*Morus* sp.	Mulberry tree	x	x
73	Földieper	*Fragaria vesca*	Strawberry	x	—
74	Málna	*Rubus idaeus*	Raspberry	x	x
75	Szeder	*Rubus occidentalis*	Blackberry	x	x
76	Ribizli	*Ribes rubrum*	Ribes, currant	x	x
77	Köszméte	*Ribes uva-crispa*	Gooseberry	x	—
78	Josta	*Ribes rubrum* × *Ribes uva-crispa*	Jostabeere	x	x
79	Áfonya	*Vaccinium myrtillus*	Blueberry	—	x
80	Som	*Cornus mas*	Cornelberry	x	—
81	Görögdinnye	*Citrullus lanatus*	Watermelon	x	x
82	Sárgadinnye	*Cucumis melo*	Sugarmelon	x	—
83	Kivi	*Actinidia deliciosa*	Kiwi	x	—
84	Naspolya	*Mespilus germanica*	Medlar	x	x
85	Dió	*Juglans regia*	Walnut	x	x
86	Mogyoró	*Corylus avellana*	Hazelnut	x	—
87	Szelíd gesztenye	*Castanea sativa*	European chestnut	x	—
88	Bodza	*Sambucus nigra*	Elder	x	—
89	Szölö	*Vitis vinifera*	Grape	x	x
90	Fekete berkenye	*Aronia melanocarpa*	Service berry	—	x
91	Mandarinfa	*Citrus reticulata*	China orange	x	—
92	Törökmogyoró	*Corylus colurna*	Turkish hazel	x	—
93	Lucerna	*Medicago sativa*	Lucerne	x	x
94	Vöröshere	*Trifolium pratense*	Red clover	x	x
95	Cirok	*Sorghum vulgare technicum*	Broomcorn	x	x
96	Takarmány répa	*Beta vulgaris* var. *crassa*	Cattle turnip	x	x
97	Facélia	*Phacelia* sp.	Phacelia	x	—
98	Tarlórépa vagy kerekrépa	*Brassica rapa* L. var. *rapa*	Turnip	x	x
99	Karórépa, csutri	*Brassica napus*	Sweedish turnip	x	x
100	Szója	*Glycine max*	Soya	x	—
101	Öszi búza	*Triticum aestivum*	Winter wheat	x	x
102	Tavaszi búza	*Triticum aestivum*	Spring wheat	x	x
103	Öszi árpa	*Hordeum hexastichon*	Winter barley	—	x
104	Tavaszi árpa	*Hordeum distichon*	Two-rowed barley	—	x
105	Rozs	*Secale cereale*	Rye	x	x
106	Tritikálé	× *Triticosecale*	Triticale	—	x
107	Zab	*Avena sativa*	Oat	—	x
108	Napraforgó	*Helianthus annuus*	Sunflower	x	x
109	Cukorrépa, burgundi répa	*Beta vulgaris* var. *altissima*	Sugarbeet	x	x
110	Mák	*Papaver somniferum*	Garden poppy	x	x
111	Köles	*Panicum miliaceum*	Millet	x	—
112	Hajdina	*Fagopyrum esculentum*	Buckwheat	—	x
113	Lóbab	*Vicia faba*	Field bean, horse bean	x	x
114	Takarmány	*Zea mays*	Fodder maize	x	x

Table 8.5. (Cont'd)

No.	Hungarian name	Latin name	English name	Home garden	Field
115	Csemege	*Zea mays* var. *saccharata*	Sweet corn	x	x
116	Pattogatni való	*Zea mays* var. *microsperma*	Popcorn	x	x
117	Bab	*Phaseolus vulgaris/ P. coccineous*	Bean	x	x
118	Fözötök	*Cucurbita pepo* subsp. *pepo*	Pumpkin	x	x
119	Sütötök	*Cucurbita maxima*	Giant pumpkin	x	x
120	Olajtök	*Cucurbita pepo*	Oil pumpkin	x	x
121	Takarmánytök	*Cucurbita pepo* var. *pepo*	Fodder pumpkin	x	x
122	Burgonya	*Solanum tuberosum*	Potato	x	x

In Szatmár-Bereg, fields have two distinct functions: (i) market sales and (ii) household consumption. Next to the fields, or at the fringes of villages, are large orchards specialized in apple, plum and sour cherry production. Even in some home gardens farmers produce fruit commercially on a fairly large scale. The most common species in gardens and fields are usually fruit trees. Cultivated crops are similar to those found in Dévaványa region (Table 8.6).

The most frequently and extensively grown species of fruit trees and crops found in home gardens are nearly the same across regions

Table 8.6. List of crops in Szatmár-Bereg. (From Hungarian Home Garden Household Survey, Hungarian On-farm Conservation of Agricultural Biodiversity Project, 2002.)

No.	Hungarian name	Latin name	English name	Home garden	Field
1	Paradicsom	*Lycopersicon esculentum*	Tomato	x	—
2	Étkezési paprika	*Capsicum annuum* var. *grossum*	Sweet pepper	x	—
3	Füszerpaprika	*Capsicum annuum* var. *longum*	Red pepper	x	—
4	Uborka	*Cucumis sativus*	Cucumber	x	—
5	Sárgarépa	*Daucus carota* subsp. *sativus*	Carrot	x	—
6	Petrezselyem	*Petroselinum crispum*	Apiaceous	x	—
7	Zeller	*Apium graveolens*	Celery	x	—
8	Pasztinák	*Pastinaca sativa*	Parsnip	x	—
9	Cékla	*Beta vulgaris* var. *conditiva*	Red beet	x	—
10	Vöröshagyma	*Allium cepa*	Onion	x	—
11	Lila hagyma	*Allium cepa*	Purple onion	X	—
12	Póréhagyma	*Allium porum*	Leek	x	—
13	Fokhagyma	*Allim sativum*	Garlic	x	—
14	Patiszon	*Cucurbita pepo* var. *patissonina*	Scallop squash	x	—
15	Padlizsán	*Solanum melongena*	Aubergine	x	—
16	Cukkini	*Cucurbita pepo* var. *giromontiina*	Vegetable marrow	x	—
17	Karfiol	*Brassica oleracea* subsp. *botrytis*	Cauliflower	x	—

(Continued)

Table 8.6. (Cont'd)

No.	Hungarian name	Latin name	English name	Home garden	Field
18	Karalábé	Brassica oleracea subsp. caulorapa var. gongyloides	Kohlrabi	x	—
19	Fejeskáposzta	Brassica oleracea	Headed cabbage	x	—
20	Lila káposzta	Brassica oleracea subsp. capitata f. rubra	Red cabbage	x	—
21	Kelkáposzta	Brassica oleracea var. sabauda	Savoy cabbage	x	—
22	Bimbós kel	Brassica oleracea subsp. acephala var. gemmifera	Brussels sprouts	x	—
23	Fejessaláta	Lactuca sativa	Cabbage head lettuce	x	—
24	Hónapos retek	Raphanus sativus	Little radish	x	—
25	Jégcsapretek	Raphanus sativus	Ice radish	x	—
26	Zöldborsó	Pisum sativum	Sugar pea	x	—
27	Sóska	Rumex acetosa	French sorrel	x	—
28	Kapor	Anethum graveolens	Dill	x	—
29	Torma	Armoracia lapathifolia	Horse radish	x	—
30	Paradicsompaprika	Capsicum annuum lycopersiciforme	Tomato paprika	x	—
31	Almapaprika	Capsicum annuum	Apple paprika	x	—
32	Citromfü	Melissa officinalis	Lemon balm	x	—
33	Borsikafü	Satureja hortensis	Summer savory	x	—
34	Kaliforniai paprika	Capsicum annuum	Californian paprika	x	—
35	Alma	Malus domestica	Apple	x	x
36	Körte	Pyrus communis	Pear	x	—
37	Birsalma	Cydonia oblonga	Quince	x	—
38	Szilva	Prunus domestica	Plum	x	x
39	Ringlószilva	Prunus domestica subsp. italica var. claudiana	Greengage	x	—
40	Meggy	Cerasus vulgaris	Sour cherry	x	x
41	Cseresznye	Cerasus avium	Cherry	x	—
42	Öszibarack	Prunus persica	Peach	x	—
43	Nektarin	Prunus persica var. nucipersica	Nectarine	x	—
44	Sárgabarack	Prunus armeniaca	Apricot	x	—
45	Eper (fa)	Morus sp.	Mulberry tree	x	—
46	Földieper	Fragaria vesca	Strawberry	x	—
47	Málna	Rubus idaeus	Raspberry	x	—
48	Szeder	Rubus occidentalis	Blackberry	x	—
49	Ribizli	Ribes rubrum	Ribes, currant	x	—
50	Köszméte	Ribes uva-crispa	Gooseberry	x	—
51	Áfonya	Vaccinium myrtillus	Blueberry	x	—
52	Som	Cornus mas	Cornelberry	x	—
53	Görögdinnye	Citrullus lanatus	Watermelon	x	—
54	Sárgadinnye	Cucumis melo	Sugarmelon	x	—
55	Naspolya	Mespilus germanica	Medlar	x	—
56	Füge	Ficus carica	Fig	x	—
57	Dió	Juglans regia	Nut	x	—
58	Mogyoró	Corylus avellana	Hazelnut	x	—
59	Bodza	Sambucus nigra	Elder	x	—

Table 8.6. (Cont'd)

No.	Hungarian name	Latin name	English name	Home garden	Field
60	Szölö	*Vitis vinifera*	Grape	x	—
61	Lucerna	*Medicago sativa*	Lucerne	x	x
62	Vöröshere	*Trifolium pratense*	Red clover	—	x
63	Takarmány répa	*Beta vulgaris* var. *crassa*	Cattle turnip	x	—
64	Öszi búza	*Triticum aestivum*	Winter wheat	x	x
65	Tavaszi búza	*Triticum aestivum*	Spring wheat	—	x
66	Öszi árpa	*Hordeum hexastichon*	Winter barley	—	x
67	Rozs	*Secale cereale*	Rye	—	x
68	Tritikálé	× *Triticosecale*	Triticale	—	x
69	Zab	*Avena sativa*	Oat	—	x
70	Napraforgó	*Helianthus annuus*	Sunflower	x	x
71	Cukorrépa, burgundi répa	*Beta vulgaris* var. *altissima*	Sugarbeet	x	—
72	Káposztarepce	*Brassica napus*	Summer rape	—	x
73	Mák	*Papaver somniferum*	Garden poppy	x	—
74	Lóbab	*Vicia faba*	Field bean, horse bean	x	—
75	Takarmány	*Zea mays*	Fodder maize	x	x
76	Csemege	*Zea mays* var. *saccharata*	Sweet corn	x	—
77	Bab	*Phaseolus vulgaris/ P. coccineous*	Bean	x	x
78	Fözötök	*Cucurbita pepo* subsp. *pepo*	Pumpkin	x	—
79	Sütötök	*Cucurbita maxima*	Giant pumpkin	x	—
80	Takarmánytök	*Cucurbita pepo* var. *pepo*	Fodder pumpkin	x	—
81	Burgonya	*Solanum tuberosum*	Potato	x	x

(Table 8.7). This confirms the role home gardens play in supplying traditional Hungarian foods and enhancing diet quality for families. Potatoes are an important starchy staple in Hungary. Beans are the basis of *gulyás*. Tomatoes, carrots and apiaceous are used for soups and in favourite Hungarian dishes. In large home gardens, farmers grow fodder crops to feed their livestock, such as maize, lucerne, wheat or fodder pumpkin. Home production of sausage and pickling of vegetables is common. When not sold, fruits are used for canning, or fresh or fermented juices. Production of oil pumpkin for home processing is a specialty in Örség-Vend.

Theoretical Model

The behavioural model employed to explain the farm households' production and consumption decisions is based on the semi-subsistence model of the farm household with missing markets (Singh *et al.*, 1986; de Janvry *et al.*, 1991; Taylor and Adelman, 2003; Chapter 5).

Although motivated by the situation of developing country farmers, the model is appropriate for analysing the case of home garden production in Hungary. Due to a combination of historical, institutional and geographical factors, home gardens are essentially small-scale farms managed with family labour and oriented towards the satisfaction of food needs. Although farm families occasionally participate in market sales of home-garden produce in some locations, profit maximization does not guide their production decisions (Swain, 2000). Even where local food markets are more plentiful, as in Dévaványa, heterogeneity of produce quality often induces families to find a 'corner' solution where they produce and consume their own output for at least some crops or varieties (Singh *et al.*, 1986; Chapter 3).

Table 8.7. Most frequently grown trees and crops in home gardens and fields. (From Household Survey, Hungarian On-farm Conservation of Agricultural Biodiversity Project, 2002.)

	Most frequently grown trees in home gardens			Most frequently grown trees in fields		
Rank	Dévaványa	Örség-Vend	Szatmár-Bereg	Dévaványa	Örség-Vend	Szatmár-Bereg
1	Plum	Apple	Plum	Plum	Apple	Apple
2	Sour cherry	Plum	Apple	Cherry	Plum	Plum
3	Pear	Walnut	Sour cherry	Sour cherry	Walnut	Sour cherry
4	Apple	Pear	Walnut	Pear	Cherry	Walnut
5	Apricot	Cherry	Pear	Apple	Pear	Pear
6	Peach	Peach	Peach	Apricot	Sour cherry	Greengage
7	Cherry	Sour cherry	Cherry	Greengage	Greengage	Peach
8	Walnut	Greengage	Quince	Walnut	Apricot	Apricot
9	Quince	Apricot	Apricot	Peach	Mulberry tree	Cherry
10	Greengage	Quince	Hazelnut	Quince	Medlar	Quince

	Most frequently grown crops in home gardens			Most frequently grown crops in fields		
Rank	Dévaványa	Örség-Vend	Szatmár-Bereg	Dévaványa	Örség-Vend	Szatmár-Bereg
1	Tomato	Tomato	Potato	Fodder maize	Potato	Fodder maize
2	Grape	Sweet pepper	Tomato	Potato	Fodder maize	Winter wheat
3	Carrot	Apiaceous	Carrot	Lucerne	Winter wheat	Sunflower
4	Apiaceous	Carrot	Sweet pepper	Apiaceous	Oil pumpkin	Oat
5	Onion	Cucumber	Kohlrabi	Winter wheat	Fodder pumpkin	Triticale
6	Potato	Potato	Apiaceous	Tomato	Winter barley	Lucerne
7	Pea	Cabbage head lettuce	Headed cabbage	Carrot	Apiaceous	Winter barley
8	Cucumber	Grape	Onion	Pea	Carrot	Potato
9	Sweet pepper	Onion	Cucumber	Cucumber	Cattle turnip	Red clover
10	Strawberry	Headed cabbage	Grape	Fodder pumpkin	Sugarbeet	Spring wheat

The model depicts a farm family that maximizes its utility over consumption of market purchased goods, C_m, and home garden outputs, C_l and home garden outputs, C_k, subscripted k for *kert*, Hungarian for home garden Eq. (8.1). The utility is maximized, subject to budget, time and production technology constraints (Eqs (8.2), (8.3) and (8.4), respectively). Household utility is influenced by Ω_{HH}, denoting a vector of household characteristics of the farm household that condition consumption preferences and choices. The utility function is assumed to be quasi-concave with positive partial derivatives. The prices of all market-purchased goods, inputs and wages are exogenous, and production is assumed to be riskless.

$$U = U(C_k, C_m, C_l; \Omega_{HH}) \qquad (8.1)$$

$$Y = w(T - H) + E - p_V V \qquad (8.2)$$

$$G(Q, H, V; \Omega_K) = 0 \qquad (8.3)$$

$$H + L_o + C_l \equiv T \qquad (8.4)$$

Full income Eq. (8.2) is composed of value of stock of total time owned by the household (T), exogenous income (E), which is non-wage, non-household production income such as direct assistance or pensions, less the values of household management input used in the home garden production (H) and other variable inputs required for production of home garden outputs (V). For cultivation of home garden plots, household management input (H) is a necessary and also sufficient input, since due to historical and institutional reasons these small farms are typically managed by family labour alone. To simplify the analysis, field production decisions are treated as predetermined or exogenous to home garden decisions, affecting them through E in full income. Time allocated to field crop production is included in the 'off-home garden employment' variable treating wages as exogenous and fixed for employment both in the field and off farm.

The household faces a production constraint, Eq. (8.3), for production technology in the home garden depicting the relationship between farm inputs (H, V) and all outputs (Q) by an implicit production function (G) that is quasi-convex, increasing in outputs and decreasing in inputs. The vector Ω_K represents the fixed agroecological features of the small farm such as soil quality. The household also faces a time constraint, depicted in Eq. (8.4), which states that the household cannot allocate more time to home garden cultivation (H), employment outside the home garden (L_O, including employment either in other forms of agricultural production, such as field production or in off-farm employment) and leisure (C_l) than the total time available to the household.

The farm family is driven towards the goal of self-sufficiency in home garden production because of thin, unreliable or missing markets. This phenomenon brings about an additional constraint that induces the household to equate home garden output demand and supply, resulting in an endogenous shadow price for home garden outputs. Consumption and production decisions cannot be separated.

$$Q_k = C_k(\Omega_M) \quad (8.5)$$

Q_k and C_k denote the quantity demanded and supplied of home garden produce and Ω_M is a vector of exogenous characteristics related to availability of and access to markets. This equality condition implicitly defines the shadow price for home garden produce, which guides production decisions. The endogenous shadow price is household-specific depending on the household characteristics that affect access to markets and consumption demand, such as wealth, education, age and age composition of the household. Agroecological features of the home garden, such as soil quality or irrigation, enter the equation through their effect on supply. Fixed factors related to market transaction costs and observed market prices also influence the shadow prices of home garden outputs. The shadow price, ρ, can be expressed as a function of all exogenous prices and household, agroecological and market characteristics:

$$\rho = \rho^*(p_m, p_v, w; \Omega_{HH}, \Omega_K, \Omega_M) \quad (8.6)$$

The solution to the household maximization with missing markets for home-garden produce results in a set of optimal choices:

$$Q_k = Q_k^*(\rho, p_v, w; \Omega_K) \quad (8.7)$$

$$H = H^*(\rho, p_v w; \Omega_K) \quad (8.8)$$

$$V = V^*(\rho, p_v, w; \Omega_K) \quad (8.9)$$

$$C_i = C_i^*(\rho, p_m, w, Y; \Omega_{HH}) \quad i = k, m, l \quad (8.10)$$

Equation (8.7) is the optimal supply of home-garden outputs; Eq. (8.8) is the optimal demand of household labour in home garden production; Eq. (8.9) is the optimal demand for all other inputs to small farm production; and Eq. (8.10) is the optimal demand for each commodity.

Substituting the solution for the shadow price into home-garden output and consumption solutions, optimal production of home-garden outputs is seen to be a function of all exogenous variables:

$$Q_k = Q_k^*(p_m, p_v, w; \Omega_{HH}, \Omega_K, \Omega_M) \quad (8.11)$$

Following Chapter 4, the observed level of agricultural biodiversity maintained on the home garden, which is a direct outcome of the production and consumption choices of the farm household, is a function of all prices and the characteristics of the families, home-garden plots and the markets where they buy or sell inputs and products.

$$ABD = ABD(Q_K^*(p_m, p_v, w; \Omega_{HH}, \Omega_K, \Omega_M)) \quad (8.12)$$

Econometric Analysis

Dependent and explanatory variables

Dependent variables are those summarized in Table 8.8, for four agrobiodiversity components. The richness of crop varieties is a count, easier to interpret with such small home-garden areas than the Margalef index. The other components are 0–1 variables. Explanatory variables used in the analysis of the survey data are divided into three sets according to the vectors denoted in the theoretical model: (i) household; (ii) farm; and (iii) market characteristics. Variable definitions and hypothesized effects are shown in Table 8.8. In the separable model of the farm households, comparative statics are ambiguous. Additional sources of ambiguity are introduced by estimating reduced form equations with dependent variables constructed as metrics over observed levels of choice variables. Hypotheses are based on economic principles as well as the findings reported in other chapters of this book and related literature.

In this model, age proxies also for experience and education level because of strong statistical correlations. Age of the home garden decision maker is positively correlated with experience and negatively correlated with their education. As in other chapters of this book and related literature, we hypothesize that age is positively related to crop biodiversity. This is especially true in Hungary where older farmers who were raised on family farms before the period of collectivization are known to be those with ancestral seed varieties and traditional practices (Chapter 15). Age probably also relates positively to traditional methods of integrated crop and livestock management

Table 8.8. Definition of explanatory variables and hypothesized effects on components of agricultural biodiversity.

Characteristics	Definition	Crop variety diversity	Landrace cultivation	Agro-diversity	Organic production
Household characteristics					
Age	The age of the main home garden decision maker	+	+	+	+
Age-squared	Age-squared	–	–	–	–
Participation	Number of household members that participate in home garden cultivation	+	+	+	+, –
Owned land	Total area of cultivated fields (in m²) that are also owned by the household in year preceding survey	+, –	+, –	+, –	+, –
Car	Household owns a car = 1, 0 else	–	–	–	+, –
Farm characteristics					
Garden size	Size of the home garden (in m²)	+, –	+, –	+, –	–
Irrigation	Percentage of the home garden area irrigated	+, –	+, –	+, –	+, –
Soil quality	Home garden soil is of good quality = 1, 0 else	+, –	+, –	+, –	+, –
Market characteristics					
Sales	Value of the sales of the home garden crop output Hungarian Forints (HUF) in preceding period, per square metre of the home garden	+, –	–	–	–
Distance	The distance of the household (in km) from the nearest food market (settlement level characteristic)	+	+	+	+, –

without the use of chemical inputs. The quadratic term for age is included since older farmers may prefer not to maintain certain practices if they require heavy investments of labour. The number of household members that participate in home garden production represents the relevant family labour stock, and its effect is hypothesized to be positive for crop and agrodiversity. However, the effect of this variable is unclear for soil microorganism diversity (i.e. organic production) as larger families might prefer to use chemicals to ensure sufficient output.

Car ownership and the total area of owned, cultivated fields account for the wealth and social status of the household. Car ownership also indicates increased market access, which could be negatively correlated with the need to maintain agricultural biodiversity in home gardens. The effect of car ownership on choice of organic production methods is ambiguous, given the luxury good property of organically produced goods in some regions. Total area of owned, cultivated fields indicates the extent to which the household is dedicated to agriculture. More 'agricultural' households may have less or more agricultural biodiversity on farms, depending on the complementarity or substitutability of inputs and outputs between the home garden and field production.

Wealth indicators are also thought to influence attitudes towards output variability or market uncertainty. Risk aversion, and hence agricultural biodiversity found on farms, is hypothesized to decrease with wealth (Meng, 1997; Chapter 3). Findings in Part III of this book do not always support this hypothesis (Chapters 5, 6, 9, 10). Although farm production is inherently uncertain because of the time lag between input choices and harvest, there is little reason to expect high degrees of output variability in home garden production in Hungary. Market sources of risk are believed to be substantial, however.

Farm physical characteristics and microecologies clearly affect the numbers and types of crops and varieties grown on farms (Brush *et al.*, 1992; other chapters of Part III and Chapter 10). Favourable production conditions such as more irrigation and good soil quality could affect agricultural biodiversity positively through increasing the productivity of labour. However, better growing conditions on farms could also influence biodiversity levels negatively by inducing specialization in production of fewer species for market sales. Farmers might also choose to maintain higher levels of agricultural biodiversity on their farms in order to raise the productivity of a microecosystem that is not very productive otherwise (Van Dusen, 2000; Di Falco and Perrings, 2002). The effect of irrigation and good quality soil on agrodiversity is hypothesized to be negative, since farmers with good crop production conditions might not choose to tend livestock. On the other hand, favourable crop production conditions on farm may allow households to supply their food crop needs on smaller areas, leaving space for livestock production. Larger home gardens may have less agricultural biodiversity found on home gardens because of families taking advantage of economies of scale to specialize; on the other hand, larger sizes may provide more space to undertake additional activities, resulting in higher levels of agricultural biodiversity. Farmers with more extensive home gardens are less likely to undertake organic production methods because of its labour costs.

Market characteristics indicate the extent to which the farm households are integrated into markets as sellers (the household-specific value of the home garden crop output sales variable), and the transaction costs the farm households face in market participation (the settlement-specific distance to the nearest food market variable). Previous studies, and most of the studies in this book, demonstrate that households located closer to markets will manage lower levels of crop biodiversity on their farms. Nevertheless, the examples from the Ethiopian highlands (Chapter 6) and Uganda (Chapter 7) illustrate that market relationships with diversity are in fact multifaceted. Households with large volume of sales on markets are expected to prefer less crop diversity in their home gardens specializing in production. Demand for agrodiversity and crop diversity is hypothesized to rise with greater distances from the nearest market. When food markets are far away farmers might prefer to ensure homegarden-produce levels by applying chemicals.

Crop variety diversity

The regression explaining the richness of all crop varieties grown, including field crops, vegetables and trees, in home gardens was estimated with a Poisson model because the dependent variable is a non-negative integer (Hellerstein and Mendelsohn, 1993). Statistical tests of both pooled and separate regressions for the three study sites revealed overdispersion (Cameron and Trivedi, 1990). The regressions were then estimated with a Negative Binomial model, an extension of the Poisson regression model that allows the distribution of the variance to differ from the distribution of the sample mean (Greene, 1997).

The results of the Negative Binomial model for crop variety richness are reported in Table 8.9. The hypothesis that parameters are constant across regions was rejected with a log likelihood ratio test at 0.5%, and separate regressions were estimated for each. Joint hypothesis tests on sets of estimated coefficients are consistent with the maintained hypothesis that production and consumption decisions cannot be separated for home garden production in any region except Dévaványa (see Chapter 4 for development of hypotheses related to model separability). In that survey region with greater market development and urbanization (Chapter 3) only the percentage of area that is irrigated, a farm characteristic, positively affects crop species

Table 8.9. Determinants of crop variety richness in home gardens, by region. (From Household Survey, Hungarian On-farm Conservation of Agricultural Biodiversity Project, 2002.)

	Marginal effects		
	Dévaványa	Örség-Vend	Szatmár-Bereg
Constant	30.1**	2***	57.4***
Age	0.51	0.023	0.32
Age-squared	−0.005	−0.0001	−0.003
Participation	1.24	0.05	0.2
Owned land	−0.000003	0.0000008	−0.00001
Car	−0.008	0.3***	−4.06**
Garden size	0.0009	0.00002*	0.0003
Irrigation	0.03*	−0.0013	0.07**
Soil quality	0.0002	0.3*	0.004
Sales	−0.04	0.0006	−0.005
Distance	—	0.005	−0.7***
Sample size	104	109	110
Log likelihood	−358.36	−409.16	−370.98
Chi squared	161.55	194.17	64.12
Significance level	0.001	0.001	0.001
Test $\Omega_{HH} = 0$ (d.o.f = 5)			
Likelihood ratio test	34.9	19.2***	6.8
Probability	0.57	0.998	0.76
Test $\Omega_M = 0$ (d.o.f. = 1 for Dévaványa; 2 for others)			
Likelihood ratio test	2.04	11.9***	9.6***
Probability	0.847	0.45	0.992
Test $\Omega_{HH} = \Omega_M = 0$ (d.o.f = 7 for all except Dévaványa; d.o.f. = 6)			
Likelihood ratio test	7.12	22.4***	15.1**
Probability	0.992	0.999	0.999

*Significant at less than 10%; **significant at less than 5%; ***significant at less than 1% with one-tailed or two-tailed tests as shown on Table 8.8.
Note: Regression is Negative Binomial; marginal effects are computed at mean values.

and variety diversity. Household, market and farm characteristics jointly determine the richness of crop varieties in each of the other regions and in all regions taken together. Crop variety richness is a metric calculated over optimal production choices that implies planting decisions through the derived demand for seed.

Greater variation across than within sites may explain why more factors are statistically significant in the pooled than in the separate regressions. Statistical tests of individual parameters confirm that older decision makers maintain more crop species and varieties, but less so as they age. The stock of family participants in home garden production also contributes to the richness of crop varieties grown in home gardens. The more extensive the home garden, the higher is the number of crop species and varieties grown. The most statistically significant variable, with the largest magnitude of effect, is the distance of the household to the nearest food market. Transaction costs induce farmers to rely on the home garden and grow a wider range of foodstuffs.

Differences emerge among tests of individual hypotheses in the more isolated regions, Örség-Vend and Szatmár-Bereg. In Örség-Vend car ownership is positively associated with the family's decision to cultivate more crop species and varieties in their garden. In other words, Örségi households that are better off cultivate more species and varieties than their poorer counterparts. An explanation for this finding is suggested by the work of Szép (2000) who investigated time allocation patterns of Hungarian home garden producer households. Szép found 'rational' labour supply behaviour with a backward-being supply curve for labour. That is, as wages of the home garden decision makers and participants increase, they choose to engage less in employment outside of home gardens, preferring to use that time for leisure activities – including the cultivation of home gardens rich in crop varieties.

Good soil quality also has a favourable effect on the richness of crop varieties. In Örség-Vend, the richness of crop varieties also rises with larger home gardens, though the effect is rather small. In Szatmár-Bereg, the influence of the most significant (and largest) factor, distance to the nearest food market, is negative. Home gardens in the more distant villages of Szatmár-Bereg are larger in size. Families cultivating these small farms tend to specialize in fewer species and varieties, especially of fruit trees, for sales to the large fruit juice industry in this region. Similarly, the coefficient on the value of sales of home garden output is negative, though not statistically significant. The size of the total farm area that is cultivated and owned also affects the richness of crop varieties negatively and significantly. Families who farm larger fields and sell their produce are more likely to have access to food markets and purchase substitutes for home garden outputs. Irrigation in the home garden contributes positively to the richness of crop varieties.

Crop genetic diversity

Estimated coefficients of the univariate probit regression for landrace cultivation in the home garden are reported in Table 8.10. Log likelihood ratio tests again support the non-separability of consumption and production decisions in each region and the dependence of parameters on region. In general, household characteristics (age, labour supply, wealth) and distances to market play an overwhelming role in the decision to plant landraces in the home garden. Stocks of family labour have both large and statistically significant effects. The importance of age and experience is particularly pronounced in Dévaványa, where it is the only significant variable. Clearly, in this more urbanized and economically developed region, the older farmers who were raised as children on farms with landraces before the collectivization period are those that retain them. Örségi families who are more agriculturally based, with larger fields and more family labour engaged on the home garden, are more likely to cultivate landraces. In this less favourable agroecology, the irrigated share of the home garden relates negatively to the prospects that a landrace is grown. Coupled with the negative sign on the soil quality variable these findings imply that in this region landraces are found in less-favoured environmental niches. Poorer families in Szatmár-Bereg, without cars and the market access they provide, are more likely to cultivate landraces. The size of the home garden counteracts this effect. In home gardens

Table 8.10. Determinants of landrace cultivation in home gardens, by region. (From Household Survey, Hungarian On-farm Conservation of Agricultural Biodiversity Project, 2002.)

	Marginal effects		
	Dévaványa	Örség-Vend	Szatmár-Bereg
Constant	−0.07	−0.15	−0.3
Age	0.0024***	0.005	0.003
Age-squared	−0.00002***	−0.00004	-0.1×10^{-4}
Participation	−0.001	0.01**	0.04***
Owned land	-0.3×10^{-7}	0.1×10^{-5}**	0.1×10^{-7}
Car	0.2	−0.01	−0.1***
Garden size	-0.5×10^{-6}	-0.3×10^{-5}	0.6×10^{-5}***
Irrigation	0.8×10^{-5}	−0.0003*	0.5×10^{-3}
Soil quality	0.4×10^{-5}	−0.01	−0.018
Sales	0.6×10^{-4}	0.02	0.13×10^{-4}
Distance	—	−0.0007	0.005
Sample size	104	109	110
Log likelihood	−49.71	−63.65	−64.70
Chi squared	21.74	23.57	22.78
Significance level	0.01	0.01	0.01
Correct predictions	73%	85%	71%
	Test $\Omega_{HH} = 0$ (d.o.f = 5)		
Likelihood ratio test	17.3***	11.7**	17.3***
Probability	0.996	0.960	0.996
	Test $\Omega_M = 0$ (d.o.f. = 1 for Dévaványa; 2 for others)		
Likelihood ratio test	2.2	9.6***	1.2
Probability	0.860	0.992	0.458
	Test $\Omega_{HH} = \Omega_M = 0$ (d.o.f = 7 for all except Dévaványa; d.o.f. = 6)		
Likelihood ratio test	20.4***	20.9***	20.6***
Probability	0.999	0.999	0.999

*Significant at less than 10%; **significant at less than 5%; ***significant at less than 1% with one-tailed or two-tailed tests as shown on Table 8.8.
Note: Regression is Probit; marginal effects are computed at mean values.

with larger sizes the likelihood that landraces are grown increases.

Agrodiversity

The dichotomous choice of whether or not to raise crops together with livestock in the home garden is estimated with a univariate probit model. Results are reported in Table 8.11. Log likelihood ratio tests confirm that production decisions are not separable from consumption decisions in any of the regions (including Dévaványa), and that regression parameters depend on region. For all regions taken together, household characteristics as a set are highly significant determinants of the decision to raise both crops and livestock, distance to market has a weaker effect and farm characteristics are of no importance. Older, more experienced and traditional decision makers are more likely to undertake both crop and livestock production in their home gardens. The effect of age declines with this labour-intensive mode of production, offset by the positive effect of the number of family

Table 8.11. Determinants of integrated crop and livestock production in home gardens, by region. (From Household Survey, Hungarian On-farm Conservation of Agricultural Biodiversity Project, 2002.)

	Marginal effects		
	Déványa	Örség-Vend	Szatmár-Bereg
Constant	−0.005**	−1.7*	−0.7 × 10^{-3}
Age	0.2 × 10^{-3}***	0.05*	0.4 × 10^{-4}
Age-squared	−0.2 × 10^{-5}***	−0.5 × 10^{-3}*	−0.6 × 10^{-6}
Participation	−0.5 × 10^{-4}	0.16***	0.3 × 10^{-3}**
Owned land	0.7 × 10^{7}*	0.3 × 10^{-5}*	0.6 × 10^{-8}**
Car	0.2 × 10^{-5}	−0.3**	−0.9 × 10^{-3}***
Garden size	0.4 × 10^{6}**	0.2 × 10^{-4}	−0.2 × 10^{6}***
Irrigation	0.1 × 10^{-5}	0.001	−0.6 × 10^{-3}
Soil quality	−0.3 × 10^{-5}	−0.14	0.4 × 10^{-3}*
Sales	0.6 × 10^{-5}	−0.003	−0.1 × 10^{-5}
Distance	—	0.017*	0.3 × 10^{-4}
Sample size	104	109	110
Log likelihood	−47.04	−58.06	−61.04
Chi squared	50.05	28.22	29.49
D.o.f	9	10	10
Significance level	0.00	0.0017	0.001
Separability test $\Omega_{HH} = 0$ (d.o.f = 5)			
Likelihood ratio test	41.1***	22.7***	20.3***
Probability	0.999	0.999	0.999
Separability test $\Omega_M = 0$ (d.o.f. = 1 for Déványa; 2 for others)			
Likelihood ratio test	2.5	5.4*	1
Probability	0.860	0.93	0.39
Separability test $\Omega_{HH} = \Omega_M = 0$ (d.o.f = 6 for Déványa; 7 for others)			
Likelihood ratio test	42.1***	27.1***	22.5***
Probability	0.999	0.999	0.999

*Significant at less than 10%; **significant at less than 5%; ***significant at less than 1%* with one-tailed or two-tailed tests as shown on Table 8.8.
Note: Regression is Probit; marginal effects are computed at mean values.

members involved. The labour requirements of livestock production are reflected in the prominent magnitudes of the coefficients on the number of family members involved in home garden production. To make space for larger animals and contribute feed and fodder, owning and cultivating larger field areas is also associated with higher prospects of undertaking integrated crop and livestock production in the home garden. Distance to the nearest food market has a less significant effect, but reflects farm family demand for self-sufficiency in consumption of pork, sausage and salami – traditional and important in the Hungarian diet.

In Déványa, where markets are prevalent, distance to the nearest market is of no consequence in the decision for integrated crop and livestock production in the home garden, though age again plays a major role. Denser settlements mean that home garden sizes are significant in the decision to raise livestock in addition to crops. In Örség-Vend, the age of the decision maker and stocks of family labour working in the home garden are also important, though garden and field areas are not in its less populated, more dispersed settlements. Owning a car, which provides access to shops in town and indicates wealth, has a large negative effect on the probability that a household raises livestock in the home garden. Distance to market has a smaller but significant effect. Similarly to Örség-Vend, the number of home garden participants, expanses of field cultivated and car ownership are significant determinants of agrodiversity management in home gardens. In Szatmár-Bereg, larger home garden areas are negatively associated with livestock production since szatmári households with larger home gardens tend to specialize in crop (especially fruit trees, as explained above) production for market sales. The negative effect of value of produce sales reinforces this finding, though the coefficient is not statistically significant.

Soil microorganism diversity

Univariate probit regressions for determinants of the decision to use organic production methods were statistically significant only for the pooled regression (Table 8.12). Econometric results are weaker statistically because of the smaller percentage of farmers engaged in organic production relative to other components of agrobiodiversity, though they are consistent with hypotheses based on economic theory. In contrast with the other components, higher numbers of family participants in home garden production imply that the household is less likely to employ organic methods. Since the stock of home garden labour is highly correlated with family size, this finding suggests that larger families may be reluctant to expose themselves to the yield risks associated with avoiding chemical inputs. Since organic techniques also require labour to substitute for chemicals in pest and disease control, larger home garden areas reduce the likelihood that they are used. Although the effects are statistically weak, good soil quality is positively associated with organic farming since it substitutes for fertilizers.

Designing Conservation Programmes

The predictions from the models estimated above enable us to identify the types of families that are most likely to sustain the four components of agrobiodiversity we have investigated on

Table 8.12. Determinants of organic production in home gardens. (From Household Survey, Hungarian On-farm Conservation of Agricultural Biodiversity Project, 2002.)

	All regions
	Marginal effects
Constant	−0.13
Age	0.0024
Age-squared	−0.00001
Participation	−0.024***
Owned land	-0.2×10^{-7}
Car	0.0002
Garden size	−0.00002**
Irrigation	−0.0002
Soil quality	0.003
Sales	0.8×10^{-5}
Distance	0.0009
Sample size	323
Log likelihood	−117.71
Chi squared	25.40
D.o.f	10
Significance level	0.0046
Separability test $\Omega_{HH} = 0$ (d.o.f = 5)	
Likelihood ratio test	13.5**
Probability	0.981
Separability test $\Omega_M = 0$ (d.o.f. = 2)	
Likelihood ratio test	1.1
Probability	0.420
Separability test $\Omega_{HH} = \Omega_M = 0$ (d.o.f = 7)	
Likelihood ratio test	14.4*
Probability	0.987

*Significant at less than 10%; **significant at less than 5%; ***significant at less than 1%* with one-tailed or two-tailed tests as shown on Table 8.8.
Note: Regression is Probit; marginal effects are computed at mean values.

the traditional small farms of Hungary. Profiles can be used to design targeted, least cost incentive mechanisms to support conservation as part of national environmental programmes. Revealed choices indicate the value farmers assign to these components, given the constraints they face.

Predicted levels and actual levels of crop variety richness are lowest in Dévaványa, as is the percentage of households choosing to grow landraces. In Dévaványa ESA only one farm family had a predicted probability, of over 75%, of growing landraces, leading to the conclusion that landrace cultivation in this ESA is not a sustainable home garden activity. The opportunity costs of maintaining crop biodiversity are clearly lower in the more remote Örség-Vend or Szatmár-Bereg, so that it would make more policy sense to consider supporting crop variety diversity and landrace cultivation in those regions. The likelihood of integrated crop and livestock production in home gardens is high and invariant to region, suggesting that there is little threat to this mode of production as long as tastes and preferences emphasize traditional Hungarian meats. Findings for organic production suggest that further research is required. Stated values placed by farmers on the components of agricultural biodiversity are similar to revealed preferences in terms of relative rank and magnitude among candidate sites (Birol et al., 2004).

In Örség-Vend, farm families with high probabilities of maintaining crop variety diversity levels above the regional average own smaller fields but cultivate larger gardens, selling more of their home garden produce compared to other farm families in the sample. They have slightly larger families and higher average income. These are active families with a high proportion of their home garden decision makers working off the farm and a smaller proportion retired. They are also less likely to cultivate their home gardens with organic methods compared to the other households in the sample (Table 8.13).

In Szatmár-Bereg, families with high probabilities of maintaining levels of crop variety diversity above the regional average also own less

Table 8.13. Comparison of households with above- and below-average predicted levels of crop variety richness.[a] (From Household Survey, Hungarian On-farm Conservation of Agricultural Biodiversity Project, 2002.)

Characteristics	Mean			
	Örség-Vend (N = 109)		Szatmár-Bereg (N = 110)	
	Above mean	Below mean	Above mean	Below mean
Family size	3.6*	3	3	2.9
Children/adult ratio	0.09	0.1	0.08	0.1
Income	94818.2**	91276.5	75117	70664.9
Owned land	6508**	11027.1	14098.5**	20653.9
Garden area	2106.9**	1502.7	2277.2*	2770.3
Sales	22.9**	2.4	26.5**	35.1
Distance	20.1	21	17.1***	18.8
	%			
Home garden decision makers with fewer than 8 years of education	0	6	30***	19
Home garden decision makers with off-farm employment	41**	32	26**	33
Home garden decision makers who are retired	63**	75	78**	72
Landrace cultivation	46	54	52	53
Agrodiversity	77	77	96*	83
Organic production	9**	20	7	8
Number of predictions	22	87	27	83

[a]Predicted with probability above 5%. Pairwise t-tests (means) or Pearson chi-squared tests (proportions) show significant differences at less than ***1% significance level; **5% significance level; *10% significance level.
Note: Regional means of crop species and variety diversity for Dévaványa, Örség-Vend and Szatmár-Bereg are reported in Table 8.3.

land. They cultivate smaller gardens as well. Although their mean income levels appear to be higher than those with lower predicated levels of crop variety diversity, differences are not statistically significant, perhaps as a reflection of the variation in sample estimates. As a result of their smaller garden sizes, they sell less of their home garden produce, even though they are close to markets by an average of 1 km. A higher percentage of these home garden decision makers have less than 8 years of education and a lower percentage work off the farm, with a higher percentage retired. They are also more likely to undertake the traditional method of integrated livestock and crop production. It is remarkable that the social and demographic profile of farm families with high predicted levels of crop variety diversity, relative to those with low predicted levels, is unique to each site.

Profiles of farm families that are most likely to cultivate landraces are reported in Table 8.14. In Örség-Vend, these families have a lower ratio of children to adults, as a reflection of their older life-cycle stage. Despite that fact, they have less income in transfers and gifts compared to those farm families that are not likely to cultivate landraces. They are more agriculturally based with less off-farm employment and larger expanses of owned land. Although their gardens are smaller, and they live farther from markets, they sell many times more garden produce. Örségi households that are more likely to cultivate landraces in their home gardens are also more likely to manage home gardens that are high in terms of other components of agricultural biodiversity (crop variety diversity, agrodiversity and use of organic techniques). These associations suggest some economies of scope in

Table 8.14. Comparison of households with high predicted probability of growing landraces and all other households.[a] (From Household Survey, Hungarian On-farm Conservation of Agricultural Biodiversity Project, 2002.)

Characteristics	Mean			
	Örség-Vend (N = 109)		Szatmár-Bereg (N = 110)	
	High probability	Others	High probability	Others
Family size	3.5	3.1	4***	2.6
Children–adult ratio	0.04***	0.12	0.1	0.09
Income	84161.8*	93750.8	82084.8	69027.6
Owned land	37374.1**	3989.3	21912.3	18286.8
Garden area	896.5**	1788.3	3684.5**	2375.5
Sales	35.7***	0.01	34.6	32.6
Distance	23.1**	20.2	19.4**	18.1
Crop variety diversity	23.5**	19.2	15.4	15.2
		%		
Home garden decision makers with fewer than 8 years of education	5	4.5	23***	21
Home garden decision makers with off-farm employment	20**	37	22***	31
Home garden decision makers who are retired	85**	70	74	74
Agrodiversity	85**	75	91*	85
Organic production	25*	16	0*	10
Number of predictions	20	89	23	87

[a]High probability is 75% or more. Pairwise t-tests (means) or Pearson chi-squared tests (proportions) show significant differences at less than ***1% significance level; **5% significance level ; *10% significance level.

designing conservation programmes in these sites. That is, households maintaining one component of agrobiodiversity also maintain another.

In Szatmár-Bereg, farm families that are predicted to cultivate landraces are also located farther from markets, in the most isolated settlements, with less off-farm employment and less education, but they have larger families and larger garden areas. Again, there is some evidence that conservation components overlap – families who are most likely to continue managing of traditional varieties of crops are also those that engage in the traditional method of integrated crop and livestock production (Table 8.14). In Chapter 15, farmers report that livestock prefer the grain of maize landraces as feed.

Conclusions

One of the salient results is the uniqueness of each region studied in terms of the level of agricultural biodiversity observed in home gardens and its determinants. In each statistical analysis conducted, whether descriptive or econometric, the hypothesis that population parameters of interest are constant across regions was rejected. Hence, any agri-environmental policy or programme that aims to support the management of current levels of agricultural biodiversity in rural Hungary will need to recognize the heterogeneity of these traditional farms and their context.

Findings are also consistent with the maintained hypothesis that for all regions, the choices farm families make in home garden production, as reflected in the components of agricultural biodiversity measured here, cannot be separated from their consumption decisions. According to the model of the agricultural household that motivates the approach, market imperfections in Hungary's transition economy continue to induce farmers to produce for themselves. Furthermore, any policy or programme that affects the wealth, education or labour participation of family members, or the formation of food markets within settlements, will influence their choices. Of the components of agricultural biodiversity investigated here, those most likely to change in a major way are observed levels of crop variety diversity and landrace cultivation.

Implications

Across regions, one of the most significant determinants of revealed preferences for agricultural biodiversity on Hungarian home gardens is the age of the home garden decision maker. Since migration of younger generation away from settlements is a common phenomenon in the more isolated regions, this finding implies that crop biodiversity, though still high today in these regions, is in jeopardy in the longer run.

National and EU level policies and programmes such as the Special Accession Programme for Rural and Agricultural Development (SAPARD) and the National Agri-Environmental Programme (NAEP), which is recently integrated into the National Rural Development Plan (NRDP), are now being implemented. The aim of these policies and programmes is to encourage economic activities in the rural areas and retain settlement populations in the countryside (Juhász et al., 2000; Weingarten et al., 2004). Ways must be found to transfer knowledge and skills to future generations of farmers in those sites.

Clearly, production on the traditional small farms of Hungary has an important role to play in promoting multifunctional agriculture through the agri-environmental measures under NRDP's extensive agricultural production schemes. Participation in these schemes, as explained in Chapter 3, is based on voluntary farmer contracts. Findings such as these can be a starting point for identifying locations and farmers to include in contracting schemes to support the sustainable management of agricultural biodiversity in home gardens. This chapter has described the type of farm household and locations that are likely to cost least in payments and whose inclusion in contracting schemes should be most effective. Some economies of scope appear in terms of the design of programmes to support components of agrobiodiversity, perhaps due to some production complementarities.

Prospects for niche markets or geographical denomination of origin might also be considered as part of the market integration that Hungary will experience with EU membership (Fischler, 2003). Numerous recent studies point to the rising demand of high-income, EU consumers for goods produced with organic methods or heirloom varieties of crop and animal species (see, e.g. Kontoleon, 2003).

Acknowledgements

This chapter is developed from the PhD research undertaken by the first author at University College London. This research is part of the Hungarian On-farm Conservation of Agricultural Biodiversity project, which is led by the Institute for Agrobotany, Tápiószele, in partnership with the Institute of Environmental and Landscape Management, Szent István University, and the International Plant Genetic Resources Institute (IPGRI), Rome, Italy. We gratefully acknowledge the funds from the BIOECON project supported by the European Commission under the Fifth Framework Programme. We would like to thank Györgyi Bela, László Holly, Phoebe Koundouri, István Már, György Pataki, David Pearce, László Podmaniczky, Timothy Swanson and Eric Van Dusen for useful comments and suggestions.

References

Birol, E., Smale, M. and Gyovai, Á. (2004) Agri-environmental policies in a transitional economy: the value of agricultural biodiversity in Hungarian home gardens. *Environment and Production Technology Division Discussion Paper No. 117*, International Food Policy Research Institute (IFPRI), Washington, DC.

Brush, S., Taylor, E. and Bellon, M. (1992) Technology adoption and biological diversity in Andean potato agriculture. *Journal of Development Economics* 39(2), 365–387.

Cameron, A. and Trivedi, P. (1990) Regression based tests for overdispersion in the Poisson regression model. *Journal of Econometrics* 46, 347–364.

de Janvry, A., Fafchamps, M. and Sadoulet, E. (1991) The peasant household behaviour with missing markets: some paradoxes explained. *The Economic Journal* 101, 1400–1417.

Di Falco, S. and Perrings, C. (2002) Cooperative production and intraspecies crop genetic diversity: the case of durum wheat in Southern Italy. Paper Presented at the First BIOECON Workshop on Property Right Mechanisms for Biodiversity Conservation, International Plant Genetic Resources Institute (IPGRI), 30–31 May 2002, Rome, Italy.

Fischler, F. (2003) Opportunities in an enlarged Europe. Address Presented at the 16th European Industry Strategy Symposium in Villach, 10 February 2003.

Greene, W.H. (1997) *Econometric Analysis*, 3rd edn. Prentice-Hall, New York.

Gyovai, Á. (2002) Site and sample selection for analysis of crop diversity on Hungarian small farms. In: Smale, M., Már, I. and Jarvis, D.I. (eds) *The Economics of Conserving Agricultural Biodiversity On-farm: Research Methods Developed from IPGRI's Global Project 'Strengthening the Scientific Basis of In Situ Conservation of Agricultural Biodiversity'*. International Plant Genetic Resources Institute, Rome, Italy.

Hellerstein, D. and Mendelsohn, R. (1993) A theoretical foundation for count data models. *American Journal of Agricultural Economics* 75, 604–611.

Juhász, I, Ángyán, J., Fesus, I., Podmaniczky, L., Tar, F. and Madarassy, A. (2000) *National Agri-Environment Programme: For the Support of Environmentally Friendly Agricultural Production Methods Ensuring the Protection of the Nature and the Preservation of the Landscape*. Ministry of Agriculture and Rural Development, Agri-Environmental Studies, Budapest.

Kontoleon, A. (2003) Essays on non-market valuation of environmental resources: policy and technical explorations. PhD thesis, University College London, University of London, London.

Meng, E.C. (1997) Land allocation decisions and *in situ* conservation of crop genetic resources: the case of wheat landraces in Turkey. PhD dissertation, University of California at Davis, California.

Singh, I., Squire, L. and Strauss, J. (1986) *Agricultural Household Models*. Johns Hopkins University Press, Baltimore, Maryland.

Swain, N. (2000) Post-socialist rural economy and society in the CEECs: the socio-economic contest for SAPARD and EU enlargement. Paper presented at the International Conference: European Rural Policy at the Crossroads, 29 June–1 July 2000, The Arkleton Centre for Rural Development Research, King's College, University of Aberdeen, UK.

Szép, K. (2000) The chance of agricultural work in the competition for time: case of household plots in Hungary. *Society and Economy in Central and Eastern Europe* 22(4), 95–106.

Taylor, E. and Adelman, I. (2003) Agricultural household models: genesis, evolution and extensions. *Review of the Economics of the Household* 1(1), 33–58.

Van Dusen, M.E. (2000) *In situ* conservation of crop genetic resources in the Mexican Milpa System. PhD dissertation. University of California at Davis, California.

Weingarten, P., Baum, S., Frohberg, K., Hartmann, M. and Matthews, A. (2004) *The Future of Rural Areas in the CEE New Member States*. Report Prepared by the Network of Independent Agricultural Experts in the CEE Candidate Countries. Available at: http://europa.eu.int/comm/agriculture/publi/reports/ccrurdev/text_en.pdf

9 Rural Development and the Diversity of Potatoes on Farms in Cajamarca, Peru

P. Winters, L.H. Hintze and O. Ortiz

Abstract

In this chapter, using household data on potato producers in Cajamarca, Peru, the relationship between rural development and potato diversity is examined. Much of the debate on *in situ* on-farm conservation of crop genetic resources has focused narrowly on the effect of the introduction of modern varieties on crop genetic diversity. The introduction of modern varieties is only one mechanism by which rural development processes may bring about a change in crop biodiversity on farms. The greatest threat to the on-farm diversity of staple crops in developing countries is that farmers will cease to grow these crops altogether. In particular, rural development is likely to bring about opportunities in new agricultural products, which cause shifts in land use patterns, to lead to alternative labour uses including employment in non-agricultural activities and to improve the ability to manage risk through more effective means than variety portfolios. The results indicate that households who are more involved in activities other than potato production, in particular dairy production and non-farm activities, have less potato diversity. This suggests that any rural development strategies that promote alternatives to potato production may reduce its diversity on farms.

Introduction

One of the key features of issues involved in the conservation of the crop biodiversity on farms is that poor farmers in developing countries are more responsible for it. The observed association between poverty and the regions where farmers continue to manage genetically diverse crops suggests that some form of relationship between the two exists. One hypothesis to explain this correlation is that the factors that lead poor farmers to maintain diverse genetic materials on-farm are the same as the factors that cause farmers to remain in poverty in the first place. If this hypothesis is correct, the implications are rather dramatic. It means that policies and programmes that are geared towards addressing the underlying causes of poverty and that seek to develop rural areas in centres of crop diversity are likely to cause genetic erosion (the loss of individual genes and gene combinations). In this chapter, we use data collected from households in the northern Peruvian Andes to examine household decision making with respect to the use of potato varieties and the effect of rural development on these decisions. The Andes are a centre of genetic diversity for potatoes and a loss of genetic diversity in this region would have global implications.

Much of the debate on *in situ* on-farm conservation of crop genetic resources has focused narrowly on the effect of the introduction of modern varieties on crop genetic diversity. The general assumption of this debate is the genetic erosion hypothesis advanced during the early phases of Asia's green revolution in wheat (see discussion in Chapter 1), repeated succinctly by FAO (1996, p.33): 'The main cause of genetic erosion in crops, as reported by almost all countries, is the replacement of local varieties by improved or exotic varieties and species'. Establishing causality has not been easy, however, in part because of data limitations (Smale, 1997). In their study of Andean potato production, Brush *et al.* (1992) found that the adoption of modern varieties leads to a reduction, but not to a complete loss, of household-level diversity and a possible loss in aggregate diversity. The analyses in Chapters 5 and 11 suggest that the introduction of modern varieties of maize and wheat has not yet replaced cereal landraces in the northern Ethiopian highlands, most likely because modern varieties have limited adaptation to local environments and farmers face economic constraints. These results, and some genetic and anthropological analyses, raise questions about the causal relationship between the planting of modern varieties and the reduction of genetic diversity, suggesting more complex relationships that are context- and crop-specific (Zimmerer, 1996; Louette *et al.*, 1997; vom Brocke, 2001).

The introduction of modern varieties, however, is only one mechanism by which rural development processes may bring about a change in crop biodiversity on farms. In particular, rural development is likely to bring about opportunities in new agricultural products, which cause shifts in land use patterns, to lead to alternative labour uses including in non-agricultural activities and to improve the ability to manage risk through more effective means than variety portfolios. It is for this reason that Dyer (Chapter 2) and Van Dusen (Chapter 5) suggest that the greatest threat to maize diversity in Mexico is that farmers will cease to grow maize altogether.

The objective of this chapter is to identify factors that influence potato diversity in the San Miguel province of the Cajamarca region of northern Peru, highlighting in particular those factors associated with rural development. The following section provides the motivation for the analysis. The agricultural economy of the study region is then characterized, with emphasis on potato production. This section also describes recent developments in Cajamarca that have affected the rural economy, and particularly the expansion of dairy production. The data used for this study are presented in the subsequent section along with summary statistics. Following the discussion of the data, the empirical approach to evaluating potato diversity and findings are reported. Conclusions are presented in the final section.

Rural Development and Potato Diversity

The study by Brush *et al.* (1992) is one of the few studies of potato diversity on farms, and perhaps the first applied economics analysis published about on-farm diversity in any crop. The authors identified factors affecting the diversity of potatoes on farms using household data from Peru. They tested a version of the 'displacement hypothesis': that modern varieties will displace native varieties and reduce diversity. They hypothesized that, in the first stages of the adoption process, diversity will be lower in farms with larger areas cultivated with improved varieties, while in the later stages of adoption, increasing area allocated to improved varieties will not affect diversity if farmers attach value to maintaining diversity. The authors conducted their study in two regions of the Peruvian Andes at different stages of adoption with different degrees of access to markets. Econometric findings supported their hypothesis. In Paucartambo, where adoption of modern varieties was relatively low and had begun more recently, an increase in the potato area farmers allocated to modern varieties was associated with lower levels of potato diversity. In Tulumayo, where cumulative adoption rates were already high, the area share in modern varieties had no statistically significant effect on the cultivation of potato landraces. Their results challenged the notion that modern varieties will inevitably replace native varieties and that only the poorest farmers will maintain crop diversity.

In fact, the greatest threat to on-farm crop genetic diversity may not be replacement by modern varieties but shifts in resource use away from the production of these crops. For example, since the 1950s, Nestle has been managing a milk processing plant in northern Peru that has provided opportunities for milk production. Recently, another dairy company offering a better price for milk has entered the market, leading to an increased interest in milk production (Godtland, 2001). Milk production has become more attractive relative to other activities, such as growing potatoes. Along with shifts away from staple production towards high value agricultural products, in general, in rural areas of Latin America there has been a shift towards non-agricultural activities (Reardon et al., 2001). In Peru, it has been long observed that small farmers engage in a range of off-farm activities. Even though a majority of peasant families still rely for most of their income on agriculture, a high percentage of them migrate seasonally and work outside of agriculture (Cotlear, 1989). As with new agricultural opportunities, expansion of non-agricultural activities is likely to affect potato production.

These changes occurring in Peru and rural areas in similar countries are often promoted by governments as part of a rural development strategy. As a guide to the policies that are considered useful for rural development, consider the strategy for rural development promoted by the World Bank (2002). Among other things, the strategy focuses on: (i) expanding rural non-farm income-producing activities; (ii) supporting agriculture diversification especially into high-value products; (iii) supporting sustainable intensification of production through the use of new technologies; (iv) encouraging, partially through demand-driven extension services, more efficient use of farm inputs and reduction of post-harvest losses; (v) strengthening farmer–market linkages; and (vi) improving rural infrastructure.

While these policies are consistent with the conventional wisdom on promoting rural development, they are bound to have implications for the conservation of crop genetic resources when carried out in centres of crop diversity. Support for diversifying agriculture or expanding non-farm income-producing activities may lead to a shift in land, labour and capital away from production of genetically diverse crops, limiting the range of varieties planted. Furthermore, access to income from other activities limits the need to use multiple varieties to manage risk. Supporting intensification and expanding extension services may lead to further promotion of modern varieties, which may be associated with some reduction in genetic diversity. Improving market linkages and rural infrastructure may present new opportunities for other activities, induce the use of modern varieties by allowing better input and output market access, or narrow the scope of varieties produced if the market demands uniformity. In general, such changes may bring about a reduction in the use of marginal lands for crop production and decrease the need to have varieties that are suited for certain agro-ecological niches. These trends are likely to put pressure on potato production, affecting farmers' choice of potato varieties and, as a consequence, the genetic diversity found at both household and aggregate levels.

Chapter 1 of this volume provides background on the previous literature regarding factors that influence crop diversity on farms. In the econometric analyses of on-farm crop diversity by Meng (1997), Van Dusen (2000) and Smale et al., (2001), followed by chapters included in Part III of this book, a number of factors known to be related to rural development have been used to explain levels of crop genetic diversity observed on farms. Explanatory factors have included, e.g. migration, wealth in livestock or housing, off-farm employment, education levels, extension contact, income from remittances and transfers, the physical infrastructure of markets, population and road densities.

In none of these cases, however, did researchers consider the possibility that farmers might invest in alternative agricultural activities in the same manner as milk production in Cajamarca. The influence of such decisions on the crop biodiversity has not been tested using comparable methods. This chapter directly addresses the relationship of diversification in agricultural income sources on the genetic diversity in a single crop. As done in the chapters noted above, it also considers how other factors related to rural development, including opportunities in non-farm activities and access to agricultural extension, affect diversity. In general, the evidence suggests that a number of rural development interventions will have significant effects on the diversity of potatoes maintained by farm households in Peru. Following a description of the site where data was collected for this analysis and a description of the data, specific hypotheses for the relationship between rural development and potato diversity are identified.

Site Description

The Cajamarca Department is located in the Northern Highlands of Peru near the border with Ecuador in an area known as the Green Andes because precipitation is high compared with the rest of the Peruvian highlands (see Fig. 9.1). The area is also characterized by steeply sloped, hilly terrain. Cajamarca is ranked the fourth poorest among the 25 departments in Peru (Godtland, 2001) and the rural communities in the department usually lack basic services such as electricity, potable water and health. Rural schools are accessible for most families. Agriculture is the main economic activity for most inhabitants although mining activities generates more overall profits. Landholdings are small, with most households owning limited amounts of arable land. A typical household in Cajamarca has an average of six members (INEI, 1996).

In the San Miguel province of Cajamarca (see Fig. 9.2), where the data for this study was collected, farmers undertake a number of income-generating activities. Godtland (2001) indicates that the two most important sources are dairy production followed by potato farming. Remaining income comes from the cultivation of other crops such as barley, wheat, Andean tubers, faba beans, maize and peas, among others, agricultural wage income and non-agricultural activities.

Cajamarca is one of the three main milk producing regions, which, taken together, account for half of Peru's milk production. In Cajamarca, the expansion of dairy production started when

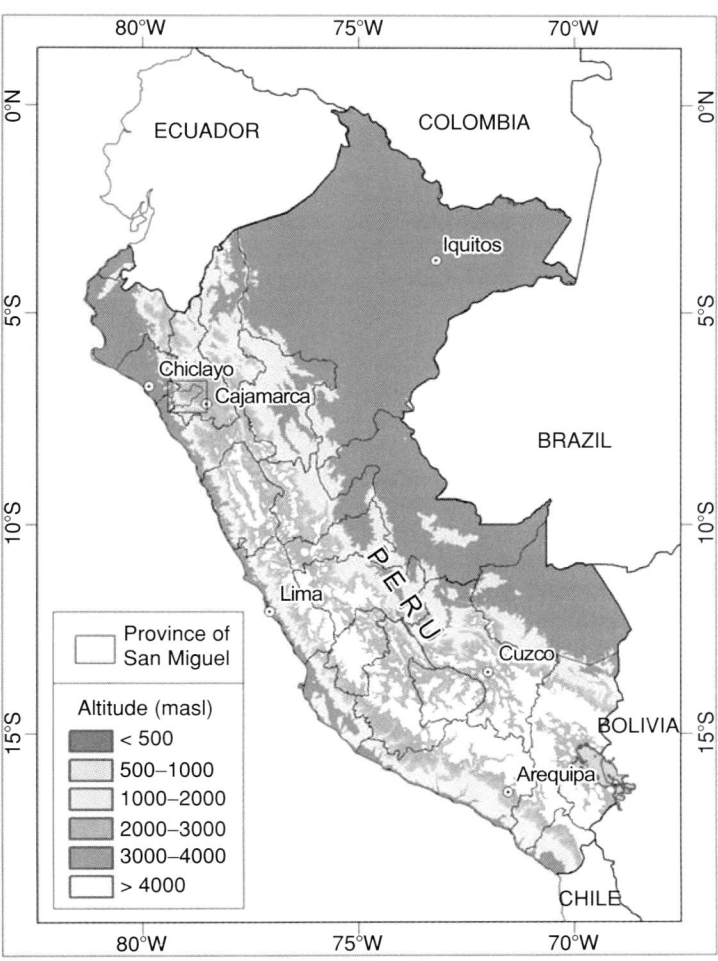

Fig. 9.1. Peruvian highlands.

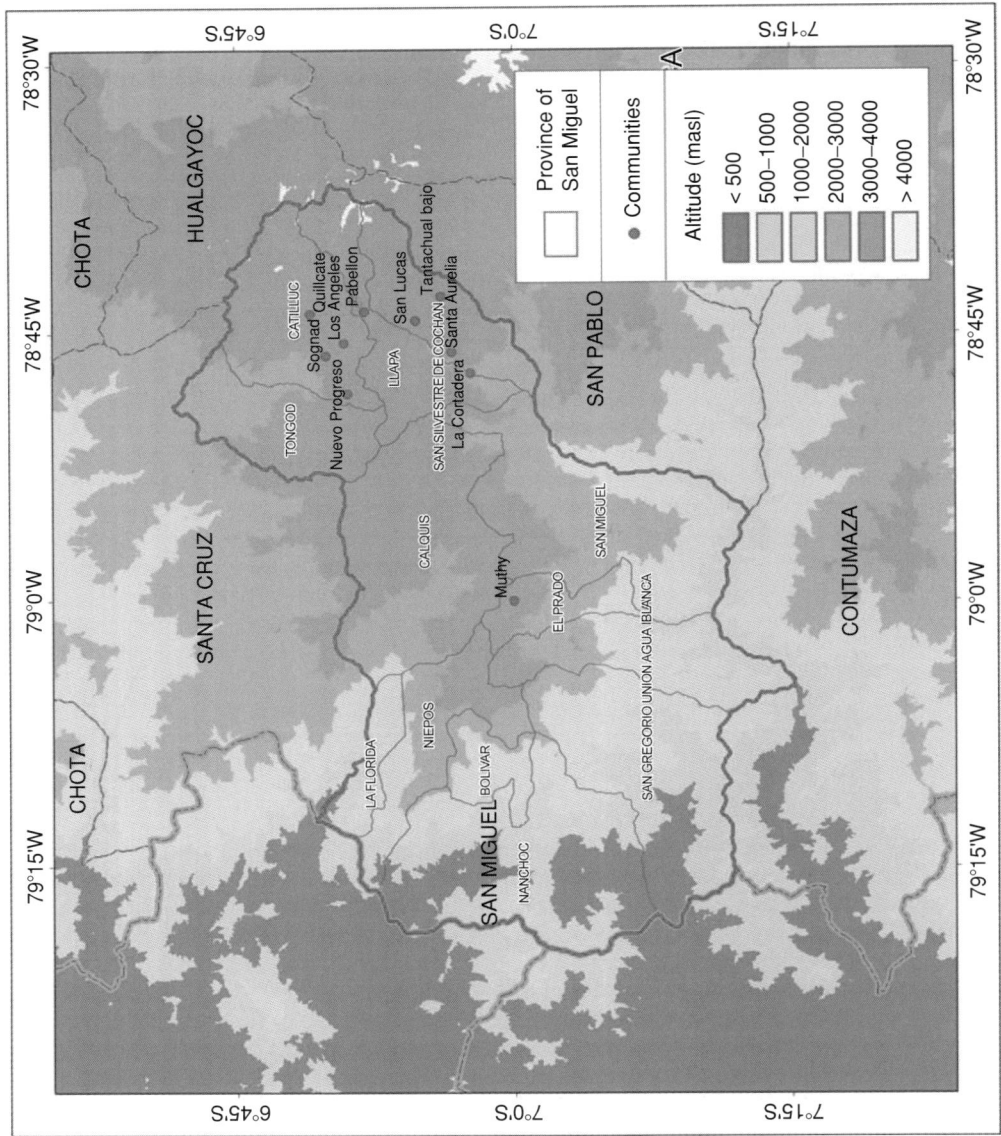

Fig. 9.2. Province of San Miguel.

the Nestlé company arrived in the mid-20th century. In the mid-1990s another processing industry was established in the region and started to buy milk from farmers, also contributing to the expansion in numbers of dairy cattle. In Cajamarca Department, total milk production per year has grown from 70,000 t in 1985 to 203,000 t in 2002, constituting a substantial increase in milk production (Webb and Fernández, 2003).

By managing livestock, farmers can earn regular income throughout the year from the sale of milk, cheese and meat. As a consequence, they have shown a tendency to shift from cultivating crops to planting permanent pastures, especially when irrigation is available. This allows them to have fodder for dairy cattle and a corresponding increase in milk production. Farmers favour milk production not only for its relatively high value, but also because of the security of market access. In dairying, purchase prices are relatively stable and payments are frequent as opposed to the uncertain markets and prices for crops. Crop income is obtained only after harvesting.

Linares (2001) indicates that 31% of households in San Miguel have at least one temporary or permanent migrant in cities such as Lima and Cajamarca. The purpose of temporary migration is to earn additional income for the household, whereas permanent migration is usually a means of pursuing access to education. Less than one in five (17%) of the families in San Miguel reported that household members worked locally in non-agricultural activities as temporary workers or in mining, though some of them also had small businesses such as craft making and small shops.

The area planted with potatoes in Cajamarca is about 23,000 ha (Linares, 2001), which accounts for 8% of the total potato area in Peru. In the Province of San Miguel, where the survey was conducted, a total of 1552 ha of potatoes are planted. The agroecosystems in Cajamarca are usually divided into two zones. The first is called the maize zone, located between 2500 and 3000 m. The second is called the potato zone, found between 3000 and 4000 m. Potatoes are an important food crop in the cold regions of the high Andes, where few other crops (some cereals and native tubers) can be planted.

Potato cultivation in San Miguel is part of a complex cropping system. Linares (2001) has indicated that, although potatoes are the most marketable crop in that system, they account for only 19% of the total area with crops. Cereals such as wheat and barley sum to 28%, maize is planted in 14% of the area, peas account for 12% and other crops such as other Andean tubers, faba beans and beans are planted in 2% of the area. Ray grass for dairy cattle is planted in 24% of the area, indicating the importance of this activity for the region.

According to Ortiz (1997), farmers usually consider three important factors when making decisions about potato cultivation. First is the rainfall pattern that divides the year into a rainy season from November to May and a dry season between June and October. The rainy season is usually associated with the presence of late blight, a serious potato disease that can lead to dramatic reductions in yields or even to total loss of the crop. Second is the risk of frosts, which depends on the location of the specific plot. Third is the availability of oxen for soil preparation.

In Cajamarca, there are three main dates for planting potatoes. Early planting happens between June and August (the coldest months of the year, thus having a high frost risk) in areas where there is enough humidity or the presence of irrigation for plant growth, but not a level of humidity that runs the risk of late blight. The second and main planting period is between September and November, at the onset of the rainy period, when there is a low risk of frost but higher risk of late blight. A third planting period occurs when, for factors such as lack of oxen for soil preparation or seed, farmers plant potatoes in December or January. During this season, there is a serious risk of late blight occurring at early stages of crop development.

Farmers from Cajamarca manage potato crops using some standard practices. Usually, soil preparation is carried out with both oxen and manual tools, depending on the soil characteristics and planting dates. They use manure or chemical fertilizers, although in most cases the amounts used are below the recommended levels for potato production. Weeding is carried out manually, and is usually combined with hilling-up (covering the base of the plant to promote tuber growth and avoid direct sun exposure on the potatoes). Pest control is another important activity. Because of the climatic conditions in San Miguel, failure to use fungicides and/or resistant varieties to control late blight leads to a very high risk of total loss. How pest control is implemented depends on farmers' economic conditions and the ability to afford pesticides. Potatoes are also harvested manually. At harvest, farmers select potatoes for sale, home consumption, seed and, sometimes, processing as 'papa seca' (dried potato). The storage process is another important activity because farmers need to keep quality potatoes for both home consumption and seed for 3–4 months. Potato plots in the San Miguel area are relatively small compared to other areas of Peru. In an analysis of 400 potato plots, Linares (2001) indicated that 66% had less than 0.5 ha and 23% had up to 1 ha. Overall, the evidence indicates that potato production serves as a complement to other crops and sources of income. Though potato production generates cash and food for the family, it is not a fully commercialized activity.

Potatoes are clonally propagated and potato tubers (which we refer to as potato seed as is common practice) are used for planting rather than true (botanic) potato seeds. Potato seed systems in Cajamarca, and in Peru in general, are characterized by the predominance of informal seed exchange. Data from the San Miguel area indicates that only 11% of farmers buy seed from external sources, but when they do so, less than 1% have access to seed from a source that can guarantee good quality (NGOs or a seed producer). The remaining 89% of farmers obtain their seed from neighbours, as wages, from their own harvest or through sharecropping (Godtland, 2001). Farmers have relatively easy access to seed of the most popular varieties in the region. Usually these are improved varieties, though the seed is mostly of doubtful quality. In general, seed is a sub-product of the cultivation of potatoes for home consumption or for the market and is not specifically produced as seed. This situation makes it harder to obtain varieties that are not well represented in the market. In particular, native varieties are planted in small plots so that harvests per household are small and therefore they do not have a strong presence in the seed system.

Farmers in the Andes usually plant several potato varieties at the same time. This is generally viewed as a strategy to reduce risk and meet multiple needs of the household that depend on the unique characteristics of different varieties. Factors that farmers take into consideration when they choose varieties for planting include resistance to late blight, good yield, early maturity, culinary quality, good market acceptance, good seed market acceptance and good appearance. A number of improved varieties are planted in the region (including Liberteña, Yungay, Canchán and Amarilis), as well as native varieties such as Chaucha, Shoga Colorada and Huagalina (Ortiz *et al.*, 1999).

Data Design and Description

The data used for this study come from a survey funded by the World Bank and administered by the International Potato Center (CIP) and CARE-Peru. The survey was designed to provide baseline data for an impact evaluation of farmer field schools (FFSs) in the province of San Miguel. Led by CIP and CARE-Peru, the FFS programme included a number of components, but focused primarily on teaching farmers about potato late blight and its management. The baseline survey was implemented in 1999. The sample included: (i) households that were already participating in the FFS programme; (ii) households that were expected to be incorporated into the FFS programme before the *ex post* evaluation surveys; and (iii) households in 'control' communities with characteristics (agroecological, social and economic, infrastructure and services) that were similar to participating communities. In communities where FFSs did not exist, households included in the survey were randomly selected for inclusion. In communities with a FFS programme, all participants were included as well as a randomly selected group of non-participants. In total, surveys were conducted in 13 communities and a total of 486 surveys were administered. All households included in the survey were potato producers.

Since the purpose of the survey was to evaluate the effect of the FFS programme on household welfare, the survey included detailed questions on household potato production as well as information on household characteristics, income-generating activities, asset ownership and participation in government programmes and local organizations. One of the key components of the FFS programme was the introduction of new late blight resistant varieties. These included Amarilis and Chata Roja, as well as fifty new clones to be evaluated with farmers using participatory methods. Because of this, detailed information on variety use was collected as part of the household survey. Along with the household survey a community survey was conducted to obtain general information regarding markets and services. Details of the survey and survey region can be found in Godtland (2001).

Table 9.1 provides an overview of the characteristics of surveyed households. The average household in the survey has a head of household that is 45 years old and has completed 5 years of education. Migration out of the Cajamarca region is significant, particularly by younger men, and the average quantity of household labour is only 3.3 members. Households own an

Table 9.1. Household characteristics.

Category	Variables	Description	All households
Human capital	Age of head	Age of household-head in years	45.0
	Head education	Years of education of household-head	5.0
	Household labour	Number of household members over 14 years of age	3.3
Natural capital	Cultivatable land owned	Hectares of land owned that can be cultivated	1.7
	Number of plots	Number of potato plots cultivated	2.4
	All black soils	Dummy variable for having only land with black soils	68.1%
	Altitude of highest plot	Altitude of highest plot in the community in metres	3452.7
	One harvest	One potato harvest possible in the community (dummy)	21.0%
	Two harvests	Two potato harvests possible in the community (dummy)	58.2%
	Three harvests	Three potato harvests possible in the community (dummy)	20.8%
Rural development	Wealth index	Wealth measure created using principle component analysis	0.0
	Milk production	Milk production in litres for previous year	5.3
	Non-farm income	Share of non-farm income in total income	14.3%
	Credit constrained	Unable to obtain credit (dummy)	20.2%
	Potato market	Kilometres to potato sales market	48.9
	FFS	Participated in Farmers Field School (dummy)	9.3%
	CARE	Participated in any CARE programme	27.0%
Region	Catilluc region	Household is in Catilluc region (dummy)	52.3%
	Cochan region	Household is in Cochan region (dummy)	24.5%
	El Prado region	Household is in El Prado region (dummy)	18.1%
	Llapa region	Household is in Llapa region (dummy)	5.1%

Note: Number of households = 486

average of 10.1 ha of land, but a major share of that amount is forest, natural pasture or other non-cultivatable land. On average, households owned 1.7 ha that could be cultivated with crops. One of the main characteristics of traditional Andean agriculture is farmer management of the high degree of environmental heterogeneity by cultivating in different altitudes and microenvironments. Farmers in Cajamarca quite often cultivate plots at various altitudes with some on hillside slopes and still others in the high altitude *puna* (high altitude hilltop meadows). Households in the survey cultivated on average 2.4 potato plots. Black soils from the high Andes are considered the best for potato production and are widespread. Over two-thirds of households (68%) surveyed had all black soils, an indicator of relatively good soil fertility. Others tended to have a mix of black and other soil types, with only 10% of households having no plots with black soils. The altitude of the highest point in the communities surveyed range from 3100 to 4000 m with the average around 3500 m. Households quite often had potato plots at the higher points in the community and some potato varieties are more suitable for these high altitude conditions.

The number of potato harvests per year in a community depends on a number of factors including the availability of irrigation. In 21% of communities only one harvest is possible, in 58% two harvests are possible and in 20% three harvests. This does not mean that all households surveyed harvest this many times, but that environmental conditions are suitable for this number of harvests.

Along with land, a number of other assets can be used as indicators of wealth. Since households may choose different assets for investment and it is difficult to get a monetary value for these assets, a wealth index was created based on a range of important (and exogenous) assets using principle components analysis. The wealth index has, by definition, a mean of 0.

As noted earlier, households in the Cajamarca region have become increasingly involved in activities other than potato production, such as dairy production. Table 9.1 provides information on litres of milk produced during the previous year, an indicator of household investment in dairy production.[1] Nearly 90% of households were involved in dairy production and on average households produced 5300 l/year, or approximately 15 l/day. Along with dairy production, over 70% were involved in rural non-farm activities. On average, these activities generate 14% of income.

Table 9.2 provides information on household income sources. Average income for households in the region was 4752 nuevo soles, which is about US$1400 at 1999 exchange rates. All surveyed households were potato producers and earned 15% of their income from potato production.[2] The most important source of income is from milk sales, which earned an average of 59% of total income and was a source of income for 88% of households. Nearly the same number of households were involved in producing other crops and earning agricultural wages, but these made up only 12% of total income. Taken together the data confirm that households are involved in multiple income-generating activities; a fact that is likely to influence potato production.

Constraints in credit access can limit the ability of households to purchase potato seeds for production, purchase other inputs for production, hire workers or invest in alternative economic activities. It can also limit the ability to smooth consumption and force households to sell productive assets during bad years, affecting their future productive possibilities. As Table 9.1 shows, one in five households are considered credit constrained.[3] As noted in other chapters of this book, gaining access to markets is likely to influence potato diversity through transaction costs. The average distance to the potato sales market is 49 km. Some households had markets in their community, and the most remote were 100 km away from the nearest market for selling potatoes.

At the time of the survey 9% of household had participated in the FFSs. In addition to the FFS programme, CARE also ran other agricultural programmes in these communities. Although other programmes were not designed to teach about late blight management and potato production, the extension agents that worked for the FFS programme and the other programme were the same. It is possible that these extension workers passed some of the same information to non-FFS participants as well as to the FFS participants, such as the existence of a new variety of potato that was resistant to late blight. For this reason, the role of CARE was also considered in this analysis. Over a quarter of

Table 9.2. Income-generating activities.

Source	Income (nuevo soles)	Share of total (%)	Share participating (%)
All sources	4752	100.0	100.0
Potato production	695	14.6	100.0
Dairy production	2827	59.5	87.5
Other agricultural activities	549	11.6	81.7
Non-agricultural activities	682	14.3	72.8

Note: Number of households = 486.

---

[1] This indicator is used instead of the number of dairy cows because there was significant variability in the productivity of the cows (litres/day/cow) based on the breed as well as feeding patterns.
[2] Note that if potatoes or other crops were consumed at home the value of that production was determined using local market prices collected at the time of the household survey.
[3] Credit constrained is defined as those households without access to credit that note that they would like but are unable to get credit and households that have access to credit that note that they would like to get more credit under the same terms and conditions but are unable to get more credit.

households surveyed participated in some CARE programme. Finally, about half the surveyed households live in the Catilluc district, with the rest spread over the three other districts.

In the survey, households identified over 50 potato varieties that they had planted in the year before the survey. Table 9.3 provides an overview of the top ten varieties planted. Only one variety, Libertaña, was used by more than 50% of households and accounted for over half of the planted area in the region. According to information collected in the survey, Libertaña is widely planted because it is considered to have a high yield, to be well accepted in the market and to be tolerant to late blight. Compared to Libertaña, no other variety represented more than 10% of the total area planted to potatoes. It is noted that 31% of farmers planted Canchan, 34% Chaucha, 24% Machala, 26% Suela Colorada and 20% Yungay. A review of the data suggested that households tended to plant one variety, normally Libertaña, for its high yields and market acceptance, planting additional varieties for culinary quality, early maturity or other desirable traits. Similar to the hypothesis presented by Bellon (1996) and the findings reported in Chapter 6, this creates a situation in which different varieties, along with serving a function of risk diversification, may meet different production and consumption needs of the household.

The data clearly show that households planted a number of distinct varieties in a given year. Figure 9.3 provides a histogram of the number of varieties grown by each household. Of the 486 households, only 15% planted one variety of potato with approximately one-quarter planting two and three varieties each. About one in every three households plants more than three varieties. As noted in Table 9.4, on average, households plant 3.1 varieties. A variety count expresses only richness, but not the relative dominance or proportional abundance of varieties on farms. The Berger–Parker and Shannon indices were also used to express these concepts, as defined and discussed in Chapter 1.

Econometric Approach

To evaluate the factors that influence on-farm potato diversity the following equation was estimated:

$$D_i = \beta_0 + \beta_1 H_i + \beta_2 L_i + \beta_3 R_i + \varepsilon_i$$

where: D_i = measure of potato diversity of household i;
H_i = human capital variables of household i (age of head, head education, household labour)

Table 9.3. Potato varieties used.

Variety	Percentage of farmers planting	Share of total planted (%)	Modern or traditional variety	Favoured characteristics noted by households
Amarilis	15.4	4.1	Modern	LB resistance, high yield
Canchan	31.3	7.7	Modern	Early maturity, culinary quality
Chaucha	33.5	2.3	Traditional	Culinary quality, early maturity
Libertena	75.5	53.6	Modern	LB resistance, high yield and market acceptance
Machala	24.3	6.5	Modern	Early maturity, culinary quality
Perricholi	12.1	3.4	Modern	LB resistance, early maturity and culinary quality
Renacimiento	10.7	2.2	Modern	Culinary quality
Shoga Colorada	7.8	2.5	Native	Culinary quality
Suela Colorada	26.1	5.4	Native	Culinary quality
Yungay	20.0	5.0	Modern	High yield, market acceptance and culinary quality
Other[a]	36.4	7.2	Modern and traditional	All of the above

[a]Includes 43 other varieties.
Note: Number of households = 486. LB, late blight.

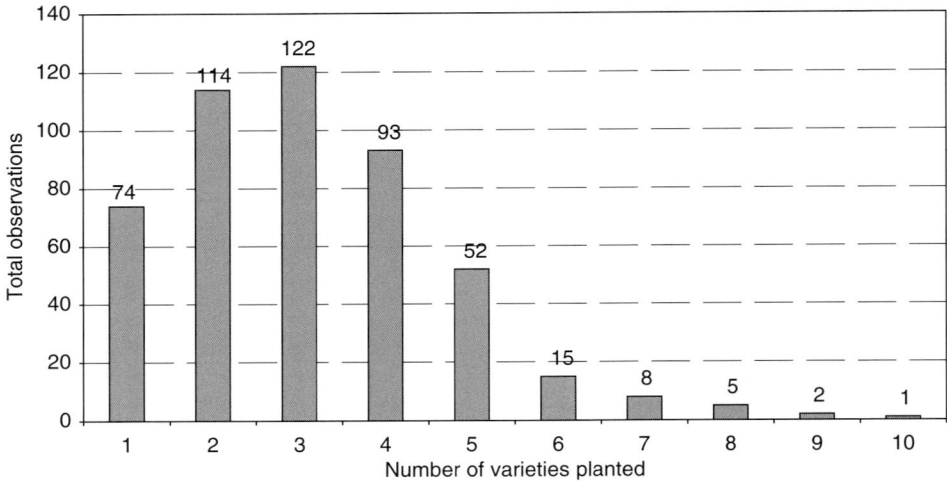

Fig. 9.3. Use of potato varieties.

Table 9.4. Diversity indices.

Diversity measure	Mean	SD	Minimum	Maximum
Number of varieties	3.12	1.58	1	10
Shannon index	0.73	0.46	0.00	1.96
Berger–Parker index	1.67	0.65	1.00	4.67

Note: Number of households = 486.

L_i = natural capital and agroecology variables of household i (land owned, land owned squares, irrigated land, number of plots, black soil, altitude, number of harvests);

R_i = variables associated with rural development of household i (wealth index, milk production, non-farm income share, credit access, potato market access, FFS participation, CARE participation); and

ε_i = error term.

Human capital variables were included in the analysis partially to control for differences in preferences across households. Older household heads may value traditional varieties more than younger household heads so the hypothesis to be tested is that age is positively associated with diversity. Maintaining a high level of diversity may also be labour-intensive. If this is the case, the expectation is that those with less labour or higher education, since the value of time may be greater, will have lower levels of diversity.

For the natural capital and agroecological variables the expectation is that greater diversity of plots will be positively associated with potato diversity. A greater number of potato plots, which presumably vary to some degree, is hypothesized to increase diversity. Similarly, having high altitude plots in the community, an indication that some of the plots are in unique agroecological settings, is expected to increase diversity. Having all black soils, which indicates a level of uniformity of plots, is presumed to decrease diversity. Cultivable land owned is expected to have a non-linear relationship at least with the number of varieties planted. An increase in land is expected to lead to a greater number of varieties planted until a point is reached that all the needs (preferences) of the household are met when this effect should be diminishing. For the other measures of diversity, the expectation is simply that more land will lead to greater diversity and not necessarily at a diminishing rate. When there are multiple harvests, specific varieties may be suitable for the different growing seasons. If this is the case, having more harvests should increase diversity and this hypothesis is tested with the data.

The primary interest of this chapter is in testing the effects of rural development on potato diversity, and the expectation as discussed earlier is that greater rural development will lead to a reduction in potato diversity. Wealth is included as a proxy for the ability to manage risk under the assumption that those with greater wealth are in a better position to manage risk through the use of their assets rather than through portfolio investment. As such, wealth is hypothesized to negatively affect diversity. Alternatively, if markets for certain varieties are incomplete, wealthy farmers may be in a better position to satisfy their preference for certain varieties through production. In such a case, wealth may be positively related to diversity. Milk production and non-farm income represent alternative paths towards improved welfare for the household instead of potato production and are hypothesized to negatively affect potato diversity since they take resources away from potato production. Potato seed is expensive compared to other staple crops and purchasing new seed, and thus accessing new varieties, is limited by an inability to obtain credit. Credit-constrained households are therefore hypothesized to have less diversity. Greater distance to market indicates that the household is less integrated into the market and faces higher transaction costs for selling output. In the absence of market forces that may push farmers to uniformity, the expectation is that greater distance to market is associated with greater diversity. Finally, since CARE promoted the use of modern late blight resistant varieties that may have displaced traditional varieties, the expectation is that FFS and CARE participation reduce diversity.

Findings

To test these hypotheses, three regressions were run for each of the indices of diversity with the identified set of dependent variables. A Poisson maximum likelihood model was estimated for the variety richness regression because the dependent variable is a non-negative count variable. Tobit models with robust standard errors were estimated for the evenness (Shannon) and inverse dominance (Berger–Parker) indices, given the range of the dependent variables. Table 9.5 presents the results of the analysis.

None of the three human capital variables – age of head, head education or household labour – has a statistically significant effect on any of the indices of potato diversity maintained by farm households in Cajamarca. This finding does not support the hypothesis that maintaining diversity is a relatively labour-intensive activity, or that younger farmers are less oriented to valuing or keeping different varieties of potatoes.

A number of natural capital variables do seem to influence potato diversity. Owned cultivatable land, as expected, appears to have a significantly positive but diminishing influence on the number of varieties planted. However, there is no significant effect of land owned or the square of land owned on any of the diversity indices. As the land owned increases, households plant more varieties (at a diminishing rate), but do not allocate large amounts of land to those varieties. This suggests a diminishing marginal utility from the traits provided by additional varieties of potato. Land ownership, however, bears no influence on the relative dominance of evenness of potato varieties individual farms.

Not surprisingly, the number of plots cultivated by a household positively and significantly influences potato diversity as measured by all the three indices, leading to more varieties, planted more evenly over a wider area. As expected, having all black soils appears to significantly reduce the number of varieties planted, since it increases the likelihood that plots are uniform and the same varieties can be used for different plots. Usually black soils are an indicator that a particular region or zone is suitable for more specialized potato production, which may lead to specialization also in terms of varieties with market purposes, thus reducing diversity. Having all black soils does not significantly affect the Berger–Parker index because the dominance of the most widely grown variety in the household is unaffected by soil type when fertility is uniformly good. Being in a community with high altitude plots appears to increase potato diversity on individual farms within that community. Higher altitudes require specialized varieties, which are quite often native cultivars. Taken together, these results suggest, as hypothesized

Table 9.5. Factors influencing potato diversity.

Category	Variable	Count (Poisson)		Shannon index		Berger–Parker index	
		Marginal effects	P-value	Marginal effects	P-value	Marginal effects	P-value
Human capital	Age of head	−0.0013	0.51	0.0003	0.89	−0.0005	0.87
	Head education	−0.0020	0.83	−0.0049	0.59	−0.0153	0.22
	Household labour	−0.0092	0.63	−0.0023	0.88	0.0053	0.83
Natural capital	Cultivatable land owned	0.0436	0.00	0.0204	0.31	0.0165	0.57
	Land owned squared	−0.0023	0.05	−0.0001	0.95	0.0007	0.68
	Number of plots	0.1190	0.00	0.1052	0.00	0.0936	0.00
	All black soils	−0.0668	0.01	−0.0820	0.06	−0.0967	0.16
	Altitude of highest plot	0.0005	0.00	0.0006	0.00	0.0009	0.00
	Two harvests	−0.3687	0.00	−0.5299	0.00	−0.8783	0.00
	Three harvests	−0.4937	0.00	−0.8214	0.00	−1.3715	0.00
Rural development	Wealth index	0.0581	0.00	0.0379	0.01	0.0243	0.34
	Milk production	−0.0147	0.00	−0.0154	0.00	−0.0176	0.01
	Non-farm income	−0.2141	0.01	−0.2928	0.01	−0.4349	0.01
	Credit constrained	−0.0567	0.26	−0.0677	0.20	−0.1324	0.09
	Potato market	0.0035	0.00	0.0042	0.00	0.0064	0.00
	FFS	0.2603	0.01	0.1681	0.03	0.0557	0.64
	CARE	0.0188	0.68	−0.0117	0.82	−0.0172	0.83
Region	Cochan region	0.0434	0.57	0.1719	0.01	0.2837	0.00
	El Prado region	−0.4753	0.00	−0.5109	0.00	−0.6720	0.01
	Llapa region	−0.0197	0.79	0.1733	0.05	0.3462	0.02
	Constant	−0.5575	0.19	−1.3244	0.00	−0.8084	0.21

Note: Number of households = 486.

here and in the literature, that owning a varied set of plots with distinct agroecologies leads to greater potato diversity. If, for some reason, households were to abandon or sell some of these plots to consolidate their holding, diversity would decline at the household level. Less diversity at the household level, however, does not necessarily imply less diversity at the community level.

Surprisingly, more potato harvests per community reduce diversity. The expectation that different varieties are more suitable to certain growing periods does not seem to hold. Examining the data more carefully reveals that farmers in communities that can harvest two or three times a year tend to plant the market-oriented variety Liberteña, or the early maturing varieties Canchan and Chaucha. Intensifying potato production in this manner could therefore lead to a reduction in potato diversity because farmers tend to specialize in particular varieties with good market opportunities.

Results for rural development variables are mixed. As found in some other chapters of this book, wealth is positively and significantly related to the richness and evenness of potato varieties grown by farmers. Households possessing more assets plant more varieties more equitably distributed on their land. However, wealth has no effect on the relative dominance of the most used variety. While keeping in mind that in general the households in this survey are poor farmers, this result lends support to the observation that richer peasants are better suited to maintain potato diversity than poor ones despite the fact that the formers are also more market oriented and take better advantage of high-yielding seed varieties (Mayer, 1992).

As hypothesized, the more intensely the household is involved in milk production, the less likely it is to maintain potato diversity. Similarly, as the share of income from non-farm sources increases, the diversity of potatoes on farms decreases by all the three measures. The more involved family members are in alternatives to potato production, the lower the level of richness and evenness they maintain in their potato varieties. Credit constraints appear to have no effect on the number of varieties planted or their evenness. This may be because seeds are obtained through relatives, neighbours and other informal channels. Potato diversity appears to be significantly and positively associated with distance to the nearest potato market, confirming, as in other chapters, that more remote households depend more directly on the diversity of the potatoes they grow for risk diversification and to satisfy other consumption needs.

Finally, though participation in the CARE programme in general had no impact on potato diversity maintained by households, contrary to the hypothesis presented earlier, participating in the FFS had a significant and positive effect on variety richness. The FFS were promoting the use of new, late blight resistant varieties, particularly Amarilis, and were encouraging farmers to experiment with them. At the time of the survey, the FFS had not been in operation a long time and it could be that participants were still at the stage of experimentation. Farmers were using Amarilis, but on small areas, and it had not yet replaced other varieties. Subsequent research by the CIP suggested that Amarilis replaced Liberteña, another modern variety, since both served the same function as high yielding, late blight resistant varieties. These two findings suggest that the FFS programme would not have reduced potato diversity in the longer term because the new variety filled a particular 'high-yield-market variety' niche thus replacing a modern variety without reducing the use of native varieties.

Of course, the coefficient on FFS could be biased by sample selection, i.e. if those households that participate may have been already inclined to experiment with a new variety or there is some unobserved difference between them and other farmers. Ideally, to deal with this problem, a two-stage estimation procedure would be used to predict FFS participation before using participation as an explanatory variable. However, good instruments for predicting participation could not be identified in the survey data. Results should therefore be treated with some caution.

Combined, the results indicate that while the wealth (permanent income) of peasant farmers does not appear to reduce diversity, and in fact may increase diversity, shifting patterns of income generation towards high value agricultural products (milk) and to non-farm activities seem to lower the level of potato diversity maintained on farms. A relatively wealthy farmer is able to enjoy the benefits of diversity: having a series of different types of preferred potatoes available for consumption (even at the risk of facing a higher risk of late blight) and to serve as presents or as payment to hired labour. Additionally, intensifying potato production through more harvests and improving access to potato markets also appear to reduce potato diversity. Given these results, if a rural development programme that follows the World Bank prescription provided in the introduction were to be promoted in an area with crop genetic diversity, the expectation is that this would lead to genetic erosion at least at the household level.

Conclusions

Policies and programmes to support rural development and reduce rural poverty are likely to seek to intensify agricultural production, to diversify household production away from staple crops, to support the production of high value crops, to enhance opportunities in non-farm activities and to promote market integration of households through improved rural infrastructure. This chapter seeks to determine how the factors that are associated with rural development are likely to affect the diversity of staple crops. To do so, data from a household survey of potato producers in the San Miguel Province of Cajamarca, Peru were analysed.

The results indicate that policies generally associated with rural development, particularly the development of alternative uses of land and labour that shift production away from staple crops, are likely to lead to a reduction in potato diversity. In other words, certain paths taken by poor households to exit poverty are not likely to lead to variety diversification.

The specific intervention of the FFS, and the introduction of the new variety Amarilis, was not associated with less potato diversity. This conclusion is tentative, however, since a more complete analysis of the data would be necessary to analyse changes over time. Ideally, a carefully constructed panel data would be collected that allows for a careful analysis of interventions. A more complete analysis would also consider how changes in the rural economy over time, such as changes in land use, shifts in allocation of labour resources, influence diversity.

Implications

If rural development is linked to a reduction in genetic diversity in centres of crop diversity like this region of Peru, the obvious question to ask is whether this is necessarily the case. It may be feasible to halt or reverse these trends by promoting the consumption and transformation of native varieties, most of which are not known in the market. This may be possible if such varieties do represent the bulk of diversity and there is a sufficient demand for them. To explore this possibility, the CIP has initiated a project called INCOPA that seeks to promote the consumption and transformation of native varieties. This project could determine the compatibility of rural development with the maintenance of crop genetic diversity on farms.

If rural development is not compatible with on-farm diversity, careful consideration must be given to how and when, in the process of rural development, to intervene to support genetic diversity. Interventions should be timed to ensure that an optimal level of diversity is maintained. This requires a much clearer understanding of the links between different types of development intervention and different types of crop biodiversity maintained on farms.

Acknowledgements

We are grateful to the World Bank, International Potato Center and CARE-Peru for funding and administering the survey and providing the data for this analysis, to Erin Godtland and Rinku Murgai for assistance with the data, the Food and Agriculture Organization for supporting Paul Winters' research in this area and Leslie Lipper, Melinda Smale and participants of the workshop *Valuing the Biological Diversity of Crops in a Development Context* in Rome, Italy for helpful comments on an earlier version of this chapter.

References

Bellon, M.R. (1996) The dynamics of crop intraspecific diversity: A conceptual framework at the farmer level. *Economic Botany* 50, 26–39.

Brush, S., Taylor, J.E. and Bellon, M.R. (1992) Technology adoption and biological diversity in Andean potato agriculture. *Journal of Development Economics* 39(2), 365–387.

Cotlear, D. (1989) *Desarrollo Campesino en los Andes*. Instituto de Estudios Peruanos, Lima, Peru.

FAO – Food and Agriculture Organization of the United Nations (1996) *The State of World's Genetic Resources for Food and Agriculture*. Food and Agriculture Organization of the United Nations, Rome.

Godtland, E. (2001) Reducing poverty in the Andes with genetically improved potato varieties: the importance of knowledge and risk. PhD dissertation. Department of Agricultural and Resource Economics, University of California at Berkeley, California.

INEI (1996) *III Censo Nacional Agropecuario*. INEI-Ministerio de Agricultura, Lima, Peru.

Linares, I. (2001) Tipificación de agricultures que cultivan papa en la provincia de San Miguel-Cajamarca: Estudio de casos de cuatro comunidades. Msc Thesis in the Faculty of Economics and Planing of the Universidad Nacional Agraria la Molina, Peru.

Louette, D., Charrier, A. and Berthaud, J. (1997) *In situ* conservation of maize in Mexico: genetic diversity and maize seed management in a traditional community. *Economic Botany* 51, 20–38.

Mayer, E. (ed.) (1992) *La chacra de papa*. Economía y Ecología, Centro Peruano de Estudios Sociales, Lima, Peru.

Meng, E. (1997) Land allocation decisions and *in situ* conservation of crop genetic resources: the case of wheat land races in Turkey. PhD dissertation, Department of Agricultural and Resource Economics, University of California at Davis, California.

Ortiz, O. (1997) The information systems for IPM in subsistence potato production in Peru: experience of introducing innovative information in Cajamarca Province. PhD thesis, Department of Agricultural Extension and Rural Development, University of Reading, UK.

Ortiz, O., Winters, P. and Fano, H. (1999) La percepción de los agricultures sobre el problema del Tizón Tardío o rancha (*Phytophthora infestans*) y su manejo: estudio de casos en Cajamarca. *Revista Latinoamericana de la Papa* 11, 97–120.

Reardon, T., Berdegue, J. and Escobar, G. (2001) Rural nonfarm employment and incomes in Latin America: overview and policy implications. *World Development* 29(3), 395–410.

Smale, M. (1997) The Green Revolution and wheat genetic diversity: some unfounded assumptions. *World Development* 25(8), 1257–1270.

Smale, M., Bellon, M.R. and Aguirre Gomez, J. (2001) Maize diversity, variety attributes, and farmers' choices in south-eastern Guanajuato, Mexico. *Economic Development and Cultural Change* 50(1), 201–225.

Van Dusen, M.E. (2000) *In situ* conservation of crop genetic resources in the Mexican milpa system. PhD dissertation, Department of Agricultural and Resource Economics, University of California at Davis, California.

vom Brocke, K. (2001) Effects of farmers' seed management on performance, adaptation, and genetic diversity of pearl millet (*Pennisetum glaucum* [L.] R.Br.) populations in Rajasthan, India. PhD dissertation, University of Hohenheim, Germany.

Webb, R.G. and Fernández, G. (2003) *Anuario Estadístico: Perú en Números 2003*. Instituto Cuanto, Lima, Peru.

World Bank (2002) *Reaching the Rural Poor: A Renewed Strategy for Rural Development*. The World Bank, Washington, DC.

Zimmerer, K.S. (1996) *Changing Fortunes: Biodiversity and Peasant Livelihood in the Peruvian Andes*. University of California Press at Berkeley, California.

10 Managing Rice Biodiversity on Farms: The Choices of Farmers and Breeders in Nepal

D. Gauchan, M. Smale, N. Maxted and M. Cole

Abstract

Farmers' choices, given the constraints they face, determine whether or not genetic resources of value to society for future crop improvement continue to be grown on farms. Plant breeders also influence crop biodiversity on farms through the supply of new genetic materials they develop. Using a conceptual approach that is similar to those presented in Part III, this chapter investigates determinants of farmers' choices and relates them to conservation goals and geneticists' preferences for landraces in Nepal, a centre of rice diversity. Factors explaining rice diversity measured at the level of the household farm include adult labour available for farm production, the ratio of expected subsistence needs to production, distance from the market and the heterogeneity of farm physical conditions. Those that enhance the richness or evenness of varieties are not always the same as those that increase the probability farmers will grow landraces that breeders consider to be of potential value for crop improvement. Isolation from markets is a major factor regardless of conservation goals. Growing rare landraces is associated with the involvement of women in farm production and landrace sales. Location in the hillside area is perhaps the single most important determinant for least cost conservation. Targeting households for landrace conservation may involve equity considerations since households in the hillsides who are most likely to continue to grow landraces with public value are better off in land, labour and household assets.

Introduction

Rice is the primary staple food in Nepal, accounting for over 50% of the total cropped area and food production (ABPSD, 2001). It supplies nearly 40% of the energy and nutrition source of the growing population in the country (FAO, 2001). Rice is also an important source of livestock feed particularly during dry winter months. With rising incomes and expanding population, the consumption and demand for rice in both urban and rural areas in Nepal continues to increase.

Nepal is an important centre of diversity for *Oryza sativa* ('Asian' rice). Asian rice was probably first cultivated in the geographically and culturally diverse region spanning from Nepal to northern Vietnam (Vaughan and Chang, 1992). Farmers' rice varieties (referred to here as landraces) still occupy over 30% of the total cultivated rice area in Nepal (ABPSD, 2001). Farmers maintain an estimated 2000 rice landraces in different parts of Nepal in association with their wild and weedy relatives (Shrestha and Vaughan, 1989; Upadhyay and Gupta, 2000). Landraces have evolved in response to wide variations in edaphic, topographic and climatic conditions, coupled with farmers' careful seed selection and management practices. In many locations, market isolation has contributed to the need for farmers to rely on their own seed sources and harvests to meet food needs, reinforcing this process.

In recent years, with economic change, the scientific community has expressed concern that as farmers gain access to markets and technologies, valuable rice genetic resources may be lost. There are indications that this is the case in Nepal (Joshi et al., 1998; Upadhyay, 1998; Chaudhary et al., 2004). Both genetic resources stored *ex situ* and those grown *in situ* are important for the crop improvement process that generates public value for society through enhanced productivity and lower food prices. Farmers' choices, given the constraints they face, determine whether or not genetic resources of public value continue to be grown *in situ*. Professional plant breeders also make decisions that affect the conservation of crop biodiversity on farms. Plant breeders select and cross materials in order to develop new varieties. The choices they make shape the range of genetic resources supplied to farmers in the form of new varieties released by commercial seed systems.

Not all landraces can be conserved on farms, and not all farmers can conserve them because of the costs involved, including direct programme costs and costs in terms of opportunities forgone. Nepal is grouped among the lowest income countries of the world in terms of Gross National Product (World Bank, 2003). The challenge for the government of Nepal is to create incentives for maintaining the rice biodiversity that benefits farmers today as well as future society. Though future needs cannot be predicted with certainty, the expert assessments of rice breeders provide us with reasonable guesses – though rice breeders, like farmers, have differing points of view.

This chapter uses detailed sample survey data from research in Nepal to investigate the determinants of farmers' revealed preferences and relate them to those stated by conservationists and breeders. Several breeders' criteria and indices for monitoring diversity levels are advanced. Similarly to the chapters of Part III, a conceptual approach drawn from a microeconomic model of farmer decision making relates the likelihood that farmers continue to grow the choice sets defined by these criteria to explanatory factors that may be influenced by public investments and policies. The relationship is then estimated econometrically. The profile of farmers who derive the greatest private value from growing those landraces thought to be publicly valuable is also presented, with implications for the design of programmes to manage rice biodiversity on farms.

The next section describes the study site and data sources. The conceptual approach is then summarized, followed by presentation of descriptive statistics, econometric methods and findings. Conclusions are drawn in the final section.

Site Description

Research was undertaken in two sites representing key rice-producing ecologies in Nepal (Fig. 10.1).[1] In most parts of Nepal rice is grown on small family-based subsistence farms with an average size ranging from less than 0.1 to 1.0 ha. The Terai lowlands are the rice bowl of Nepal producing 75% of the national rice crop; hill and mountain regions produce the remaining 25% (ABPSD, 2001). The Kaski site is located in a lake watershed in the hill region and is composed of a cluster of communities with moderate-to-high population density (155 persons/km^2). The agroecosystem is mid-altitude (600–1600 metres above sea level (masl)) and warm temperate to subtropical, with a wide range in altitude and ecological features including upper and lower hill terraces. Precipitation per annum is about 3900 mm. Rice production

---

[1] These sites were selected in the process of implementing a globally coordinated project aimed at the *in situ* conservation of on-farm agrobiodiversity in Nepal. The Nepal Agricultural Research Council (NARC) and the Local Initiative for Biodiversity Research and Development (LIBIRD), both based in Nepal, implement the project. The International Plant Genetic Resource Institute (IPGRI), based in Rome, Italy, coordinates the project.

Fig. 10.1. Map of Nepal showing Kaski (hill) and Bara (lowland Terai) sites.

is semi-subsistence and dominated by landraces that are grown in microniches, often in close association with their wild relatives found in the periphery of the two major lakes, Begnas and Rupa.

The Bara site is a lowland river watershed with higher population density (210 people/km^2). Located on the flat and fertile Indo-Gangetic plain (Terai region) on the southern border with India, this agroecosystem is low altitude (80–150 masl) and subtropical with an average rainfall of 886 mm/annum. Rice production is semi-commercial and dominated by modern varieties, with few farmers growing landraces. Bara farmers have easy access to high yielding modern varieties and modern technologies as well as market information from both local and external sources (Sthapit et al., 2000; Gauchan and Smale, 2003). Recent evidence documents the rapid displacement of landraces in this site by popular modern varieties with stronger market demand (Chaudhary et al., 2004).

Here, the term agroecosystem refers to an ecological landscape that covers farms and communities at the watershed level. These are hybrid ecosystems that are used for agriculture and have strong interactions with human components (Conway, 1985). Rice is cultivated across a range of microecological conditions. Upland, lowland and swamp environments are often found within the same farm. Farmers typically plant several varieties to match land types, soils, moisture conditions and cropping sequences. Normally a wheat crop, maize, lentils, potatoes and vegetables follow a rice crop planted in a single season. Some farmers also practise double cropping of rice in lowland areas. Farmers have their own way of classifying rice land types or *khet*[2] with different elevation, soil types and moisture level. The hill agroecosystem has five rice land types, compared to four in the lowland.

Rice land types are a major determinant of the variety adaptation within the agroecosystem (Table 10.1). High quality aromatic and fine landraces (e.g. *Basmati*, *Jetho Budho*, etc.) generally have longer duration and are adapted to lower wetlands (high moisture, good soils) and better fertility conditions. Coarse-grained, short-

---

[2] *Khet* is a bunded field (irrigated or rainfed) well suited for wet rice cultivation.

Table 10.1. Landraces cultivated in microecological niches in agroecosystems.

Agroecosystems	Land types (microecological niche)	Local name for land types	Local name of adapted landrace
Lowland (Bara)	Upper wet land	Uchha khet	Mutmur
	Mid-wet land	Samtal khet	Mansara
	Lower wet land	Nicha khet	Basmati
	Deep water	Ghol khet	Bhathi
Hill (Kaski)	Lower river basin	Sinchit khet/Phant	Jetho Budho, Anadi
	Hillside terrace/upper river basin	Tari khet	Kathe Gurdi, Mansara
	Seasonal stream irrigated terrace	Kulo khet	Ekle
	Swampy land	Dhab khet	Anadi, Jarneli Dhave
	Unbunded terrace (upland)	Ghaiya bari	Seto Ghaiya, Rato Ghaiya, Kunchhale Ghaiya

Note: Only landraces grown by farmers surveyed are listed here.

Table 10.2. Diversity of cultivated rice varieties at agroecosystem level, Nepal. (From Computed Project Survey Data, 2002.)

Study sites	Total cultivars	Total landraces	Total modern varieties	Per cent landraces in all cultivars	Per cent area in landraces	Per cent area in modern varieties
Hills	50	39	11	78	72.5	27.5
Lowland	23	5	18	21	4.0	96.0

maturing landraces (*Mut Mur, Ghaiya, Kathe Gurdi, Mansara*) are generally adapted to areas with moisture stress and poor fertility conditions in upland or upper wetlands. Some rice landraces in lowland such as *Bhathi* are adapted to deep water or flooded conditions where modern varieties perform poorly during heavy monsoon seasons.

The greater the range of land types or niches found within an agroecosystem, the more likely it is that farmers grow distinct cultivars with different growth habits (Brush *et al.*, 1992a,b; Bellon and Taylor, 1993; Meng, 1997; Rana *et al.*, 2000; Van Dusen, 2000). Heterogeneity tends to decrease when farms and agroecosystems are located in uniform slopes (flat lowland) with less altitudinal and climatic variation, or when more favourable moisture conditions are present (irrigation or rainfall). In the Philippines, Bellon *et al.* (1998) reported that only modern varieties of rice were grown in the irrigated lowland ecosystem where the environment is homogenous, while farmers' varieties persisted in the upland heterogeneous ecosystem.

Survey data confirm this to be the case in the study sites. Sample farmers in the study agroecosystems maintain a total of 50 and 23 rice cultivars in the hill and lowland agroecosystems, respectively (Table 10.2). As expected, the highest number of rice landraces (39) and percentage of area allocated to landraces (72.5) was found in the hill agroecosystem. Although both modern varieties and landraces coexist in both agroecosystems, almost all of the area in the lowlands is allocated to modern varieties (96%). Sample farmers in the lowland agro-ecoytem also cultivate a relatively higher number of modern varieties (n = 18) compared to those in the hills.

Data

Sample survey of rice-growing households

The sample survey research and analysis reported here builds on several years of intensive, participatory research with farmers as part of the national project. Initially the survey team led by the principal investigator listed all 1856 households in both sites. Through local contacts, the team learned that some of the households were no longer engaged in farming, some were no longer located in the original settlement and a few did not grow rice. A random sample representing 17.25% of actively farming, rice-growing households was drawn, numbering 159 in Kaski and 148 in Bara.

The survey instrument was a structured questionnaire administered in personal interviews. Questions covered social, demographic and economic characteristics of farmers and their households as well as physical characteristics of their farms, economic aspects of rice production and market access. The principal researcher coordinated the survey with the support of experienced local staff. Both men and women involved in rice production and consumption decisions were interviewed. To enhance data quality and uniformity, peer review of the questionnaires was undertaken in regular intervals to check for measurement errors, ambiguities and missing information. Households were revisited immediately for missing information and inappropriate responses immediately during the survey period. To ensure uniformity in units of measurement and consistent terminology, the researcher and enumerators edited the questionnaires at the survey site.

Key informant survey of rice breeders

A structured key informant survey of plant breeders and conservationists was implemented. First, a total of 16 scientists (both plant breeders and conservationists) were asked to identify criteria breeders use to select landraces as potentially useful. Scientists were chosen based on their active involvement in on-farm crop genetic conservation and national rice breeding programmes. Members of the national agricultural research system, the NARC and a local NGO, the LIBIRD, participated. Criteria included: diversity (expressed as a non-uniform, heterogeneous population); rarity (embodying unique or uncommon traits) and adaptability (exhibiting wide adaptation). Then, scientists were supplied with a list of rice landraces cultivated in the project site and asked to classify them according to selection criteria. Categories are not mutually exclusive. That is, the same rice landrace may be classified under more than one criterion. Table 10.3 reports breeders' stated preferences for rice landraces grown in the study sites. Their preferences reflect their perception of the potential value of the cultivars for crop improvement.

Conceptual Approach

Some empirical studies have investigated trade-offs in one type of diversity compared to another when policies promote changes in an explanatory variable, such as investments in education and infrastructure, using indices of diversity adapted from the ecological literature (Van Dusen, 2000; Chapters 6 and 11; Smale et al., 2003). This chapter investigates these as well as the explicit relationship between the stated preferences of rice breeders and the revealed preferences of farmers. Referring to the values of crop genetic resources described in Chapter 1, stated preferences of rice geneticists and breeders represent public value for future crop improvement, while the revealed preferences of farmers represent private value.

The conceptual approach is based on the theory of the agricultural household (Singh et al., 1986) as applied to variety choice and crop diversity outcomes (Part III, especially Chapter 5). When markets are not functioning well for a crop or its trade is associated with substantial costs of transaction, then production and consumption decisions cannot be separated and a shadow price for the crop guides decision making rather than its market price. Incomplete markets for rice landraces have been documented for the study area (Gauchan et al., 2005). The reduced form equation from the non-separable model expresses optimal area allocations among crops and varieties as functions of a vector of prices (p), farm size (A^0), exogenous income (Y^0) and vectors of farm household (Ω_{HH}), farm physical (Ω_F) and market characteristics (Ω_M):

$$\alpha^* = \alpha^*(p, A^0, Y^0, \Omega_{HH}, \Omega_F, \Omega_M) \quad (10.1)$$

Diversity indices are metrics constructed from area shares, as described in the next section. Diversity D on household farms is an outcome of choices made in a constrained optimization prob-

Table 10.3. Breeder perceptions of the potential value of rice landraces in crop improvement.

Variety name	Diverse	Rare	Adaptive	Variety name	Diverse	Rare	Adaptive
Anadi Rato	0	0	1	Jhinuwa Ghaiya	0	1	0
Anadi Seto	0	0	1	Jhinuwa Kalo	0	1	0
Anga	0	1	0	Jhinuwa Pakhe	0	1	0
Badahari	0	1	0	Jhinuwa Seto	0	1	0
Basmati	0	0	0	Jhinuwa Tarkaya	0	1	0
Basmati	0	0	0	Juwari	0	1	0
Bayerni	0	1	0	Kathe Gurdi	1	0	0
Bayerni Jhinuwa	0	1	0	Kaude 1 (NL + KG)	0	0	0
Bhathi	0	1	0	Kaude 2 (Md + Mn)	0	0	0
Bichara Ghaiya	0	1	0	Kunchhale Ghaiya	0	1	0
Ekle	0	0	0	Madhese	0	0	1
Faram Ialka	0	0	1	Mala	0	0	1
Gajale Jhinuwa	0	1	0	Mansara	0	0	1
Gauriya	0	1	0	Mansuli Ghaiya	0	1	0
Gurdi	1	0	0	Mut Mur	1	0	0
Gurdi Sano	1	0	0	Naulo Madhese	0	0	1
Gurdi Thulo	1	0	0	Pahenle	0	0	0
Jarneli	1	0	0	Ramani	0	1	0
Jarneli Dhave	0	1	0	Rato Ghaiya	0	1	0
Jarneli Pakhe	0	1	0	Sathhi	0	1	0
Jetho Budho	1	0	0	Seto Ghaiya	0	1	0
Jhinuwa	1	0	0	Tunde	0	1	0

Note: 1 = of high potential value, 0 otherwise.

lem rather than an explicit choice. Equations estimated econometrically take the following conceptual form, as in Chapter 5:

$$D = D(\alpha *(p, A^0, \Upsilon^0, \Omega_{HH}, \Omega_F, \Omega_M)) \quad (10.2)$$

Rice land allocation among rice varieties is not affected by the choice of other crops unless earlier maturity or growth habit reduces the cost of a rotation or intercrop. Although this may be the case in many land-intensive production systems of Asia, it was not observed in the study area. Therefore, decisions regarding other crops are treated as independent of decisions about how much rice area is to be allocated to each variety.

Econometric Methods

Equation (10.2) is the basis of econometric analysis and hypothesis tests. The econometric analysis investigates two types of policy trade-offs that may occur in public investments to support conservation of rice biodiversity on farms in Nepal. A public investment that promotes the mainte- nance on farms of one form of diversity on farms (such as richness levels) at the expense of another (such as evenness levels) results in a diversity trade-off. A conservation trade-off occurs when a public investment targets the maintenance of a set of rice landraces defined by one choice criterion (such as genetic diversity) at the expense of a set defined by another competing criterion (such as rarity or adaptive potential).

In the first set of econometric regressions, explanatory factors specified in the farm household decision-making model enhance or detract from rice diversity (richness, dominance and evenness) measured at the household level. In the second, explanatory factors specified in the farm household decision-making model influence the likelihood that landraces in various core subsets (selected with different choice criteria, such as rarity or diversity) are grown. If the effect of an explanatory factor is the same regardless of the diversity index or choice set, policy or public investments related to that factor are 'neutral'. If they differ, enhancing the prospects for maintaining one type of diversity level or one choice set may diminish prospects for another, entailing a policy trade-off.

Dependent variables

The dependent variables for the first set of tests are spatial diversity indices measured at the level of the household farm (Table 10.4), as in most of the chapters of this book. A simple count of distinct rice varieties expresses richness. Although standardizing by area may be preferred in principle to a simple count (as in the Margalef index), the small areas farmed in Nepal led to problems of variable definition. The Berger–Parker index represents relative abundance or the inverse dominance of the major variety grown by the household in terms of its share of rice area. The Shannon index reflects both richness and relative abundance, or evenness in the allocation of rice area among varieties on the farm. The dependent variables for the second set of tests are defined according to breeder scores for diversity, rarity and adaptability, as shown in Table 10.3.

Explanatory variables

Explanatory variables and hypothesized effects are shown in Table 10.4, grouped according to the sets of observed variables that represent the variables in Eq. (10.2). Market prices are not included because they are fixed for all households in each site. Moreover, since rice markets are incomplete, shadow prices govern the decisions of farmers.

Hypothesized effects are similar to those presented in the chapters of Part III, supported by previous research in the study site. Household characteristics affect crop biodiversity both

Table 10.4. Definitions of variables and hypothesized effects on diversity.

Variable name	Definition	Hypothesized effect
Dependent variable		
Richness	Count of distinct rice varieties	—
Dominance^{-1}	Berger–Parker index	—
Evenness	Shannon index	—
Household characteristic		
Age (PDM)	Age of production decision maker (years), typically a man	(+)
Education (PDM)	Education of production decision maker, typically a man (years)	(+, −)
Education (CDM)	Education of consumption decision maker, typically a woman (years)	(+, −)
Farm labour	Active adults working on farm (number)	(+)
Female labour	Per cent female of actively working adults	(+)
Livestock value	Value of large animals (bullocks, dairy animals)	(+)
Expenditures	Average monthly household expenditure since last harvest preceding this season (exogenous cash income)	(+, −)
Subsistence ratio	Ratio of 5-year average of kilograms of rice produced/rice consumed	(+, −)
Farm physical characteristic		
Irrigation	Per cent rice area under irrigation	(+, −)
Land types	Number of rice land types	(+)
Farm dispersion	Total walking distances (min) from house to rice plots, divided by cultivated hectares	(+)
Market characteristic		
Market distance	Total walking distance from house to local market (min)	(+)
Landrace sales	Landrace grain sold by household, past season (kg)	(+)
Modern variety sales	Grain of modern variety sold by household in preceding season (kg)	(−)

Note: Expenditures and livestock value variables are measured in Nepalese rupees.

through preferences and household-specific costs of market transaction as well as through labour stocks and opportunity costs. Age, education and the gender composition of households influence the set of crop varieties chosen for cultivation through the preferences of household members and their farming experience.

The age of production decision makers may be positively related to rice diversity since older farmers are more likely to have experience and knowledge about cultivating a range of varieties, and particularly landraces. Similarly, active adult labour on-farm is hypothesized to have a positive effect on rice diversity since more labour allows households to engage in the cultivation of a larger set of rice varieties with differing management requirements. The educational levels of decision makers (production and consumption) may have either a positive or a negative influence on rice diversity. More years of schooling may increase farmers' ability to acquire information and experiment with diverse varieties, or it may be associated with a preference for modern varieties and specialization. The proportion of active working females is thought to relate positively to rice diversity through variety preferences for consumption attributes. An earlier study by the project team revealed a greater role of women on rice seed maintenance and cultivation (Subedi et al., 2000).

Livestock assets, exogenous income and subsistence ratio are wealth-related variables hypothesized to affect variation in crop diversity levels through their association with the ability to bear production risk. Households owning a larger number and value of draft (bullock) and dairy (buffalo, cows) animals are expected to grow more diverse rice varieties through increased access to inputs and information and capacity to experiment, or because of greater demand for fodder. Ownership of bullocks (draft power) also induces farmers to maintain diversity because these allow for timely land preparation, threshing and transportation of inputs and harvested products. Cash income might be either positively or negatively associated with diversity. On one hand, cash income enhances farmers' capacity to hire labour and purchase inputs in order to engage in a wider range of activities. On the other hand, it may imply that households are allocating household labour to non-farm activities or specializing in the production of a few

modern varieties for the market. Farmers producing rice in excess of their expected consumption needs may maintain either more or less diverse rice varieties.

Farm physical characteristics include farm fragmentation and land heterogeneity measured by the number of land types, distances among rice plots and the percentage of rice area irrigated. The more heterogeneous the conditions in which farmers' cultivate the crop, the higher the expected level of diversity since such heterogeneity leads farmers to choose a broader set of varieties to suit multiple classes of farm land and seasonal niches. Thus, farmers are expected to maintain more diversity when they own and cultivate different land types. Since total farm plot distance was highly correlated with area cultivated, the two variables were combined into one to capture the effect of scattered plots while controlling for total hectares cultivated. The ratio of total rice plot distance to total cultivated hectare is a measure of dispersion of rice plots around homesteads, or fragmentation. The percentage of rice area that is irrigated affects rice production potential by improving moisture availability and may have either negative or positive effects on the rice diversity on individual farms. Better availability of water may enhance specialization or dominance of few varieties by making the production process more uniform; it may also enable cultivation of diverse varieties with different moisture requirements and maturity periods.

Market-related variables affect diversity through the extent to which households trade their rice crop and purchase inputs, foods and other household needs in the market. The distance of the farm from the market is a major component of the cost of engaging in market transactions. The more removed a household farm from a local market centre, the more likely it will be to rely on its own production to meet its consumption needs. Consumption needs may include a range of food products as well as fodder. Grain sales from production of landraces are expected to relate positively to incentives for cultivating them. Grain sales from production of modern varieties may relate to specialization in fewer, uniform modern varieties. Past rather than current sales were used because these are not choice variables in the survey season. Sales amounts also express more variation than would a 0–1 variable.

Regression models

Regression models used to investigate the two types of trade-offs differed because of the definitions and measurement of the dependent variables. A Poisson regression model was used to relate explanatory factors to richness (a count) in rice varieties. Since some households grow only one variety of rice, the Shannon evenness index takes on a value of 0 and the Berger–Parker index assumes a value of 1 in some cases, both lower limits. A tobit model for censored regressions was used to estimate these relationships. A probit model was used in regressions about conservation subsets. All regressions were estimated using LIMDEP (version 7.0).

Findings

Diversity trade-offs

Agroecosystem differences suggest that the underlying parameters of the multivariate regression models relating these factors to rice diversity should be estimated separately for the hill and lowland sites. The hypothesis that parameters are equal between the sites was rejected with a log-likelihood ratio test (Greene, 2000, pp. 152–153) confirming the statistical significance of site-specific factors both in terms of levels of diversity and the marginal effects of explanatory factors on these levels. The Poisson model assumes equality between the conditional mean and variance and is usually tested for fitness against the Negative Binomial regression model. The test resulted in failure to reject the Poisson model.

Factors explaining variation in the richness, evenness and dominance among rice varieties grown by farmers in Bara and Kaski sites are presented in Table 10.5. The age and education of decision makers is a significant factor explaining rice diversity in Bara ecosite, but not in Kaski ecosite. Older farmers in the plains are more likely to allocate rice area more equally among varieties, perhaps due to their experience and because they are not as receptive to adopting and specializing in a single modern variety. More education among production decision makers (usually men) is positively related to both evenness and inverse dominance, though the opposite is the case for consumption decision makers (typically women). Education may expand the variety choice options for men through access to information, while enabling women to substitute rice products that are available on local markets for home-produced items. More active labour on farms generally contributes positively to rice diversity, and the marginal effects are particularly large in the hills where there are fewer non-farm opportunities and rice production requires more labour time. The gender composition of active labour is of no apparent importance, however.

Although neither outside sources of cash income nor livestock assets are significantly related to the diversity of rice varieties grown in these sites, households who expect to produce more rice relative to their consumption requirements also grow more varieties that are more equitably distributed. The statistical significance of household characteristics and market characteristics in explaining rice diversity supports other evidence that rice markets are incomplete in both the sites. As expected, distance from market centre, or the degree of isolation of the household farm, is positively related to rice diversity because purchased rice is not so easily substituted for own farm production. While sales of the grain of landraces is of no importance, sales of the grain of modern varieties is associated with less evenly distributed varieties in either site and in the lowland, with greater dominance by any single variety – presumably through market-driven incentives for its production. Sales of landrace grain were limited to a few farmers.

As hypothesized, the more heterogeneous the conditions in which farmers cultivate the crop, the greater the numbers of rice varieties grown and evenness in their area distribution. Rice plots are more widely dispersed per unit area in Kaski than in Bara, and within the Kaski ecosite their dispersion is positively related to the richness and equality among rice varieties on farms. In the Bara ecosite, but not in Kaski, the percentage of rice area irrigated is positively related to evenness and inverse dominance.

No trade-offs appear to be associated with public investments to promote richness, evenness or equality in the distribution of rice varieties on farms in either site. The direction of any statistically significant effect is the same across diversity concepts. Though different factors are significant

Table 10.5. Factors explaining variation in the diversity of rice varieties grown by farmers in two agroecosystems of Nepal.

	Bara site (N = 148)			Kaski site (N = 159)		
Variables	Richness	Evenness	Inverse dominance	Richness	Evenness	Inverse dominance
Constant	−0.5533	−0.8159***	0.3927	0.3247	−0.1531	0.7915
Household characteristics						
Age (PDM)	0.0038	0.0052**	0.0064	−0.0059	−0.00107	−0.0049
Education (PDM)	0.0405	0.0193*	0.0331*	0.0257	0.000729	−0.0088
Education (CDM)	−0.0479	−0.0420*	−0.0861**	0.0331	0.00739	0.0063
Farm labour	0.1896	0.0781***	0.0214	0.4914***	0.14438***	0.1300*
Female labour	−0.6165	−0.2571	−0.4569	0.5353	−0.10769	−0.6442
Livestock value	0.000007	0.000005	0.000004	0.00001	0.000004	0.00002
Expenditures	0.00003	0.00002	0.00008	0.00002	−0.00001	−0.00003
Subsistence ratio	0.2940	0.2163***	0.2584*	1.2439*	0.31228**	0.46832
Farm characteristics						
Irrigation	0.2537	0.1203*	0.2436**	0.4308	0.04773	−0.1370
Land types	0.4198*	0.1937***	0.1233	0.4094	0.1142*	−0.1301
Farm dispersion	−0.0003	−0.0002	−0.0003	0.0022**	0.00034*	0.0007
Market characteristics						
Market distance	0.0012	0.0009***	0.0012***	0.0013***	0.00032***	0.0003
Landrace sales	−0.0002	−0.0002	−0.0003	−0.0006	0.00003	−0.0002
MV sales	−0.0005	−0.00009*	−0.00014*	−0.00058	−0.0004*	−0.0006
Value of Log-likelihood function	−204.6	−78.63	−157.15	−264.0219	−88.61	−252.56

Note: Regression model for richness is Poisson; for evenness, Tobit censored at 0; and for inverse dominance, Tobit censored at 1. One-tailed Z-tests significant at $P < 0.01$ (***), 0.05 (**), 0.1(*) per cent level. Z-statistic is relevant for maximum likelihood estimation. The values reported are marginal effects that are computed at the means of explanatory variables. The significance of marginal effects in some cases is not supported by the significance of coefficients. See Table 10.4 for variable definitions.

in each site, the directions of effects are the same between sites.

Conservation trade-offs

Probit regressions investigating the trade-offs among conservation subsets could not be estimated separately for the two ecosites since the regression for the Bara lowlands did not converge, perhaps because few farmers there grow landraces. Instead, a 0–1 variable for site was included in the pooled regressions, and results confirm the positive and significant effect of hill location on the probability that landraces thought to be of public value will be cultivated (Table 10.6).

Education, labour composition and livestock assets are statistically significant predictors that households will grow landraces that are considered important for future crop improvement. Human capital appears to be a critical factor, in more ways than one. The more educated the decision maker in rice consumption (typically a woman), the greater the likelihood that that household grows a landrace that is genetically heterogeneous. More adult labour engaged in agriculture has a large effect on the probability that adaptive landraces are grown, also contributing significantly to cultivation of genetically diverse landraces. A higher percentage of women among active adults in the households means that a rare landrace is more likely to be grown. The more endowed with livestock assets (buffalo, cattle and bullocks), the more likely the household is to grow genetically diverse landraces. Cash income levels are of no apparent significance, since growing landraces does not cost money. The number of rice land types increases

Table 10.6. Factors predicting that farmers will grow landraces breeders identify as potentially valuable, in two agroecosystems of Nepal, by choice criterion.

	Choice criterion of rice breeders		
Explanatory variables	Diversity	Rarity	Adaptability
Constant	−0.6221***	−0.4289***	−2.6499***
Site	0.2792***	0.1074***	1.0596***
Household characteristics			
Age (PDM)	−0.000029	−0.00058	0.000387
Education (PDM)	−0.0101	0.00212	0.00931
Education (CDM)	0.0218**	−0.00483	−0.00679
Farm labour	0.04315**	0.01702	0.14948***
Female labour	−0.03892	0.13687*	−0.05048
Livestock value	0.000005*	−0.0000019	−0.000002
Expenditures	−0.000023	−0.000018	0.0000003
Subsistence ratio	−0.09510	−0.02833	0.05185
Farm characteristics			
Irrigation	0.080216	0.005799	0.1390
Land types	−0.05990	0.06588***	0.03843
Farm dispersion	0.000029	0.000056	0.001112**
Market characteristics			
Market distance	0.00040**	0.000137**	0.000665*
Landrace sales	0.00021	0.000111*	−0.000094
MV sales	−0.00004	−0.000005	−0.0001188
Value of Log-likelihood function	−93.79	−75.50	−54.65

Note: N = 307. The regression model used in all the cases is a probit. One-tailed Z-tests significant at $P < 0.01$ (***), 0.05 (**), 0.1(*) per cent level. See Table 10.4 for variable definitions. Z-statistic is relevant for maximum likelihood estimation. The values reported are marginal effects that are computed at the means of explanatory variables. The significance of marginal effects is not always supported by the significance of coefficients in small samples (Greene, 2000).

the chances that a rare landrace is grown, and the dispersion of rice plots relative to the total area cultivated contributes positively to growing adaptive landraces. Isolation from markets is associated with higher probabilities of growing any landrace that is identified as potentially valuable by rice breeders.

The factors predicting that farmers will grow landraces of potential value to society are the same in sign for subsets defined according to either genetic diversity or adaptive potential, though different factors are significant and the magnitude of effects depends on the choice set. One policy-relevant factor that is significantly associated with growing rare landraces, but not diverse or adaptive landraces, is past sales of the grain from landraces. This finding is consistent with other project evidence that the development of specialized markets through specific policy interventions might provide incentives for farmers to continue cultivating rare landraces (Gauchan et al., 2005).

Targeting households for conservation

Table 10.7 presents the profile of farm households with high and low likelihood of growing landraces that breeders identify as potentially important in Kaski (hill) ecosite, by choice criterion. Profiles were constructed by selecting households with high predicted probabilities of growing landraces in each group, according to probit regression results. For diverse and adaptive landraces, predicted probabilities for these households were above 90%, although for rare landraces, they were above 50% because of smaller subgroup sizes. Low probability households are those with less than 10% of probabil-

Table 10.7. Profile of households with high and low likelihood of growing landraces breeders identify as potentially valuable in Kaski agroecosystems, Nepal, by choice criterion.

Characteristics	High predicted probability			Low predicted probability of growing any choice landrace (N = 81)
	Grow diverse landraces (N = 20)	Grow rare landraces (N = 17)	Grow adaptive landraces (N = 76)	
Family size	6.15	5.8	6.36	5.86
Per cent men of active working adults	0.34	0.31	0.33	0.27
Ratio inactive/active persons	0.88	0.85	0.85	1.07
Number of persons working off-farm	1.3	1.35	1.35	1.71
Share of adults working on-farm	0.91**	0.98**	0.83**	0.50
Total value of household assets (NRs)	40043**	39877**	31366**	23408
Total land cultivated (ha)	0.92**	0.91**	0.76**	0.42
Rice land cultivated (ha)	0.75**	0.75**	0.62**	0.32
Landrace share of cultivated rice area	0.91**	0.88*	0.82	0.64
Rice landraces (number)	5.5**	5.35**	4.0**	1.59

Note: Since the subsets of farmers with high predicted probabilities of growing diverse, rare or adaptive landraces are not mutually exclusive, statistical tests on differences of means among them could not be conducted. Tests compare means for any one of these groups to the mean for households with low predicted probabilities of belonging to any group. Further explanation is provided in the text.
**Significant at 5% level; *Significant at 10% level.

ity of cultivating landraces classified in any of the groups. Means were compared statistically between households with high and low predicted probabilities.

To the extent that public value is expressed by any of the three criteria identified by rice breeders, targeting a location like Kaski has a major impact on the likelihood that landraces with public value would be conserved on farms. Within this hill agroecosystem, however, there are clear differences between the households with a high and low likelihood of growing such landraces. Tests comparing any one of the groups with high likelihood of growing publicly valuable landraces to the group with low predicted probabilities show significant differences in almost all means, except those related to social and demographic characteristics. For example, the family size, proportion of adults who are men and the dependency ratio (inactive to active persons) are of similar order of magnitude regardless of whether or not a household is likely to grow a publicly valuable landrace.

Although the number of persons working off-farm is also invariant to the likelihood of growing any of these landraces, the involvement of adults in farm production is an important defining feature. Any policy that would draw additional labour off the farm would therefore have a deleterious effect on the chances that particular landraces would continue to be grown. Another salient feature of households with high likelihood of growing rare, diverse or adaptive landraces is their overall wealth relative to those with low likelihood of growing them. They have much higher total asset values, their land areas are nearly twice as extensive and their rice areas are more than twice as great. At the same time, landraces occupy a greater share of their rice area and they grow on average 5–6 distinct rice landraces on their farms. These families ascribe more private value to rice landraces relative to modern varieties than do their poorer counterparts. Finally, meaningful differences are not apparent when examining means among conservation choice subsets in Table 10.7, and statistical differences cannot be tested because subsets overlap.

Conclusions

Farmers determine the survival of crop varieties or the maintenance of specific gene complexes in any given reference area by choosing whether or not to grow them and in what proportions. The choices they make today not only affect their welfare but that of future society. Farmers choose which varieties of a crop to grow according to their private value, and this value depends in semi-subsistence agriculture on farmer-specific characteristics and market conditions as well as the physical features of their farms.

Plant scientists employ decision making criteria when they select materials for breeding or conservation purposes, and these differ from those of farmers. They also differ among scientists. For example, they may identify varieties that are genetically diverse, those that have rare traits or express wide adaptation as potentially important for breeding programmes, and hence for genetic resource conservation. These are 'best guesses' regarding the public value of landraces.

This chapter has focused on types of policy trade-offs associated with the choice of criteria for conservation. One type of trade-off may occur when policy instruments promote one form of diversity (such as richness) at the expense of another (such as evenness). Richness refers to the count of varieties and evenness to the equitability in the area shares distribution among varieties. Similar to the findings reported for Ethiopia in Chapters 6 and 11, results for Nepal suggest few trade-offs.

As constructed, the diversity indices treat all material as equally important for conservation. That is, these goals are related to the numbers, evenness or equitability of varieties grown in communities without regard to the nature of the varieties. The second type of trade-off involves differences in the materials targeted for conservation, according to the criteria of rice breeders. Increasing the likelihood that farmers will maintain varieties that are members of one choice set may decrease the prospects that varieties in other sets continue to be grown. If so, policies designed to attain one objective might have serious consequences for another. Again, results show no trade-offs among conservation choice criteria. They do suggest, however, that while the policies designed to support the conservation of diverse and adaptive landraces are similar, these are different from those required to support the conservation of rare landraces. Findings also indicate that the factors determining variation in rice diversity levels on farms are sometimes distinct from those that influence the prospects that farmers grow particular landraces identified as important for crop improvement.

Implications

Regression results of this type could potentially be used to identify sites and households for programmes aimed at local conservation of rice biodiversity. Clearly, any rice-growing household in hill agroecosystem of Kaski is more likely to grow genetically diverse, rare or adaptable landraces. However, households with a high likelihood of growing more diverse combinations of rice varieties on their farms are those who have higher ratios of expected production to subsistence in both the lowland and hills. This finding suggests that there may be food policy reasons to be concerned about maintaining diversity in both agroecosystems – though this would be accomplished through means that are system-specific. Rice diversity in the lowlands encompasses both modern varieties and landraces, while the agroecosystem to target for managing socially valuable landraces on farms is the hills.

Not all households in Kaski, and not all landraces in Kaski, are equally promising candidates for conservation. Households with more active adults engaged in agriculture are more likely to maintain landraces of public value, so that increasing opportunities for competing sources of off-farm employment may have a negative impact on prospects for conservation. Targeting households for landrace conservation may involve other trade-offs in terms of equity considerations. Those most likely to grow landraces identified as important for future crop improvement are significantly richer in total value of assets, with more extensive farms, larger rice areas and higher share of adult family labour working on farm. Although most farmers on the hillsides of Nepal are ranked as poor by global standards, targeting the households relatively more likely to maintain valuable landraces is by no means equivalent to targeting the poor within those locations.

Findings indicate that it will be cost-effective to target conservation of valuable landraces in isolated locations of high crop biodiversity where farm households have more access to land, labour and capital assets. Policies and programmes aimed at supporting the economic viability of these communities could have positive effects on crop biodiversity conservation if targeted appropriately. Market linkages and opportunities for landraces should be further investigated. Support from scientists (plant breeders and conservationists) in the formal seed system is fundamental. Existing initiatives in Nepal include participatory plant breeding, community biodiversity registration, public awareness campaigns, diversity fairs and other relatively low cost activities. These efforts, undertaken by Nepal's on-farm conservation project, have been endorsed by the national government.

Acknowledgements

This chapter draws from the doctoral dissertation of the first author (Gauchan, 2004). It is based on research conducted as part of the *In Situ* Conservation of Agrobiodiversity On-farm Project, Nepal (NARC/LIBIRD/IPGRI). We are grateful to senior scientists T. Hodgkin, D. Jarvis, P. Eyzaguirre and B. Sthapit (International Plant Genetic Resources Institute) for their insights, and to Eric M. Van Dusen, University of California, Berkeley. This research was supported by the International Development Research Centre of Canada (IDRC) and the European Union.

References

ABPSD (Agri-Business Promotion and Statistics Division) (2001) Statistical Information on Nepalese Agriculture. Ministry of Agriculture, HMG, Nepal.
Bellon, M.R. and Taylor, J.E. (1993) 'Folk' soil taxonomy and the partial adoption of new seed varieties. *Economic Development and Cultural Change* 41, 763–786.
Bellon, M.R., Pham, J.L., Sebastian, S., Francisco, S.R., Loresto, G.C., Erasgo, D., Sanchez, P., Calibo, M., Abrigo, G. and Quiloy, S. (1998) Farmers' perception of varietal diversity: implications for on-farm conservation of rice. In: Smale, M. (ed.) *Farmers, GeneBanks and Crop Breeding: Economic Analysis of Diversity in Wheat, Maize, and Rice*. CIMMYT and Kluwer Academic Publishers, Boston, Massachusetts/Dordrecht, The Netherlands/London, pp. 95–108.
Brush, S.B., Taylor, J.E. and Bellon, M.R. (1992a) Technology adoption and biological diversity in Andean potato agriculture. *Journal of Development Economics* 39, 365–387.
Brush, S.B., Taylor, J.E. and Bellon, M.R. (1992b) Biological diversity and technology adoption in Andean potato agriculture. *Journal of Development Economics* 39, 365–387.
Chaudhary, P., Gauchan, D., Rana, R.B., Sthapit, B.R. and Jarvis, D.I. (2004) Potential loss of rice landraces from a Terai community in Nepal: a case study from Kachorwa, Bara. *Plant Genetic Resources Newsletter* 137, 14–21.
Conway, G.R. (1985) Agroecosystem analysis. *Agricultural Administration* 20, 31–55.
de Janvry, A., Fafchamps, M. and Sadoulet, E. (1991) Peasant household behaviour with missing markets: some paradoxes explained. *Economic Journal* 101, 1400–1417.
FAO (2001) FAO Database 2001, Food and Agricultural Organization (FAO), Rome, Italy.
Gauchan, D. (2004) Conserving crop genetic resources on-farm: the case of rice in Nepal. PhD dissertation, University of Birmingham, Birmingham, UK.
Gauchan, D. and Smale, M. (2003) Choosing the 'right tools' to assess the economic costs and benefits of growing landraces: an illustrative example from Bara District, Central Terai, Nepal. *Plant Genetic Resources Newsletter* 134, 41–44.
Gauchan, D., Smale, M. and Chaudhary, P. (2005) Market-based incentives for conserving diversity on farms: the case of rice landraces in Central Tarai, Nepal. *Genetic Resources and Crop Evolution* (in press).
Greene, W.H. (2000) *Econometric Analysis*, 4th edn. Prentice Hall, New York.
Joshi, K.D., Subedi, M., Kadayat, K.B. and Sthapit, B.R. (1998) Factors and process behind the erosion of crop genetic diversity in Nepal. In: Pratap, T. and Sthapit, B.R. (eds) *Managing Agrobiodiversity. Farmers' Changing Perspectives and Institutional Responses in the Hindu-Kush-Himalayan Region*. International Center for Integrated Mountain Development (ICIMOD), Kathmandu and International Plant Genetic Resource Institute (IPGRI), Rome, Italy.
Meng, E. (1997) Land allocation decisions and *in situ* conservation of crop genetic resources: the case of wheat landraces in Turkey. PhD dissertation, University of California at Davis, California.
Rana, R.B., Gauchan, D., Rijal, D.K., Khatiwada, S.P., Paudel, C.P., Chaudhary, P. and Tiwari, P.R. (2000) Socio-economic data collection and analysis: Nepal. In: Jarvis, D., Sthapit, B. and Sears, L. (eds) *Conserving Agricultural Biodiversity In Situ: A Scientific Basis for Sustainable Agriculture*. IPGRI, Rome, Italy, pp. 51–53.

Shrestha, G.L. and Vaughan, D.A. (1989) Wild rice in Nepal. Paper presented in the Third Summer Crop Working Group Meeting, National Maize Research Programme, Rampur, Chitwan, National Agricultural Research Centre (NARC), Nepal.

Singh, I., Squire, L. and Strauss, J. (eds) (1986) *Agricultural Household Models: Extensions, Applications, and Policy*. The World Bank and Johns Hopkins University Press, Washington, DC and Baltimore, Maryland.

Smale, M., Meng, E., Brennan, J.P. and Hu, R. (2003) Determinants of spatial diversity in modern wheat: examples from Australia and China. *Agricultural Economics* 28(1), 13–26.

Sthapit, B., Upadhyay, M.P. and Subedi, A. (2000) A scientific basis of *in situ* conservation of agrobiodiversity on-farm: Nepal's contribution to the global project. NP Working Paper No. 1/99. NARC/LIBIRD/IPGRI, Nepal.

Subedi, A., Gauchan, D., Rana, R.B., Vaidya, S.N., Tiwari, P.R. and Chaudhary, P. (2000) Gender: methods for increased access and decision making in Nepal. In: Jarvis, D., Sthapit, B. and Sears, L. (eds) *Conserving Agricultural Biodiversity In situ: A Scientific Basis for Sustainable Agriculture*. IPGRI, Rome, Italy.

Upadhyay, M.P. (1998) Gene pools of crop landraces and threats. In: Pratap, T. and Sthapit, B.R. (eds) *Managing Agrobiodiversity. Farmers' Changing Perspectives and Institutional Responses in the Hindu-Kush-Himalayan Region*. International Center for Integrated Mountain Development (ICIMOD), Kathmandu and International Plant Genetic Resource Institute (IPGRI), Rome, Italy.

Upadhyay, M.P. and Gupta, S.R. (2000) The wild relatives of rice in Nepal. In: Jha, P.K., Karmacharya, S.B., Baral, S.R. and Lacoul, P. (eds) *Environment and Agriculture: At the Crossroad of the New Millennium*. Ecological Society, Kathmandu, Nepal, pp.182–195.

Van Dusen, E. (2000) *In situ* conservation of crop genetic resources in Mexican milpa systems. PhD thesis, University of California at Davis, California.

Vaughan, D. and Chang, T.T. (1992) *In situ* conservation of rice genetic resources. *Economic Botany* 46, 369–383.

World Bank (2003) *World Development Indicators 2003*. The World Bank, Washington, DC.

11 Determinants of Cereal Diversity in Villages of Northern Ethiopia

B. Gebremedhin, M. Smale and J. Pender

Abstract

This chapter compares factors explaining the inter- (interspecific) and intracrop (infraspecific diversity) of cereals grown by villages in Peasant Associations (PAs) of the northern Ethiopian highlands of Amhara and Tigray, building on the household-level analysis presented for the same sites in Chapter 6. The village is the smallest social unit for policy interventions targeted at sustainable management of crop biodiversity on farms. Villages have the capacity to govern the utilization and conservation of genetic resources, reconciling private and social objectives. Econometric analysis indicates that a combination of factors related to the agroecology of a village, proximity to markets and the characteristics of households and farms within the village influence the level of inter- and intracrop diversity of cereals. Determinants differ between Amhara and Tigray regions. There are no apparent trade-offs between policies seeking to enhance the richness or the equitability among cereal crops or within any single crop grown in villages. Trade-offs may occur among crops, however. Growing modern varieties of maize has a positive effect on the evenness of maize types grown within communities, although modern varieties of wheat have no observable effect. At the village level, markets appear to introduce cereal crop diversity in some cases, while in others, they reduce it. Ambiguity of market effects could reflect local demand relative to local seed supply, or different phases of seed and product market development.

Introduction

Crop diversity can be observed and measured at any one of several levels, including those of the individual farm family and the village (Almekinders and Struik, 2000). The studies presented in Parts II and III of this book, as well as several of those in Part IV, are based on sample surveys of household members and their farms. In several of these chapters, factors measured at the village, settlement or community level are introduced as variables that influence the behaviour of individual farmers. This chapter explores the variation in crop biodiversity measured at the level of the village. The labour, literacy and asset ownership profiles of household farms within villages are explanatory factors, along with regional and village characteristics. In other words, this chapter shifts the geographical scale of observation from the household to the village.

There are at least two reasons why a village, as compared with a household, is an appropriate unit of analysis for considering interventions to support the sustainable management of crop biodiversity on farms. First, a village is the smallest social unit that has the capacity to govern the utilization and conservation of genetic resources. Even if targeting is to occur within villages, as described in several of the chapters in this book, programmes will need to be managed by a larger collection of individuals. This collection itself, and others within a region of the same country, are likely to have been approached initially by a rural development and conservation institution.

Second, genetic diversity is a public good, and in locations where it is clearly a 'good' or a positive (as opposed to negative) externality, the village would be the focus of any policy incentives designed to reconcile private objectives with social objectives. Seed has both private and public good attributes (Chapter 1), especially for cross-pollinated species. As a result, the structure of genetic variation may most closely reflect the combined practices of farmers in a village rather than that of any single household farm (vom Brocke, 2001; Berthaud et al., 2002). The combination of private seed choices made by individual farmers each cropping season generates the spatial distribution of distinct types and genetic diversity across the village and higher levels of aggregation.

Empirical research can help identify the types of social trade-offs that may occur in designing institutional mechanisms or strategies to sustain the biodiversity of crops on farms. For example, if the introduction of modern varieties and crop genetic diversity has an inverse relationship, then social trade-offs would result from policies aimed at promoting either, unless interactions of specific genotypes and cumulative adoption ceilings are taken into account. To the extent that the determinants of crop biodiversity differ among crops, policies designed to enhance the diversity in one crop may have adverse consequences for the diversity of another crop. Similarly to Chapter 10, this chapter compares policy trade-offs among conservation objectives, such as promoting higher numbers of distinct types versus greater evenness in the spatial distribution of those types. As in a number of other chapters in this book, both intra- and intercrop diversity are assessed. Complementing the approach developed in Chapter 5, the analysis is conducted with data collected during group interviews at the village level, rather than in individual interviews with household members.

The chapter is organized as follows. The following section describes the survey sites in the highlands of Tigray and Amhara. The conceptual approach is then presented, followed by the data design. Subsequent sections describe the econometric approach and report the findings. Policy implications are drawn in the final section.

Context

The highlands of northern Ethiopia (the regions of Tigray and Amhara) are a suitable empirical context for testing hypotheses about the determinants of cereal crop diversity. Ethiopia is a centre of diversity for barley, wheat, faba bean and some forage crops, among others, and is often referred to as one of the eight Vavilovian gene centres of the world. In recognition of this importance, national activities to conserve genetic resources on farms and in genebanks have been undertaken systematically in Ethiopia over the past two decades (Worede et al., 2000). A brief synopsis of the history of cultivation of the cereal crops grown in the Ethiopian highlands, and a description of the farm households that grow them, is provided in Chapter 6. A map showing the site location is also provided in Chapter 6.

The highlands of northern Ethiopia are relatively less favoured than other areas of the country in terms of both growing environment and market infrastructure, two of the generic factors hypothesized to determine the extent of diversity maintained on farms. There are important social, economic and physical differences between the two northern highland regions (Tigray and Amhara), from which this data set was collected, as indicated by the statistics in Table 11.1.

The physical environment in Tigray is more degraded and the area has lower agricultural potential than Amhara. The average annual rainfall in Amhara is estimated at 1189 mm, compared with only 652 mm in Tigray. Soils are also generally deeper and more fertile in Amhara. Since 1991, concerted efforts have been made to rehabilitate the environment, especially in Tigray (Gebremedhin, 1998; Gebremedhin et al., 2002). The average size of landholding per household is larger in Amhara (1.72 ha) compared with Tigray (1.05 ha). Larger landholdings, combined with better soil fertility, are reasons for higher agricultural production per family in Amhara, as compared with Tigray. The average distance from the village to the nearest market town is much lower in Amhara (58 walking min) than in Tigray (212 walking min). The difference in access to the nearest market is very small, however. Towns are the major market loci used by farmers in Tigray, while other (usually community-based, smaller) markets are commonly used in the Amhara highlands. Cooperative marketing is minimal in either region.

Table 11.1. Social, economic and farm physical characteristics of villages in the highlands of Tigray and Amhara.

Characteristic	Tigray	Amhara
Population density (persons/km^2)	133	133
Average annual rainfall (mm)	652	1189
Proportion of female headed households	22	12
Distance to nearest market town (walking hours)	3.40	3.63
Distance to nearest market (walking hours)	3.53	1.00
Distance to nearest all-weather road (walking hours)	2.5	3.9
Villages with marketing cooperatives (per cent)	1.8	2.2
Average landholding (ha/family)	1.05	1.72

About 80% of the population depend on subsistence mixed crop–livestock farming. The average population density in the highlands in 1998/9 was about 133 persons/km^2, with population growth rate of above 2.9%. Cereal crops are the most widely grown crops, covering about 85% and 65% of cultivated land in Tigray and Amhara, respectively. Seven major cereal crops are grown in the highlands: teff, barley, maize, sorghum, wheat, millet and finger millet. Households typically produce cereal crops on different plots dispersed in their villages. Perennial crop production is limited in both regions, although farmers in the Amhara highlands engage in some.

Livestock, especially cattle, sheep and goats, are an integral part of the farming system in the highlands. Oxen power is used for land preparation and threshing. Households typically maintain livestock herds in order to sustain the supply of oxen power for crop cultivation. Due to imperfections in the markets for oxen power and its critical role in crop production, households prefer to keep their own oxen year-round to ensure supply during the cropping season. Since cereal crop residues are an important source of livestock feed, crop residue yield is an important consideration for households when they select from among local crop varieties or decide to adopt improved ones.

Conceptual Framework

The regression models used in this chapter are based on the model of the agricultural household model as presented in Chapter 6 and applied in other chapters of Part III, as well as several of the chapters in Part IV. Rather than representing the decision making of an individual farm household, the regression models in this chapter represent aggregated, village-level associations. Estimated marginal effects are village-level parameters.

The theoretic framework of the agricultural household is appropriate for analysing on-farm conservation of crop diversity in the Ethiopian highlands. Farmers in the highlands both produce and consume their cereal harvests. Under the assumption of perfect input and product markets, the production decisions of farm households are separate from their consumption decisions. Households are assumed to make production decisions that maximize profit and then allocate net profits to consumption decisions at a second stage. In this case, only farm characteristics are important in production decisions. However, when factor and/or product markets are imperfect, production and consumption decisions are made jointly.

Since market imperfections abound in the highlands of northern Ethiopia, household characteristics become important in influencing production decisions of farm households in these areas. Hence, it is assumed that the diversity within a particular crop i (intracrop) in a village c (D_{ic}) is determined by village-level fixed factors (C_c); regional-level fixed factors (R_c); aggregate household characteristics in a village (H_c); aggregate farm-level characteristics in a village (F_c) and random factors (u_c). In addition, adoption of modern varieties of a particular crop within a village (MV_{ic}) is hypothesized to influence the pattern of cultivation and intracrop diversity. These relationships can be expressed as

$$D_{ic} = D(C_c, R_c, H_c, F_c, MV_{ic}, u_c) \quad (11.1)$$

The intercrop diversity of a cereal j in a given village c (D_{jc}) is determined by village-level fixed factors (C_c), regional-level fixed factors (R_c), aggregate household characteristics in a given village (H_c), aggregate farm-level characteristics in a village (F_c) and random factors (u_c):

$$D_{jc} = D(C_c, R_c, H_c, F_c, u_c) \quad (11.2)$$

All of the vectors of determinants indicated in Eq. (11.1) are exogenous, except the adoption of modern varieties, which is single-dimensioned. Adoption of a modern variety (MV_{ic}) may be partly or wholly determined during the current period. Hence, adoption of modern varieties is potentially endogenous to current decisions about crop diversity. In the econometric analysis, predicted probabilities of the adoption are entered rather than observed adoption rates. Adoption is predicted using village-level fixed factors (C_c), regional-level fixed factors (R_c), aggregate household characteristics in a village (H_c) and aggregate farm characteristics in a village (F_c). Individual variables included in these blocks of factors are distinct from those included in the diversity equations.

$$MV_{ic} = MV(C_c, R_c, H_c, F_c, u_c) \quad (11.3)$$

Equations (11.1)–(11.3) are the basis for the econometric estimation.

Data

Results are based on data collected from a sample of 198 villages in Tigray and Amhara regions of northern Ethiopia between 1998 and 2001. The PA is the lowest administrative unit in Ethiopia. A stratified random sample of 99 PAs, usually consisting of four or five villages, was selected from highland areas (above 1500 masl) of the two regions. Strata were defined based on variables associated with the comparative advantages of areas. Comparative advantages are based on moisture availability, a major factor affecting agricultural productivity, as well as market access and population densities.

In Amhara region, secondary data were used to classify the *woredas* (districts) according to access to an all-weather road, the 1994 rural population density (greater or less than 100 persons/km^2) and whether the area is drought-prone (following the definition of the Ethiopian Disaster Prevention and Preparedness Committee). Two additional strata were defined for PAs where irrigation projects are found. In each of the ten strata, four to five PAs were randomly selected. From each sample PA, two villages were randomly selected, for a total of 98 villages.

In Tigray region, PAs were stratified by whether an irrigation project was present or not, and for those without irrigation, by distance to the *woreda* town (greater or less than 10 km). A total of three strata were defined in Tigray. PAs closer to towns and in irrigated areas were selected with a higher sampling fraction to assure adequate representation. A total of 54 PAs were selected. Four of the PAs in northern Tigray could not be studied due to the war with Eritrea. From each of the remaining PAs, two villages were randomly selected, for a total of 100 villages.

Information collected at the PA, village and household levels includes agricultural and natural resource conditions, household composition and assets, access to markets and infrastructure and agricultural practices in 1991 and 1998/99. The data were supplemented by secondary geographic information.

Data were collected by group interviews, both at the PA and at the village levels. Each interview involved up to ten respondents, selected to represent different age groups (below 30 years of age and older), primary occupation (farming or off-farm), gender, literacy and administrative responsibility. At the PA and village levels, information on changes in agricultural and natural resource conditions between 1991 and 1998/99 was collected.

Econometric Approach

Regression models

Poisson regression models were used to predict inter- and intracrop (variety) counts of richness across the seven commonly grown cereals (barley, wheat, sorghum, finger millet, pear millet, maize and teff). Villages that did not grow a particular crop were assigned zero values. Poisson regression models are appropriate for count data that take on non-negative integer values and where the outcome is zero for at least some members of the population (Wooldridge, 2002). The Poisson model assumes equality between the conditional mean and variance. To check for over- or underdispersion, the estimated Poisson model was tested against the Negative Binomial regression models, resulting in failure to reject the Poisson model. Since all villages grew more than one cereal, the intercrop Shannon and Berger–Parker diversity indices were computed for all villages at values greater than the lower limit (0 and 1, respectively), and regressions run with ordinary least squares (OLS).

Several estimation problems were encountered in estimating the equations with respect to the Shannon and Berger–Parker indices of intracrop diversity. First, when a village did not cultivate a cereal, a sample selection problem occurred in the variety diversity index for that cereal. Second, even when the cereal was cultivated, if a large proportion of the sample grew only one variety, the diversity index is censored because many of its values cluster at the limit (i.e. 0 for Margalef and Shannon indices and 1 for the Berger–Parker index). A standard OLS or seemingly unrelated regression (SUR) of the diversity indices will yield biased and inconsistent estimates in this situation. In principle, a maximum likelihood approach such as a tobit model may be employed to address the censoring and account for correlations in error terms across equations by specifying a multivariate density function for the error terms. This approach is difficult to implement with more than two equations. Consequently, although a systems approach was originally envisaged, single regression equations were estimated.

The general approach most often used to address selectivity bias is to employ a technique similar to that advanced by Heckman: the probability that the cereal is grown and inverse mills ratio (IMR) are predicted in the first stage, and the IMR is then used to estimate a second-stage censored regression. Even if the explanatory variables in the first- and second-stage regressions are identical, because the predicted IMRs or probabilities from the first-stage regressions are non-linear functions of the explanatory variables, the second-stage regressions are identified under normality of the probit models. However, since the second stage is a censored regression, the IMR correction introduces heteroskedasticity (Maddala, 1983). The errors in the predicted IMR depend on values of the explanatory variables, which, unlike in a linear model, causes the estimator to be inconsistent (Maddala and Nelson, 1975; Maddala, 1983). In addition, there is a problem in obtaining the correct standard errors, since the predicted rather than the actual IMR is used. As in Chapter 6, the censored least absolute deviations (CLAD) estimator, which is robust to heteroskedasticity (Deaton, 1997), could be used. With CLAD, bootstrapping is used to compute the standard errors. However, due to a relatively small number of observations with the village-level data, the CLAD regression failed to converge. An interval regression, with probability weights to correct for the standard errors, was used to estimate the intracrop Berger–Parker and Shannon indices at the village level.

Third, a problem with an endogenous explanatory variable also occurs in investigating the effects of choosing to grow modern varieties on intracrop variety. Problems of this type are typically addressed through regressions with treatment effects or self-selectivity. Including a dummy variable expressing whether or not at least one household in the village has adopted an improved variety will give inconsistent estimates (Barnow et al., 1981; Maddala, 1983; Greene, 1993). Instead, predicted probabilities from a probit regression of whether or not an improved variety is cultivated have been included in the second-stage regression (Barnow et al., 1981).

As in many two-stage estimation approaches, identification of the second-stage regression is an important issue. In general, it is difficult to find variables that are correlated with the decision to grow a cereal crop or an improved variety, but not correlated with the associated diversity index (which is constructed from area shares). At the village level, mean altitude in a village was a strong predictor of whether or not a crop was grown.

Dependent variables

The dependent variables used in this analysis are counts and diversity indices constructed using area shares allocated to crops or varieties. The most widely cultivated cereal crops that were included in the analysis are barley, maize, sorghum, teff, wheat, finger millet and pearl millet. Within these cereal crops, 'variety' is simply understood as a crop population recognized by farmers. Usually 'named' by farmers, varieties have agromorphological characters that farmers use to distinguish among them and that are an expression of their genetic diversity. Although the relationship between variety names and genetic variation may not be well defined, names that are reported at the village level are likely to coincide with genetic distinctions to the extent that genetic diversity is determined at the village level.

Many indices are available to represent diversity based on crop and variety units. Three indices are adapted from ecological indices of spatial diversity in species (Magurran, 1988) to represent either inter- or intracrop diversity. The Margalef index of richness could not be constructed at the village level because, although proportions of area allocated to crop and variety were reported, total area was not. The Berger–Parker index was used as an indicator of relative abundance, and the Shannon index was used to express proportional abundance, or evenness. The properties of these indices and formulas used to construct them are discussed in Chapter 1.

Independent variables and hypotheses

Definitions, hypothesis and summary statistics for explanatory variables are presented in Tables 11.2. The explanatory variables used in this analysis are village-level aggregates representing those tested in the applications presented in this book and in related literature. Aggregates are categorized into household characteristics, farm physical characteristics and village or regional characteristics. Variables were constructed from group interviews conducted at the PA and village level. The comparative statics of the household model are for the most part ambiguous in sign. Since the independent variables are aggregates and the dependent variables are metrics over choice variables, economics principles and previous empirical research provide the only guidance concerning expected directions of effects.

Household characteristics include the education (the proportion of households that are literate), credit use (the proportion of households using formal credit), the extent of landlessness and ownership of oxen. About 50% of household heads in the highlands are literate, and due to the recent expansion of rural credit services, about 60% of households have access to formal credit.

In the average village surveyed, 50% of households have literate members. Literacy may be positively or negatively associated with crop diversity, since access to information may lead to specialization or diversification. By raising the opportunity cost of labour, education may take labour away from farm production and diverse cropping activities.

Farmers have usufruct right to land, but cannot sell, buy or mortgage land. The most common means of land acquisition is land redistribution. Due to population pressure and land scarcity, not all households have land allocated to them through redistribution. When farmers know the soil characteristics of their land through having used it, they are in a better position to match crops and varieties to specific niches for better performance. A higher proportion of landless households in the community would be associated with lower cereal crop diversity. In these villages, the proportion of landless households ranges from none to over half (52%).

Oxen power is used predominantly for ploughing and threshing. Since oxen power supplies the only draft power and the oxen power market is not well developed, oxen ownership is a critical determinant of crop intensification in the highlands of northern Ethiopia. Only 40% of households had one ox or more in the survey year. The effect of oxen ownership on crop diversity in the highlands has no predicted direction. On one hand, a large proportion of households owning oxen are expected to enhance diversity since it increases the capacity of farmers to grow more crops. On the other hand, greater oxen ownership may lead to specialization, allowing more intensive cultivation of high-value cereal crops such as teff.

Formal credit in the highlands of northern Ethiopia has been associated with a certain fixity in the type and amounts of modern seed extended through the extension system. Access to formal credit may lead to reduced numbers of

Table 11.2. Definition of explanatory variables, summary statistics and hypothesized effects on cereal (inter- and intracrop) diversity in villages of the highlands of Amhara and Tigray regions, Ethiopia.

Variable name	Description	Hypothesized effect		Mean	Standard error	Minimum	Maximum
		Intercrop	Intracrop				
Household characteristics							
Education	Proportion of literate households in 1998	(+,−)	(+,−)	0.50	0.03	0.03	0.9
Credit	Proportion of households who use formal credit in 1998	(+,−)	(+,−)	0.60	0.25	0.00	0.9
Landlessness	Proportion of landless households in 1998	(−)	(−)	0.17	0.15	0.00	0.5
Oxen ownership	Proportion of households owning oxen in 1998	(+,−)	(+,−)	0.40	0.02	0.05	1.0
Farm characteristics							
Extent of erosion	Proportion of cultivated land under severe erosion in 1998	(+)	(+)	0.30	0.03	0.00	0.8
Extent of good soils	Proportion of soil considered good by village in 1998	(−)	(−)	0.40	0.03	0.00	0.9
Village and regional characteristics							
Range in altitude	Range of altitude of topography	(+)	(+)	274.20	32.90	3.00	1524.0
Mean rainfall	Average annual rainfall (mm)	(+,−)	(+,−)	1753.00	87.40	501.40	3389.0
Distance to market	Walking time in minutes to nearest market	(+,−)	(+,−)	145.70	13.10	10.00	720.0
Distance to road	Walking time in minutes to nearest all-weather road	(+,−)	(+,−)	208.50	33.90	0.00	1236.0
Population density	Population per square kilometres in village	(+)	(+,−)	143.10	11.90	15.00	397.0
Location in Tigray	Administrative region of PA (Amhara = 0; Tigray = 1)	(+,−)	(+,−)	0.174	0.01	0.00	1.0

Note: Means and standard errors are adjusted for stratification, weighting and clustering of sample.

varieties and less evenness in their distribution if it results in specialization. If the size of the packages promoted is small relative to the holdings of the household, formal credit might enhance rather than detract from intracrop diversity. Credit use in the surveyed villages ranges from nil to 90%.

Farm physical characteristics are measured in: (i) the extent of soil erosion and (ii) the extent of good-quality soils in the village. Erosion is a serious physical production constraint in the northern Ethiopian highlands. The proportion of cultivated land under severe erosion is used to measure the extent of erosion in a village. For various purposes, such as for land redistribution, communities typically classify their cultivated lands into three categories representing good quality, medium quality and poor quality. The extent of good soils ranges to 90% of all lands in the villages surveyed, although the proportion of eroded lands reaches a maximum that is nearly that high (80%). When land degradation is severe, the risk of crop failure is likely to be higher. In such conditions farmers would be expected to diversify in order to reduce the risk of loss. When soils are good, farmers would tend to specialize in order to benefit from crops with higher productivity and net returns. This chapter hypothesizes that while the extent of erosion increases farmer demand for cereal crop diversity, good-quality land detracts from it.

Village or regional characteristics are those that are fixed to all households in villages but vary among villages, such as the range in altitude in the village, average annual rainfall, walking time to nearest market or road, population density and location in Tigray or Amhara. The working hypothesis that the degree of environmental heterogeneity encountered on farms, in villages or regions affects the diversity of crop populations has been borne out in previous studies as well as in those collected in this book (Marshall and Brown, 1975; Van Dusen, 2000; Part III). This chapter uses the range in altitude as an indicator of environmental heterogeneity. The average range in altitude in the highlands is 208 m.

Studies conducted in the Peruvian Andes, Turkey and Mexico demonstrated a positive relationship between marginal growing conditions for the crop and farmers' decisions to continue to grow landraces (Brush, 1995), although a regional study of maize landraces conducted by Aguirre Gómez (2000) in the state of Guanajuato failed to support it. Here, the effect of mean annual rainfall is hypothesized to be indeterminate. Higher rainfall levels could increase the possibility of growing more diverse crops; less rainfall could increase farmer demand for diversification given the risk of crop failure.

A second major working hypothesis advanced in previous studies and in other chapters of this book is that isolation from markets drives farmer diversification of crops and varieties. When villages are removed from market centres and roads, farmers face higher transaction cost of buying and selling, inducing them to rely primarily on their own production for subsistence. In line with this argument, Van Dusen (2000) found that higher number of maize, beans and squash varieties were grown by farmers who were more distant from markets. On the other hand, when possibilities of trade open up, new crops and production possibilities may be added to the portfolio of economic activities available to farmers. Brush et al. (1992) found that access proximity to markets in the Andean potato agriculture was positively associated with the adoption of modern varieties, but this adoption was not associated with a decrease in the number of potato types grown. Farmers surveyed in the highlands of northern Ethiopia walk for over 2 h on average to reach the nearest market and for 1 h to the nearest all-weather road. The range in market and road access is wide. Farmers sell most of their produce or purchase farm inputs at district markets. Better access to markets and roads could serve to introduce or detract from the breadth of cereals and varieties grown.

Population density may induce land-saving technical change or agricultural intensification. Such intensification may arise from the use of modern varieties or from an increased number of production activities. The effect of population density on infra- or intercrop diversity is therefore an empirical question for which the answer depends on the development path or phase.

Results

Seven cereal crops (sorghum, barley, wheat, maize, teff, pearl millet and finger millet) are grown in the communities in the highlands of Tigray and Amhara. On average, each village grows four cereals. The number of cereals grown per village ranges from one to seven. The range in numbers of varieties per cereal is from three to ten. Barley, maize, wheat and teff are grown by the largest numbers of communities, as compared with sorghum, pearl and finger millet. Villages grow more than two varieties of wheat and teff on average, and between one and two varieties of maize and barley on average. Mean numbers of varieties grown per cereal are less than one per village for sorghum, pearl and finger millet (Table 11.3).

Higher numbers of varieties should not be taken to imply that one cereal crop exhibits higher levels of richness than another for two reasons. First, numbers cannot be directly compared across crops because of differences in crop reproduction systems. That is, diversity may be partitioned more within varieties than among varieties for cross-pollinating as compared with self-pollinating species. Second, farmers' names may either overstate or understate genetic distinctions.

Intercrop diversity of cereals

Table 11.4 presents regression results for the determinants of intercrop cereal diversity. Separate regressions reveal important differences in factors related to the intercrop diversity of cereal crops between communities located in the highlands of Amhara and those found in the highlands of Tigray, although the results for Amhara are relatively weaker statistically. For example, while population density and severity of erosion are important in explaining cereal diversity in Tigray, their effect in Amhara is insignificant. Similarly, market access appears to have a stronger effect in Amhara than in Tigray. Aside from regional distinctions, however, the signs of statistically significant factors are consistent across indices, suggesting minimal trade-off in maintaining different types of cereal diversity.

Regional- and village-level factors are jointly significant in explaining variation of intercrop diversity in cereals among villages, although not all individual factors are. Range in altitude generally has no effect on the intercrop diversity of cereals grown, except for the richness of cereals grown in villages located in Amhara. Level of rainfall has no significant effect on cereal diversity in either Tigray or Amhara. Villages in Amhara may concentrate more on fewer crops to take advantage of higher yield potential as well as commercial benefits, given their relative proximity to markets.

Controlling for region, however, the relationship of market access to intercrop diversity of cereals remains ambiguous, as hypothesized. The larger the average distance of households in the village to all-weather roads, the greater the intercrop diversity of cereals they grow, by any of the three indicators. The further the village is from the district market the less diverse the mix of cereals grown in the more remote Tigray, but the more diverse the cereals grown in Amhara. Longer distances to the all-weather road, however, are positively related to intercrop diversity.

Table 11.3. Numbers of cereal varieties grown in villages in the highlands of Tigray and Amhara regions of northern Ethiopia.

	Barley	Maize	Wheat	Teff	Sorghum	Finger millet	Pearl millet
Number of varieties planted							
Mean	1.66	1.39	2.22	2.07	0.55	0.42	0.29
Standard error	0.28	0.13	0.26	0.14	0.10	0.07	0.06
Minimum	0	0	0	0	0	0	0
Maximum	9	6	10	8	8	3	3
Number of communities planting	166	149	139	178	75	64	49
Sample size	198	198	198	198	198	198	198

Note: Mean and standard errors are adjusted for stratification, weighting and clustering of sample. Data on named varieties of finger and pearl millet were not collected in the Amhara region survey.

Table 11.4. Regression results, factors affecting the intercrop diversity of cereals in villages of the highlands of Amhara and Tigray regions, Ethiopia.

	Tigray			Amhara		
Explanatory variable	Richness (Poisson regression)	Inverse dominance (OLS)	Evenness (OLS)	Richness (Poisson regression)	Inverse dominance (OLS)	Evenness (OLS)
Village and regional characteristics						
Range in altitude	−0.00008	−0.00071	−0.000058	−0.0005***	−0.000076	−0.00016
Mean rainfall	0.00014	−0.000046	0.00075	−0.0002	−0.0002	−0.00017
Distance to market	−0.0007*	−0.00201**	−0.00135***	0.0005	0.0019	0.00122*
Distance to road	0.00076**	0.001744*	0.001***	0.0001	0.00032*	0.00025**
Population density	0.0015***	0.00321**	0.0015***	0.0002	0.00056	0.00038
Household characteristics						
Education	0.2606**	0.18098	0.2214	0.3303*	0.6598	0.27174
Credit	−0.0029	−0.40922*	−0.02523	0.03371	0.1672**	0.0746*
Landlessness	−1.11e−07	−0.000368	−0.00003	0.000021	−0.00014	0.00009
Oxen ownership	0.2397**	0.5729	0.19972	0.3285*	0.7692*	0.3154*
Farm characteristics						
Extent of erosion	0.3769***	1.0718**	0.60489***	0.0244	−0.2763	−0.2671
Extent of good soils	0.0608	0.3171	0.14244	−0.2479	−0.2017	−0.1315
Constant	0.9611***	1.3457*	1.2828***	1.4036***	1.4758**	0.9011**
Number of observations	85	85	85	69	69	69
F	7.58	4.72	8.77	3.04	1.56	1.93
Probability > F	0.000	0.000	0.000	0.012	0.017	0.019
R^2		0.3551	0.4395		0.2508	0.2706

Note: Indices are defined in Table 11.1. Coefficients and standard errors are adjusted for stratification, weighting and clustering of sample. *Statistically significant at the 10% level; **statistically significant at the 5% level; ***statistically significant at the 1% level.

Population density is positively associated with the richness, evenness and inverse dominance of cereals in Tigray, and is of no significance in Amhara.

The characteristics of households within villages are also important determinants of variation in the levels of the intercrop diversity of cereals observed among villages. Education is positively associated with the diversity of cereals grown in both Tigray and Amhara, suggesting that human capital and access to information are favourable for growing a wider range of cereal crops. In both Tigray and Amhara, the greater the proportion of households owning oxen within the village, the higher the intercrop diversity of cereals they grow. The statistical significance, positive direction of the effects and large magnitude of the effects of human capital and assets are consistent and evident across diversity indices and regions. The higher the proportion of households with access to formal credit in the communities of Amhara, the greater the intercrop diversity of the cereals they grow, although this same factor has a negative effect or is of no significance in Tigray. The proportion of landless households has no effect on variation in levels of cereal crop diversity among communities in either region.

While higher proportions of land in good soils have no effect on lower cereal diversity in Tigray, the proportion of land that is eroded is strongly and positively related. Neither of these factors is significant among villages in the highlands of Amhara. Soil-related factors appear more important in explaining patterns of cereal crop cultivation in the more environmentally degraded region of Tigray than in Amhara.

Intracrop Diversity of Cereals

Regressions explaining the intracrop diversity of all cereals except teff are shown in Tables 11.5

Table 11.5. Regression results, factors affecting intracrop diversity of barley, wheat and maize in villages of the highlands of Amhara and Tigray regions, Ethiopia.

	Barley			Wheat			Maize	
	Tigray and Amhara	Tigray		Tigray and Amhara	Tigray		Tigray and Amhara	Tigray
Explanatory variable	Richness (Poisson regression)	Inverse dominance (interval regression)	Evenness (interval regression)	Richness (Poisson regression)	Inverse dominance (interval regression)	Evenness (interval regression)	Richness (Poisson regression)	Evenness (interval regression)
Village and regional characteristics								
Range in altitude	−0.00018	0.00004	0.0001	0.0003	0.00067	0.0005**	0.00028*	0.00045**
Mean rainfall	0.00067***	−0.00038	−0.0005	0.0013	0.0022	0.0003	−0.00058***	−0.00118
Distance to market	−0.00096	0.0033**	0.0015	−0.00059	−0.0001	−0.00035	0.00024	0.00017
Distance to road	0.00018	−0.004***	−0.0022**	0.00035***	−0.00034	−0.00006	−0.00069	−0.00011
Population density	0.00148	0.0014	0.0012	−0.00072	−0.0024	−0.0012	0.001222*	−0.00018
Location in Tigray	1.0753***			−0.32183			−0.92876***	
Household characteristics								
Education	−0.9163**	0.1183	0.0864	−0.30173	1.0534	0.2721	0.4474	−0.0254
Credit	0.00706	0.0313	0.2977	−0.01958	0.2977	0.1591	0.1046**	0.1452
Landlessness	−0.000026	0.0004	0.00026	−0.00005	0.0012	0.0011*	−0.00031	0.00039
Oxen ownership	−0.70953**	0.0553	−0.7859	−0.33622	1.2656	0.5634	1.4691***	0.6286
Farm characteristics								
Extent of erosion	0.33072	0.1269	0.1155	0.07597	0.04915	−0.2305	−0.6792**	0.3605
Extent of good soils	−0.21296	0.0009***	0.0007***	−0.6639	−0.0018***	−0.0019***	−0.6792**	−0.7947**
Inverse Mills ratio, growing cereal		−0.1295	−0.0688		−0.4894***	−0.3782***		
Probability of growing modern variety					1.2333	0.478		
Constant	−0.06865	0.9869	0.06191	1.2263*	−2.11696	−0.6853	0.71415	0.5082*
Number of observations	154	71	72	154	56	56	154	75
F	5.57	7.7	5.34	4.12	6.08	6.5	4.12	1.99
Probability > *F*	0.000	0.000	0.000	0.000	0.000	0.000	0.000	0.039

Note: Indices are defined in Table 11.1. Coefficients and standard errors are adjusted for stratification, weighting and clustering of sample. *Statistically significant at the 10% level; **statistically significant at the 5% level; ***statistically significant at the 1% level.

Table 11.6. Regression results, factors affecting intracrop diversity of sorghum, finger millet and pearl millet in villages of the highlands of Amhara and Tigray regions, Ethiopia.

Explanatory variable	Sorghum	Finger millet		Pearl millet
	Tigray and Amhara	Tigray and Amhara	Tigray	Tigray and Amhara
	Richness (Poisson regression)	Richness (Poisson regression)	Inverse dominance (interval regression)	Richness (Poisson regression)
Village and regional characteristics				
Range in altitude	−0.00043	−0.00092*	−0.00041	−0.00112**
Mean rainfall	−0.00225***	0.0002	−0.002447**	−0.000312
Distance to market	0.00257**	0.00107	0.001975	0.00088
Distance to road	0.00053*	−0.0012	−0.00167*	0.00042**
Population density	−0.00186	0.00051	−0.00015	0.0021
Location in Tigray	−1.5024**	1.4711**		0.9644
Household characteristics				
Education	0.34497	0.84854	−0.55735	1.3301*
Credit	−0.5395	0.13126	−0.03787	−0.64834
Landlessness	0.00043	−0.00063	−0.000049	0.00059
Oxen ownership	0.43478	2.16221***	0.58891	−0.1209
Farm characteristics				
Extent of erosion	1.1559	−0.61815	−0.26976	0.8567
Extent of good soil	−0.22947	−0.57056	0.3107	−0.3097
Inverse Mills ratio, growing cereal			−0.61982*	
Constant	2.18422	−2.7673**	3.6071***	−1.8825
Number of observations	154	154	53	154
F	4.13	7.09	2.35	3.54
Probability > F	0.000	0.000	0.022	0.000

Note: Indices are defined in Table 11.1. Coefficients and standard errors are adjusted for stratification, weighting and clustering of sample. *Statistically significant at the 10% level; **statistically significant at the 5% level; ***statistically significant at the 1% level.

and 11.6. The factors explaining variation in intracrop diversity clearly differ from those explaining variation in intercrop diversity, and they also differ among cereal crops. Findings for teff were not statistically significant. Although richness (variety count) regressions could be estimated for both Tigray and Amhara, inverse dominance and evenness regressions could be estimated only for Tigray, due to absence of area share information at the village level in the Amhara survey. The Berger–Parker index of inverse dominance was not statistically significant in the regression explaining sorghum diversity, while the evenness regression was not significant for finger millet.

Regional and village characteristics influence variation in the intracrop diversity of cereal crops among villages. A wider range in altitudes is associated with more evenness among wheat and maize varieties, although it is negatively associated with richness in pearl and finger millet. Pearl and finger millet are crops grown at lower altitude and farmers may diversify to other crops (as suggested by the findings for intercrop diversity) and their varieties with increasing altitudes. Specific wheat or maize varieties may grow better in some altitude niches. Higher mean rainfall implies greater barley richness, but fewer numbers of maize and sorghum varieties. Higher rainfall also implies higher dominance of finger millet varieties in Tigray.

As is the case for intercrop diversity in cereals, market and road access have mixed impacts on patterns of variety cultivation across cereal crops. In Tigray, while market access implies higher dominance in barley varieties, access to roads implies less dominance and evenness. Road access is also associated with less richness in wheat, sorghum and pearl millet varieties in both regions, as was market access for richness of sorghum varieties. More densely populated communities grow more varieties of maize, but this factor is not related to variation in patterns of intracrop diversity for other cereals. When controlling for other factors, communities located in Tigray grow more varieties of barley and finger millet, and fewer varieties of maize and sorghum, but there are no significant differences for wheat and pearl millet.

The structure of household characteristics within villages has an impact on variation in intracrop diversity of cereal crops among villages. Access to credit in communities is positively associated only with intracrop diversity in maize. In Tigray, the higher the proportion of landless households in the village, the more diverse are its wheat varieties. Although this result appears to contradict the negative relationship of population density to wheat diversity, landlessness is higher in low population density areas perhaps due to less cultivatable land (Gebremedhin et al., 2002). Education is positively associated with the richness of pearl millet varieties, although negatively associated with the richness of barley varieties. The greater the proportion of households that own oxen, the more diverse their maize and finger millet varieties, but the fewer are the number of barley varieties grown in the village.

Farm physical characteristics are also important, as hypothesized. In the Tigray region, communities with better quality of land grow more diverse barley, perhaps because barley is grown on relatively better soils in the region. The higher the proportion of good-quality land, the lower is the diversity of wheat and maize varieties. It may be that households concentrate on fewer wheat or maize varieties on good soils in order to take advantage of higher yields. Maize richness is associated negatively with both the extent of eroded land and the extent of good-quality soils. Maize may be grown on soils with intermediate quality that are less eroded.

Adoption of modern varieties of maize is associated with greater evenness in the distribution of varieties across communities in Tigray. This finding is consistent with the notion that in environments that are less favoured with respect to either market infrastructure or productivity potential, modern varieties that are suited to some production niches can provide traits that complement (rather than substitute for) local varieties. Interestingly, the effect of adoption of modern varieties is insignificant for wheat, a relatively old crop, compared with maize. The effect of modern varieties on diversity of either wheat or maize at the household level was insignificant (see Chapter 5).

The IMR was associated with lower diversity in wheat and finger millet in Tigray, suggesting that correcting for sample selection is important. This means that using only the observations on communities that cultivated wheat or finger millet in a Tobit model, without the correction, would have yielded inconsistent estimates. The coefficient on the IMR for barley was statistically insignificant.

Policy Implications

Scale of policy or programme

In the highlands of both Amhara and Tigray, as hypothesized, a combination of agroecological variables, market access factors and the characteristics of farms predicts variation in the inter- and intracrop diversity of cereal crops when measured at the village level. Factors that are significant differ markedly between the highlands of Amhara and those of Tigray, the more environmentally degraded region.

These findings reveal the location-specific nature of any policies or programmes that are designed to encourage the maintenance of diversity, and the dangers of drawing generalizations from any single case study. They also suggest that the cost of assembling the information required to design programmes for local conservation of crop diversity is high.

Trade-offs between richness and equitability of cereal crops and varieties

The direction of the effect of statistically significant factors is the same for indices of richness, evenness and inverse dominance among cereals. Results therefore suggest that a policy whose goal is to augment the richness of cereals grown would not entail trade-offs in terms of 'equitability' or dominance among crops.

The same appears to be true for the intracrop diversity of any given cereal crop grown in villages. Different factors are significant in explaining the richness and equitability among varieties grown for any single cereal crop but they are consistent in sign. A programme designed to conserve the richness of varieties of any single crop is not likely to have a negative impact on the evenness among them at the village level.

However, the set of factors that determines the pattern of intracrop diversity varies among cereal crops and some are clearly more important for one crop than another. Policies designed to encourage the intracrop diversity in one cereal crop at a village level might have the opposite effect on that of another crop. Conserving the richness or equitability among varieties of one cereal crop might lead to less richness or equitability among those of another.

These findings indicate the 'partial' nature of most empirical research conducted so far concerning the on-farm conservation of crop genetic resources. Crop genetic resources evolve within a farming system and agroecosystem. Other tools must be brought to bear on analyses if system interrelationships involved in agrobiodiversity conservation are to be adequately understood. For example, in these communities, the relationship between animal husbandry and cereal diversity is evident.

Trade-offs in conserving inter- versus intracrop diversity of cereals

Policies related to oxen ownership will affect both the intercrop diversity and the intracrop diversity of cereals at the village level, but in different ways and differentially among cereal crops. Owning more oxen is generally associated with more diversity among cereals in villages, and more among maize and finger millet varieties, but less diversity among barley varieties. Similarly, farm physical characteristics, agroecological conditions and market access are related in various ways to both inter- and intracrop diversity of cereals at the village level. Therefore, the incidence of related policies would be differential and difficult to predict. These findings illustrate that programmes designed to influence the intracrop diversity of cereals are not likely to be neutral to their intercrop diversity, and vice versa, as was found for the household-level analysis conducted in Chapter 5.

Development and diversity

In the northern Ethiopian highlands there appears to be no trade-off between seeking to enhance productivity of wheat and barley through the use of modern varieties and the spatial diversity among named varieties of these two cereal crops. So far, introduction of modern varieties has not meant that any single variety dominates or that modern varieties have displaced landraces, most likely because they have limited adaptation and farmers face many economic constraints in this environment.

Instead, as hypothesized, it is just as likely that small amounts of seed of improved varieties diversify the seed set of these farmers by meeting a particular purpose or filling a particular niche, rather than contributing to uniformity. The obvious reason is that neither the physical terrain nor the market network is particularly favourable for specialized, commercial agriculture. This is not to say that the improved varieties introduced in such areas are themselves genetically diverse, but that the traits they add to those of the other varieties grown enables farmers to better meet their production and consumption objectives in this difficult and uncertain growing and marketing situation.

In villages of the northern Ethiopian highlands, there seems to be little trade-off at present between the needs of development and maintaining complex combinations of crops and varieties. On the contrary, access to credit and oxen and education are more likely to have positive rather than negative relationships with the cereal crop diversity observed at the village level. Use of formal credit is in general positively related to the infra- and intercrop diversity of cereals. Currently, in this resource-poor system, modern varieties appear to contribute to rather than threaten wheat and maize diversity. These findings confirm that opportunities to pursue development while enhancing cereal crop diversity do occur in areas of the world that are less favoured in terms of environmental conditions and economic infrastructure.

Acknowledgements

The financial support of the Ministry of Foreign Affairs of Norway and the Swiss Agency for Development and Cooperation is gratefully acknowledged. The Food and Agriculture Organization of the United Nations (FAO) supported this analysis. Special appreciation goes to the many officials, village leaders and farmers who graciously and patiently participated in the research and responded to the numerous questions.

References

Aguirre Gómez, J.A., Bellon, M.R. and Smale, M. (2000) A regional analysis of maize biological diversity in southeastern Guanajuato, Mexico. *Economic Botany* 54(1), 60–72.

Almekinders, C. and Struik, P.C. (2000) Diversity in different components and at different scales. In: Almekinders, C.A. and de Boef, W. (eds) *Encouraging Diversity: The Conservation and Development of Plant Genetic Resources*. Intermediate Technology Publications, London.

Barnow, B.S., Cain, G.S. and Goldberger, A.S. (1981) Issues in the analysis of selectivity bias. *Evaluation Studies Review Annual* 5(5), 43–59.

Berthaud, J., Pressoir, G., Ramirez-Corona, F. and Bellon, M.R. (2002) Farmers' management of maize landrace diversity: a case study in Oaxaca and beyond. Paper presented at the 7th International Symposium on the Biosafety of Genetically Modified Organisms, Beijing, 10–17 October.

Brush, S.B. (1995) *In situ* conservation of landraces in centers of crop diversity. *Crop Science* 35, 346–354.

Brush, S.B., Taylor, J.E. and Bellon, M.R. (1992) Biological diversity and technology adoption in Andean potato agriculture. *Journal of Development Economics* 39, 365–387.

Deaton, A. (1997) *The Analysis of Household Surveys: A Microeconomic Approach to Development Policy.* Johns Hopkins University Press, Baltimore, Maryland.

Gebremedhin, B. (1998) The economics of soil conservation investments in the Tigray region of Ethiopia. PhD dissertation. Department of Agricultural Economics, Michigan State University, East Lansing, Michigan.

Gebremedhin, B., Pender, J. and Tesfay, G. (2002) Village natural resources management: the case of woodlots in northern Ethiopia. *Environment and Development Economics* 8, 35–54.

Greene, W.H. (1993) *Econometric Analysis.* Macmillan Publishing, New York.

Maddala, G.S. (1983) *Limited Dependent and Qualitative Variables in Econometrics.* Cambridge University Press, New York.

Maddala, G.S. and Nelson, F. (1975) Switching regression models with exogenous and endogenous switching. Proceedings of the American Statistical Association, Business and Economic Section, pp. 423–426.

Magurran, A. (1988) *Ecological Diversity and Its Measurement.* Princeton University Press, Princeton, New Jersey.

Marshall, D.R. and Brown, A.H.D. (1975) Optimum sampling strategies in genetic conservation. In: Frankel, O.H. and Hawkes, J.G. (eds) *Crop Genetic Resources for Today and Tomorrow.* Cambridge University Press, Cambridge, Massachusetts.

Van Dusen, E. (2000) *In situ* conservation of crop genetic resources in the Mexican *milpa* system. PhD dissertation. University of California at Davis, California.

vom Brocke, K. (2001) Effects of farmers' seed management on performance, adaptation, and genetic diversity of pearl millet (*Pennisetum glaucum* [L.] R.Br.) populations in Rajasthan, India. PhD dissertation. University of Hohenheim, Germany.

Wooldridge, J.M. (2002) *Econometric Analysis of Cross Section and Panel Data.* Massachusetts Institute of Technology, Cambridge, Massachusetts.

Worede, M., Tesemma, T. and Feyissa, R. (2000) Keeping diversity alive: an Ethiopian perspective. In: Brush, S.B. (ed.) *Genes in the Fields: On Farm Conservation of Crop Diversity.* International Development Research Center, International Plant Genetic Resources Institute, and Lewis Publishers, Ottawa, Rome and Boca Raton, Florida.

12 Social Institutions and Seed Systems: The Diversity of Fruits and Nuts in Uzbekistan

M.E. Van Dusen, E. Dennis, J. Ilyasov, M. Lee, S. Treshkin and M. Smale

Abstract

This chapter builds on the household model presented in Chapter 5 by exploring the role of social institutions in household access to planting material and use of crop biodiversity. Two types of institutions are analysed: (i) local community groups to which a household may belong and (ii) sources of planting material and agricultural information. Findings from a household survey implemented in two districts of Samarqand, Uzbekistan, provide a description of the diversity of fruits and nut trees, as well as the nature of systems for planting material. Econometric analysis reveals a linkage between participation in community groups and the levels of fruit and nut tree diversity managed by households. No relationship is found between the type of institution used to obtain genetic material and the level of diversity in orchards. Household participation in community groups influences the type of institution used for access to genetic material, however.

Introduction

From the time of the Silk Road until the present, home gardens in Central Asia have served as repositories of agricultural genetic resources, reflecting cultural traditions and contributing to the local economy. Uzbekistan is a hotspot of both agrobiological and cultural diversity. Over 60 distinct cultural and linguistic groups of people exist in Uzbekistan alone, and the country is located in the centre of origin for over 40 crops including apple, peach, plum, almond, walnut, pistachio, grapes and such horticultural crops as garlic, melon and spinach. Vavilov (1930) visited the area in his early collecting missions, and described the combinations of wild forests of fruit and nut trees next to ancient oasis cities inhabited for millennia. Although the Soviet modernization of agriculture led to centralized planning of widespread monocultures on vast irrigated acreages, significant diversity was maintained in household garden plots where traditional agricultural practices and inherited varieties were cultivated with no interventions from state planning institutions. The favourable climate of Uzbekistan led it to be used by the Soviets to supply fruits and vegetables to the northern cities, and there was a small but significant infrastructure and trade in these crops.

The Soviet agricultural research system was well developed, and Soviet botanists pioneered the study of wild and farmer varieties of crop plants. As in most areas, the state sector contracted in the 1990s. The Shreder Institute and its regional research stations are in charge of fruit and nut genetic resources for Uzbekistan. The stations are faced by limited funding for activities necessary to keep their collections alive, and have largely limited other activities, despite the huge benefits to local farmers. These stations play a dual role, collecting and conserving traditional varieties, and making selections and providing both improved and local genetic materials for planting.

The issue of agricultural biodiversity is also related to other important environmental issues in Uzbekistan. One pressing issue is the diversification of the overall economic system away from cotton monoculture. The cotton production system is notoriously reliant on intensive chemical inputs and inefficient irrigation infrastructure. The diversification into different fruit, nut and vegetable crops, especially ones that draw on local genetic resources, has the potential to reduce polluting inputs, and provide more economic options than the single dominant cash crop. From the point of view of the individual farmer, given that cotton is subject to production orders, and most of fruit and vegetable production is not, when given the option farmers seek to move out of cotton into these other activities.

The economic role of household production has become even more pronounced since independence, as Uzbekistan has initiated changes in land tenure and households have diversified income-earning activities as a strategy for surviving the crisis of economic transition. In the villages of rural Uzbekistan, a range of local organizations and social groups interlink households, supporting their access to goods and information in a rapidly changing society.

This chapter describes the biodiversity of fruit trees, grapes and nuts in a rural economy in transition, exploring relationships between household production to seed systems and social institutions. The research on which it is based was conducted through international collaboration between the International Plant Genetic Resources Institute (IPGRI) and national partners in Uzbekistan, under the auspices of the system-wide project on Collective Action and Property Rights (CAPRi). The institutions involved contributed in several ways to the direction of the research and methods applied. First, while prominent agricultural issues in Uzbekistan involve cotton production, the management of water resources and the related environmental issues, the focus of this research on genetic resources guided the analysis away from state farms and collective enterprises towards households and village institutions. Second, the overall research project emphasizes local (household and village level) effects of changes in social institutions on the management of genetic resources. Finally, analysis of garden diversity calls for a synthesis of methodologies, combining conservation efforts that targeted varieties and populations within a single crop species, with research tools from studies of home gardens and agroforestry, where many species are grown together.

The first section describes the importance of home gardens to rural households, and the role of local social organizations in rural communities. Utilizing an original household survey from a series of rural villages near Samarqand, Uzbekistan, a quantitative portrait of fruit, grape and nut diversity within household production is then presented. Next, elements of the seed system are documented to illustrate sources of genetic material. Descriptive statistics and econometric analysis explore the relationship between diversity and social institutions.

Home Gardens, Local Institutions and Rural Poverty in Uzbekistan

Home gardens are the focus of this study not only because of their role in the conservation of crop genetic resources, but because they are critical components of household income and represent a key sector in the agricultural economy of Uzbekistan. Home gardens have been the major source of fruit and vegetable crops in Uzbekistan, and this general tendency holds for Central Asia, and in Russia and the Commonwealth of Independent States (Seeth *et al.*, 2003; Lerman *et al.*, 2004). In Uzbekistan, home gardens are oriented not only towards home production but also satisfy more than half of national demand in fruits and nuts, furnishing a significant share of the export market in those commodities (Lerman, 1998; Thurman and Lundell, 2001). Despite the collectivization of the vast majority of land during the Soviet period, individual initiative was always permitted in household gardens. In the perestroika period of economic opening at the end of the Soviet Union (1986–1991), initiative in garden production was encouraged, and the Soviet Union used this resilient sector of the economy to face falling production on the state-run farms (Seeth *et al.*, 2003). Growing fruits and tending vegetables is a fundamental feature of culture in rural Uzbekistan.

Furthermore, recent studies have demonstrated that gardens are central to household strategies for combating poverty and economic collapse in the post-Soviet era (Kandiyoti, 1999). In an assessment of rural poverty carried out by the Expert Centre (1999), access to garden lands buffered the household income shocks of the early 1990s. Seeth *et al.* (2003) found that gardens in rural Russian were the second most important source of income in rural households, providing from 10% to 50% of real earnings. In Uzbekistan this is exaggerated by the fact that many parts of the rural economy remain without cash; households may work for the collective farm (*shirkat*) for access to benefits or payment of gas and electric utilities, but sales of horticultural goods are a rare source of cash income (Bloch, 2002).

Rules regulating the process of obtaining garden plots and their size have undergone multiple changes since Uzbek independence in 1991. In rural areas each household is allotted a parcel of land for use as a garden (*hovli*), typically located around the home. Many households include three to four generations and garden work is a shared enterprise. Households usually manage 10–15 *sotkas* of land (0.10–0.15 ha) in the garden, combining fruit and nut trees, grape vines, staple crops such as potatoes and cabbage and vegetables for household consumption or markets sales. They also raise livestock and corresponding forage crops such as maize. The manager of the shirkat, formerly known as the collective farm, allocates *hovli* lands to households. A new family wishing to establish a household petitions the shirkat for a grant of land. Receiving the grant may take up to 4 years and may require political connections or the equivalent in cash.

Institutions can be viewed at a number of different levels, from 'the humanly devised constraints that shape human interaction' (North, 1990, p. 4) to 'complexes of norms and behaviours … serving a collectively valued purpose' (Uphoff, 1986, p. 8). For the purpose of this study, there are two levels at which institutions shape the levels of agricultural diversity used by a household. One level comprises the local community groups and forms of interaction where households meet and form social networks. Another level is the seed system, the different places and organizations that a household may use to access planting material and information.

In this study, institutions are defined as a map of the different social networks that a household may use to acquire genetic material or related genetic information for fruit trees, grapes or nuts. Institutions that act indirectly are the community groups and social networks that exist within communities for reasons that are distinct from circulation of crop genetic material or related information, although households may use them for that purpose. Institutions that act directly are established specifically in order to provide crop genetic materials and corresponding information.

Planting material is the physical element of plant genetic resources – the seed, saplings in the case of trees or the rootstock in the case of grapes. Agricultural information describes the knowledge required to properly cultivate the plant, such as which varieties are pest- or drought-resistant, the watering schedule of a variety or the maturation date of a variety. It can also include social information such as plant uses, market price or transportation characteristics. Knowledge of the attributes unique to a variety and necessary for its proper cultivation passes among individuals, through generations, and over geographies via a myriad of channels. Agricultural information may be conveyed as custom, tradition or ritual in a context that emphasizes the particular meanings and significance associated with that variety. In the local-level seed system agricultural information is conveyed through individuals, and consequently the norms regulating the conditions under which people meet can influence what and how much agricultural information passes between farmers.

Community-level organizations for the collective management of natural resources have been shown to responsibly and rationally manage scarce resources under a variety of circumstances (Ostrom, 1990). The most successful examples are organizations managing a distinct resource from which users directly obtain tangible and soon-realized benefits. Crop genetic resources, unlike other natural resources such as forests or water, do not derive value solely from the physical planting material (Chapter 1). Much of the value of crop genetic resources is determined when engaged in a relationship with farmers, scientists and ecosystems. The benefits of crop genetic resources are neither soon realized nor particularly tangible; and no examples of natural resource management organizations like those described by Ostrom (1990) exist for managing fruit, grape and nut genetic resources in villages of rural Uzbekistan. Although all of the farmers we interviewed in Uzbekistan shared a common interest in, and lamented the loss of, local plant diversity, they have not developed organizations for managing it.

Data Design

The data were collected from May to July of 2003 in the districts of Urgut and Bulungur in the province of Samarqand, Uzbekistan. These two districts were selected because they offered a range of agricultural contexts within each district. The location of Samarqand within Uzbekistan and Central Asia is presented in Fig. 12.1. The location of Urgut to the northwest of Samarqand, and Bulungur to the southwest of Samarqand is presented in Fig. 12.2.

The Urgut district is located at the foothills of the Pamir Alay mountains, and the high glaciated peaks loom above the city of Urgut. Urgut is known as an important market centre, and Urgut traders continue to travel throughout the former Soviet Union. Many of the villages sampled in Urgut were in semi-mountainous foothill regions, although villages in the irrigated lowlands were found to be contract-farming tobacco for a joint venture factory in Urgut city. The Bulungur district is located along the main highway connecting Samarqand to Tashkent, and has some industrial areas along this corridor. The topography of the Bulungur district also ranges from the mountain foothills to flat irrigated plains, but Bulungur had a larger percentage of arable land and a more extensive road network.

The first step in the sampling process was to define the sampling domain, using secondary data and expert advice from preliminary site visits. The survey team visited with the district leader (hokimiyat) and acquired maps, summary data on agriculture (available at the shirkat level) and lists of villages. At this stage the domain was determined by eliminating shirkats based on a set of criteria that sought to screen out dominating effects such as if a village would be too urban, too dominated by livestock or under extreme ecological conditions. The geographical

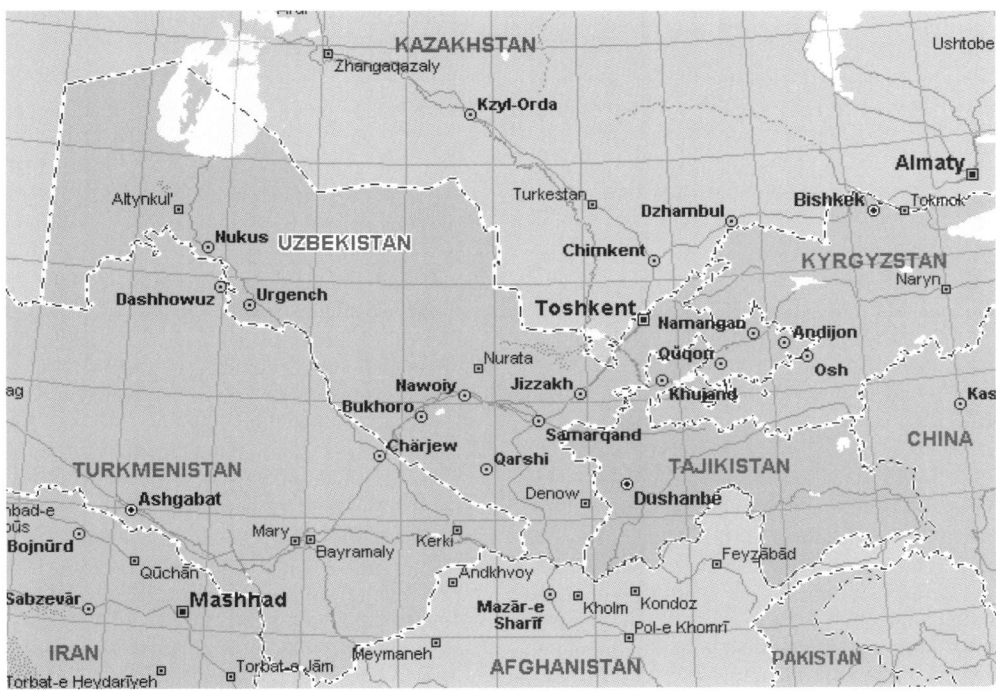

Fig. 12.1. Map of Samarqand in the Central Asian region.

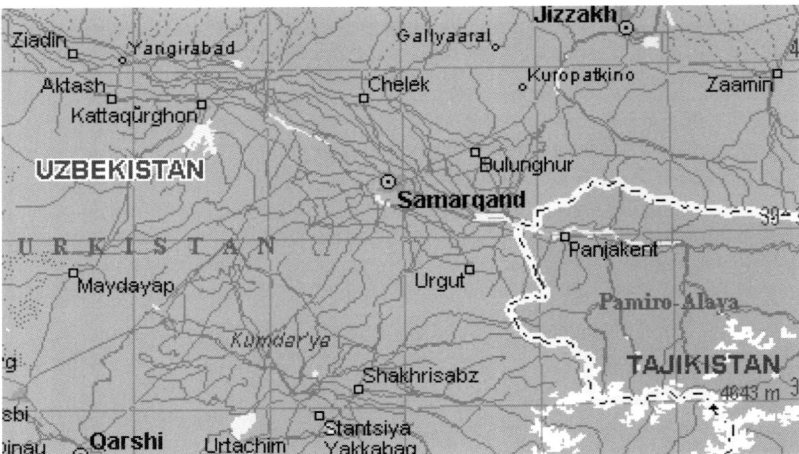

Fig. 12.2. Map of Urgut and Bulungur districts in Samarqand province.

and historical heterogeneity of the region was also studied, taking into account the age of villages, proximity of markets and location in relation to the Pamir Alay mountain range.

The second step was the selection of ten villages in each district. A screen was applied to remove outliers, and a series of villages was selected to capture regional heterogeneity in distance to markets, agroecological conditions and other conditions. From this set of villages ten were selected in each district. In the Urgut district these ten villages were located within seven different shirkats (the administrative unit corresponding to agricultural lands) and four different selsoviets (the administrative unit corresponding to social services such as schools). In the Bulungur district the ten villages corresponded to six shirkats and four selsoviets. The surveyors visited the villages and obtained a list of households from the village committee.

Households were then sampled at random within the village with a list of 20 households and 5 alternates generated. A household survey was implemented that had six sections: demographic, land, production characteristics, crop diversity, social and institutional participation and markets and sales. In the household survey data analysed there are 368 households with data complete enough for statistical purposes, 182 in Urgut and 186 in Bulungur. This analysis focuses on the crop diversity and the institution modules. Group surveys were also carried out with groups of local expert farmers, old men, young men and women. An agricultural expert from the Shreder Institute visited the villages to record the history of local varieties.

Description of Households, Farms and Diversity

The average age of the household head is just under 50 years. The average household has 4.5 adults, although the number of adults is as high as 10 in some extended households. Cash income from pensions, off-farm work and other transfers was recorded, although the information was occasionally incomplete and, under the circumstances, difficult to collect consistently. Another indicator of income from activities outside the household farm, the number of workers with an off-farm job, averages 1.9 per household.

An index for household wealth was also created using a combination of the ownership of key durable goods and the characterization of household construction. Fifty per cent of households report owning a radio, and 90% report owning a television. Only 19% report owning a car, and in a large majority of cases these are Soviet-era cars. The average number of rooms in a home is just over four rooms, and just under half of the households have a concrete floor. The majority

Table 12.1. Household characteristics.

	Per cent or mean
Age household head	47.57
Number of adults	4.54
Cash income (10^3 UZ Soum)	368.0
Number of off-farm workers	1.85
Wealth (durable goods and housing)	
Radio	48%
TV	90%
Refrigerator	22%
Bicycle	13%
Motorcycle	7%
Car	19%
Rooms (average number)	4.48
Concrete floor	46%
Concrete walls	2%
Paint	7%
Barn	86%

N = 386 households.

of households in rural villages are constructed from mud or mud brick, only 7% of households have painted walls and only 2% are constructed from brick or concrete block. A summary of household characteristics and key wealth variables is presented in Table 12.1.

Table 12.2 presents summary statistics by district for the household plot, and for an additional garden plot if the household has one. While almost all households have a garden plot, only a few have a secondary plot. Part of the agricultural reform in Uzbekistan has been to give additional lands to households. This is not common in the survey sample – only 10% and 13% of households in Urgut and Bulungur, respectively, had a second garden plot.

Land in gardens is measured in *sotkas*, equivalent to 100 m² or one-hundredth of a hectare. The average garden sizes are 17 *sotkas* in Bulungur and 15 *sotkas* in Urgut, close to the medians. The total area cultivated is significantly higher in Bulungur (0.69 ha) than in Urgut (0.34 ha). The age distribution of parcels reveals that households have held most of them since the Soviet era. In Bulungur, 55% of households with an additional garden plot received it in the last 10 years as a consequence of recent land reforms. In contrast, only 23% and 26% of households in Bulungur and Urqut have received their household plots during that period. When households were asked how long they would officially own the land, almost all households responded that they believed the garden plots to be inheritable.

Crop biodiversity in perennial tree crops has some features that differentiate it from that found in annual crops, such as the cereals (maize, rice, sorghum, millet, wheat) studied in a number of chapters of this book. Like bananas (Chapter 7) and potatoes (Chapter 9) and some coffee plants (Chapter 4), most fruit trees and grapes are clonally propagated, which is more difficult than seed reproduction but produces perfect genetic similarity. There are fewer individual plants in each garden, but in total for a household there are often more varieties and species than in the case of annual crops. Instead of living for one season, individual perennial plants can live 20–30 years. This has several implications. First, genetic resource decisions are not made annually and may not be made very

Table 12.2. Characteristics of home garden parcels.

	Bulungur		Urgut	
	Household plot	Additional plot	Household plot	Additional plot
N	185	26	177	19
Mean parcel area (hectares)	0.17	0.17	0.15	0.13
Mean total area cultivated	0.69**		0.34	
How long you have farmed it				
0–10 years	23%	55%	26%	18%
10–20 years	34%	18%	24%	27%
>20 years	43%	27%	51%	55%

**Significantly higher using a chi-square test.

Table 12.3. Diversity indices.

	All households
N	386
Variety count	7.14
Species count	4.89
Shannon index (by variety)	1.30
Simpson index (by species)	0.61

frequently. Second, trees can be inherited – without a crop choice being made. Furthermore, someone leasing or renting lands with trees or vines already planted may not make genetic resource decisions, although they might make them for annual crops.

Fruit crops in Uzbekistan do not provide subsistence for a household, and even in rural villages, commercial markets may be relatively developed for these fruits. Information gathered from preliminary research, market surveys and group interviews suggests that markets may value traits demanded on export markets, or processing characteristics. This is similar to the case of coffee in Ethiopia (Chapter 4).

Table 12.3 presents summary statistics for variety counts, species counts and diversity indices for each household. The average number of fruit and nut species grown by each household is 4.9, and counting total varieties across species the average number was 7.1. A household that grew (apple, grape, apricot) received a count of three. A household that grew (apple varieties 1 and 2, grape varieties 1 and 2, apricot variety 1) received a count of five. This distinction is important for the interpretation of the regression analysis, where the count by varieties is used as the independent variable. The count by varieties expresses both inter- and intracrop richness. No significant difference was found between the two districts in the level of overall fruit and nut diversity. Both the Shannon and Simpson indices are presented as alternative ways of representing relative richness or evenness considering the area share allocation among crop varieties. The fruit and nut crops grown in the sample are grape, apple, apricot, walnut, peach, plum, sweet cherry, mulberry, fig, quince, almond, pomegranate, pear and prune. Summary statistics for grape, apple, apricot, walnut, sweet cherry, peach and mulberry are shown in Table 12.4. The first line shows the number of households growing each crop. There is no crop that all households grow, and there is a range in frequencies across households for any single crop. Apple is the most commonly grown, followed by grape, walnut and apricot.

For households growing grapes, the average number of varieties is 2.3. Households with apples trees grow an average of 1.7, but for most other crops, the average number of crops is close to 1. The number of individuals is the number of vines or trees, and this is much higher for grapes because there are some households with leased grape plantations and large numbers of plants. In general these distributions are highly skewed, even for minor crops principally grown in gardens, by many households with one or two and fewer households with several. 'New' trees are those that do not yet bear fruit. Fruit trees may take 5–10 years until they bear fruit, and the percentage of new trees is taken as an indication of the level of turnover of trees – the rate at which households need to replace genetic material and how frequently they can decide whether to change species or

Table 12.4. Crop descriptive statistics.

	Grape	Apple	Apricot	Walnut	Cherry	Peach	Mulberry
Number of households	288	330	245	281	182	159	42
Mean number of varieties	2.32	1.68	1.08	1.11	1.29	1.07	1.19
Mean per household							
Number of trees, vines	161.84	15.65	2.98	4.72	26.29	3.48	3.67
Per cent new trees	0.17	0.23	0.27	0.26	0.25	0.26	0.16
Per cent households that sell	0.26	0.16	0.04	0.17	0.08	0.04	0.10

varieties. Across all crops a quarter to a fifth of all vines and trees are recently planted. The data on sales show that a small percentage of households are selling their production and this is concentrated in certain crops like grapes, apples and walnuts.

To understand the underlying distribution in both crops and varieties, it is necessary to look at the distribution within as well as between crops. In Fig. 12.3, a histogram of the count of the number of varieties grown is shown for grape, walnut, apple and apricot. Including all 368 households, it is clear that for each crop some households grow none. Most households (63%) have just one variety of apricot, and 70% have only one variety of walnut. Nearly half of grape growers (47%) have two or more varieties, and 41% have two or more varieties of apple.

Seed Systems

Planting material

An essential feature of the seed system is *from whom* the planting material is obtained. Farmers were asked from whom they obtained sapling or seed material of the trees and vines currently growing in their garden. In Table 12.5, the sources of genetic material are presented for each of seven crops. At the bottom of the table the sources are presented by percentage of total, and aggregated: parents, siblings, children, neighbour, friend, other relatives are grouped as unofficial sources, and Shreder and shirkat are grouped as official. The official sources are those with a mandate from the government to distribute genetic material, the unofficial are all other sources.

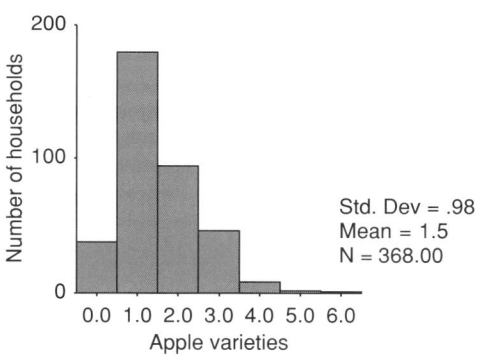
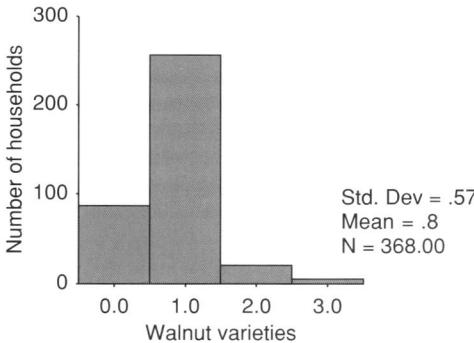
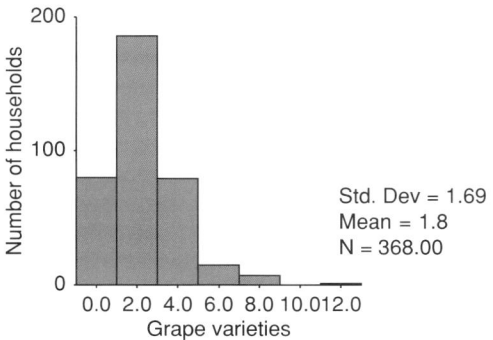
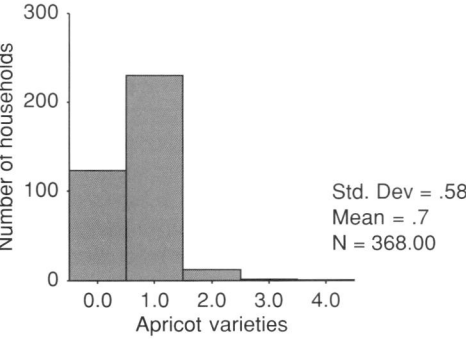

Fig. 12.3. Number of varieties grown.

Table 12.5. Source of genetic material – social relation.

	Grapes	Apples	Apricots	Walnuts	Peach	Cherry	Mulberry
No answer	10	7	13	12	5	2	2
Parents	40	24	13	28	3	8	4
Sibling	7	3	4	6	2	2	
Children	3	1					
Neighbour	20	5	6	9	5	14	3
Friend	6	5	1	3	3	5	1
Relative	16	12	6	12	4	12	1
Own source	85	34	54	83	45	44	9
Bazaar	65	205	124	120	82	79	16
Shirkat	26	7	4	2	2	3	3
Shreder	10	25	18	4	7	10	2
Other		2	2	2	1	3	1
Total	288	330	245	281	159	182	42
Own plants	30%	10%	22%	30%	28%	24%	21%
Other local	32%	15%	12%	21%	11%	23%	21%
Bazaar	23%	62%	51%	43%	52%	43%	38%
Official	13%	10%	9%	2%	6%	7%	12%

There is no single source that dominates the seed system. The bazaar is the most common, providing over 50% of genetic material for apple, apricot and peach. Grape, one of the most commonly grown crops, is also the most evenly sourced. This may reflect the fact that grape is relatively easier to propagate by using cuttings from vines than are the tree crops. For all crops, official sources never represent more than 12% of planting material, although the materials from the bazaar may originally be released by the Shreder Institute. The high percentage of own and unofficial sources shows that the unofficial channels are incredibly important for these households, despite 70 years of Soviet rule and continued centralization of the agricultural sector. Certainly for cotton and many other crops the government retains the Soviet monopoly on seed supply. While the government retains a technical monopoly on fruit variety release and breeding, other means of access to planting material are fundamental.

A second feature of the seed system is *where* planting materials are sourced. For each crop, households were asked the geographical origin of the saplings. Table 12.6 presents a picture of the seed system by geographical location across crops, with each household referencing their answer to their own household. A small number, 2–5%, are sourced from immediate neighbours, while the mode is to source from the same village. The categories of the same shirkat and same selsoviet (an administrative unit) are not necessarily mutually exclusive (but only one

Table 12.6. Source of genetic material – location.

	Grapes	Apples	Apricots	Walnuts	Peach	Cherry	Mulberry
No answer	11%	13%	13%	12%	14%	9%	10%
Next door	5%	2%	2%	2%	3%	4%	5%
Same village	62%	38%	42%	59%	52%	57%	52%
Same shirkat	10%	12%	12%	6%	9%	12%	12%
Same selsoviet	2%	9%	9%	6%	6%	6%	10%
Other	10%	26%	22%	14%	16%	13%	12%
Total	288	330	245	281	159	182	42

answer is recorded) and basically correspond to receiving the variety from the nearby villages.

In every species except apricots and apples, the majority of households reported that they obtained planting material within the same village. In general, the relatively constant proportions across crops are convincing. There are contrasts, such as that grapes are only sourced outside of the local area 10% of the time, while apples are the most frequently sourced from outside, at 26%. There is also a contrast between Table 12.6, where households reported receiving 40–60% of the material from the bazaar, and Table 12.7, where only 22–41% of households reported obtaining material from outside the village. Because the bazaar is usually at the district level this appears to be a contradiction, although there may be more subdistrict bazaars that cater specifically to genetic resources, or households may seek sellers from their local area when conducting transactions in a district bazaar.

Agricultural information

Knowledge of the attributes unique to a variety – like disease resistance and watering schedule or plant uses and market prices – is necessary for its proper cultivation. In the local-level seed system, agricultural information is conveyed through individuals, and consequently the norms regulating the conditions under which people meet influence what and how much agricultural information passes between farmers. For this aspect of the study, the places and organizations where villagers meet in the local community were used as the unit of analysis. From preliminary fieldwork, the survey was designed to ask about social institutions and types of groups in which the head of the household interacted with other community members. The survey then led to a series of questions about the intensity and frequency of participation in these groups. The principal groups and places where household heads met are shown in Table 12.7, presented in order of their popularity according to the data.

Weddings are the most important single social gathering in rural life. Almost the entire village and many young people from surrounding villages attend festivities lasting for several days. A range of social functions occur at weddings including reciprocal exchange, meeting between young people, and for the sake of this study, the general exchange of information between relatives and friends.

The *bazaar* or marketplace is the principal place of commerce and where villagers from an entire region trade with each other as well as with merchants from outside the region. The bazaar is therefore an important node of social communication, where key agricultural information such as prices or demand for different varieties is communicated. (The bazaar will also be examined later as a direct institution because it is a source of genetic materials for households.)

Hashar is a form of reciprocal labour and either it can be publicly organized by authorities for the construction of a community good (e.g. the cleaning of drainage ditches or the building of a communal football pitch) or it can be privately organized by individuals (e.g. building a house, harvesting, planting and sowing). Participants in *hashar* receive no payment for their services but are traditionally fed for the day or receive a share of the day's harvest. Respondents revealed that feeding *hashar* helpers was becoming more expensive, and in many communities people were hiring day labourers instead of organizing private *hashars*. However, *hashars* remain the second most important source of labour (15% according to group interviews in villages with men) after the family unit. In some villages *hashars* are necessary for the cultivation of grapes that require covering with mud during the winter seasons and uncovering during the spring.

In Uzbekistan, villagers celebrate a mix of patriotic holidays, Muslim rituals and Zoroastrian traditions. Holidays include Navruz (21 March), Independence Day (1 September), Eid ul-Fitr (after Ramadan) and Eid ul-Adha (after the Pilgrimage). Other festivals include Mustaqilik, Children's day, New Year's Eve, Yilboshi (beginning of a year) and Darveshona. The most prominent Muslim custom is the *hudoyi*, meaning thanksgivings to Allah. The *hudoyi* involves giving food to the community or poorer community members as a public act of sacrifice. The *hudoyi* is commonly celebrated multiple times a year.

Mahalla is an administrative unit defined by a neighbourhood block and sometimes centred around a teahouse (*chaykhana*) where social and ritual functions are performed. More frequently it

Table 12.7. Community-level institutions.

	Wedding	Bazaar	Hashar	Other festivals	Mahalla	Guzar	Other	Solkbak 1	Work brigade	Solkbak 2	Chaykhana
Per cent of households participating	98	71	61	57	51	38	35	30	30	6	6
Average number of meetings per month	2.18	2.9	2.08	1.44	4.27	9.55	5.25	1.12	18.02	1.58	6.25
Average number of people in a group	154	9	11	135	13	41	20	50	15	10	11
Per cent of participants reporting a financial obligation	59	5.7	1.3	5.3	0.5	2.9	58.6	58.6	3.7	36.4	13.6
Per cent of participants reporting talk about fruit	83	79	88	75	85	91	92	91	90	86	91
Average number of times agricultural information is received from this group or a member	1.88	1.59	1.79	1.22	1.68	2.41	1.68	2.16	3.26	5.5	1.75

refers to a unit of community defined by the respondent himself. Each *mahalla* has a *mahalla committee* that provides a forum for male elders, farmers and community leaders to discuss community problems and take voluntary collective action (*hashar*). As a forum to discuss problems and share solutions, the *mahalla* facilitates awareness about local varieties and the movement of information necessary for their proper cultivation.

Guzar refers to meeting community members in the centre of the village. Similar to the *mahalla*, the term implies both the meeting place and the convention of sharing news, information or gossip with other community members.

Solkbak is a very common organization in rural Uzbekistan, involving a semi-formalized meeting of close friends and colleagues. Groups of a dozen acquaintances and friends gather bi-weekly for entertainment and socializing at a different member's house for each meeting. The *solkbak* is a Central Asian tradition that regained popularity in the 1970s when the old custom incorporated a ritual of reciprocal exchange of consumer goods. The *solkbak* provides members with access to capital using a simple rotating credit mechanism. At each meeting, participants contribute a small and equal sum of money, which is then given to the host. The honour of hosting the *solkbak* rotates among the group members. In many cases the amount contributed at each meeting is indexed to the price of a kilogram of meat in order to control for inflation. Some respondents belonged to more than one *solkbak*.

Work brigade or *pudrat* is the way grouped labour is organized on the collective farms. Work brigades are often based on extended families and neighbours. Currently, *pudrats* have a range of contracting arrangements with the collective farm manager. For the villagers involved, it is both a shared social obligation and an agricultural workplace and where crop activities are done jointly.

Chaykhana translates as 'teahouse' and is another local community meeting place. Most neighbourhoods have a principal teahouse where men meet and socialize. This is another important place for the exchange of informal information and building of social networks.

Some organizations are better conduits than others for agricultural information. Table 12.7 presents a set of indices about the intensity of household participation, the numbers of households participating in each kind of group and the approximate size of each meeting. Table 12.7 also presents data about the rate at which participants discuss the cultivation of fruit trees. Of these organizations the best sources of information are those groups in which participants speak more often about fruit and from which they receive agricultural information more times per year. Although the primary purpose of all of the organizations in this group is not to facilitate the movement of agricultural information, they each accomplished this end with varying degrees of success.

The per cent of households participating in an institution or organization does not indicate how important the source is for agricultural information. While almost all households participate in weddings and the bazaar these are not very common sources of agricultural information. In all groups, the per cent of households that discuss fruit within the group is above 75%. The frequency with which households meet in these groups ranges from monthly in the case of the *solkbak* to 9.5 times for the *guzar* or 18 times for a work brigade. Similarly, the number of people that one meets in each group can range from under 20 in the case of the *hashar* or the work brigade to over 150 in a wedding. Differences are apparent among groups in the number of times that a household has received agricultural information. For example, the *guzar* and work brigade are the highest in terms of agricultural information. Although these are not the largest or most common organizations, they are more frequently attended than other groups.

The common attribute of most of the four important organizations (work brigade, *solkbak*, *hashar*, *guzar*) is that membership is restricted to a well-defined group of individuals. Neither the size of the group nor the frequency of meetings fully explains how frequently members obtain agricultural information and talk about fruit. Although 30–60% of village populations participate in *solkbak*, *hashar* and work brigade, membership is restricted by invitation, and multiple groups operate in each village. The ability of a group to invite or expel individuals develops trust among members, and prestige may even be associated with participation in certain groups. These data indicate that trust built through community groups can facilitate the sharing of agricultural information.

An exception to this pattern is the *guzar*, which appears as the second most frequently utilized for agricultural information but grouped in the last category for frequency of discussions about fruit. One hypothesis is that the *guzar* is an important source for agricultural information not because it establishes a basis of trust, but because it provides a forum for sharing agricultural information specific to the local environmental conditions of the village. The *guzar*, literally translated as 'the centre of the village', is where local news of both dependable and dubious quality is disseminated. Thirty-eight per cent of male heads of household spend time there on average once every 3 days. In the *guzar*, information may be learned more quickly, and may be more relevant to the village conditions than if it is obtained at the local bazaar, but it is not necessarily highly trusted. Consequently, individuals do not speak about fruit as 'often' as in more trusted organizations like the *hashar*, but because the topic is a common concern, it is spoken about more consistently. Few participants speak about fruit only 'rarely'. As a result, men who frequent the *guzar* obtain agricultural information relatively frequently.

Within the survey, a separate set of questions asked about the institutions where households directly acquired agricultural information and foreign varieties. Table 12.8 reports the per cent of the households using each institution as a source of agricultural information and an index (from 1 to 5) ranking the effectiveness. While the data do not show much variation, a couple of patterns are evident. First, the local breeder is a relatively uncommon source of information. Relatives and neighbours are far more common. Second, the Shreder Institute, or most official institution, received the highest ranking for effectiveness. Third, the bazaar, which focus groups described as an unreliable source of information, does not score significantly differently than the community institutions.

However, when the data are grouped by the number of institutions that each household reported using, different institutions appear to be used with different intensities (Table 12.9). Almost half of the households reported using only one institution, and of these, the majority, 64%, used only the bazaar. A pattern emerges showing that as households use more sources of information,

Table 12.8. Direct institutions – sources of information and new varieties.

	Shreder	Bazaar	Breeder	Relative	Neighbour	*Oxacol*
Source of agricultural information						
Per cent households using	36	45	20	44	43	40
Mean effectiveness ranking	4.58	3.82	3.65	3.99	3.82	3.57
Source of new varieties						
Per cent households using	31	53	20	32	31	23
Mean effectiveness ranking	4.4	3.57	3.69	3.77	3.48	3.36

Table 12.9. Use of direct institutions – agricultural information.

		Per cent using each institution					
Number used	Total	Shreder	Bazaar	Breeder	Relative	Neighbour	*Oxacol*
0	67	0	0	0	0	0	0
1	148	13.5	63.5	2.7	6.1	5.4	8.8
2	26	34.6	30.8	15.4	46.2	38.5	23.1
3	37	40.5	13.5	10.8	89.2	83.8	56.8
4	34	67.6	11.8	26.5	97.1	94.1	97.1
5	27	81.5	22.2	88.9	96.3	100.0	96.3
6	14	100.0	92.9	92.9	100.0	100.0	100.0
Total	353	29.2	36.8	16.4	36.0	34.6	32.0

the per cent accessing the bazaar for agricultural information declines sharply. Households using a greater number of institutions are less likely to use the bazaar and more likely to use a combination of institutions. For households using two sources, 40% use relatives and neighbours. For households using three sources, relatives are used by 90%, neighbours by 85% and the *oxacol* (village leaders) by 55% of respondents. Among these households only 15% use the bazaar. Households using four and five institutions follow a similar pattern. No discernible pattern is obvious for households using the local breeder.

Households appear to be divided into three categories according to where they obtained agricultural information: (i) households using one institution primarily use the bazaar; (ii) households using two or more institutions rely upon relatives and neighbours, and sometimes the Shreder Institute; and (iii) households using multiple institutions use relatives, neighbours, *oxacols*, and the Shreder Institute.

Econometric Analysis

Regression analysis is used to test the effect of institutions on the level of fruit and nut diversity in a household's garden, controlling for other confounding effects. The theoretical motivations for this econometric model are provided in Chapter 5. The household is taken as a unit of analysis, and a combination of household and institutional variables are regressed against the total number of fruit and nut varieties planted by each household. A Poisson regression was used because of the nature of the dependent variable, and a likelihood ratio test applied to test for the joint significance of sets of independent variables. A list of the variables used and definitions of each variable is provided in Table 12.10.

Because there is such a large number of institutions that each household can participate in and a range of possible answers, a set of indices was created to summarize the data about household participation in the informal community institutions, described and discussed above. The first variable is a compound index that combines the number of institutions and the household's subjective ranking of the importance of this institution.

The second variable is the number of institutions in which the household reported participating, with a mean of 5 and a maximum of 10. Two intensity variables are used: the number of meetings per month and the number of institutions to which household reports some financial commitment. Finally, there is an index of the household's ranking of the institutions as sources of agricultural information.

Information already reported in Table 12.9 was introduced to represent the institutions from which farmers obtain materials or information directly, including the Shreder Institute, use of the bazaar and a combination of the remaining categories (relative, neighbour, *oxacal*) as indicators of informal institutions. These variables can be understood as 'stated preference' variables, in contrast to the 'revealed preference' data on seed sources presented in Table 12.6; they are the answer to the question 'where would you source new materials' rather than 'where have you sourced existing materials'. The stated preference nature makes them more acceptable as exogenous instruments. There is potential endogeneity between observed fruit and nut diversity levels and choice of the system for obtaining planting materials.

A set of regressions were run on subgroups of the variables, first on the household characteristics alone, second on the household characteristics and the institutions that operate indirectly, third on the household characteristics and the institutions that operate directly, and fourth on all of the variables.

Regression results are presented in Table 12.11. Households in Bulungur have lower variety counts in their gardens, holding other effects constant. Older farmers have a greater number of fruit and nut varieties. Households with more adults are associated with more richness in fruits and nuts, which could be related to the labour intensity of cultivating a higher number of fruit and nut varieties. More household members working off-farm relates negatively to variety richness, suggesting that labour is displaced by off-farm work. The wealth index is almost significant in the regression containing only household variables, and positive and significant in the regression containing only the variables that operate indirectly. Wealthier households grow the largest number of fruit varieties. The coefficient on off-farm cash income is

Table 12.10. Definition of explanatory variables.

Variable	Definition
Bulungur region	= 1 if household is in Bulungur region
Household characteristics	
Age of household head	Age of household head in years
Number of adults	Number of members of household older than 15 years of age
Cash income	Sum of reported cash income from wages, salaries, pensions
Number of off-farm workers	Number of household members with off-farm employment in the past year
Durable assets	Sum of key durable goods and characteristics of household construction
Farm characteristics	
Garden size	Area of household garden plot in 100 m^2
Community institutions	
Number of institutions	Total number of local community groups that a household reports participating in
Number of sources of agricultural information	Total number of local institutions that are reported to be sources of agricultural information
Number of meetings/month	Total number of meetings attended per month summed across all groups
Financial commitment	Total number of groups to which the household reports a financial commitment
Index ranking of agricultural information	Sum across groups of household ranking of institutions for quality as sources of agricultural information
Seed institutions	
Use Shreder	= 1 if household would look to Shreder Institute to access new material
Use bazaar	= 1 if household would use bazaar to access new material
Use informal	= 1 if household would use informal institutions to access new material

Note: Since the sources and opportunities for official farm income are extremely limited in Uzbekistan and there is no 'labour market' in rural areas, cash income from pensions and salaries largely reflects predetermined employment decisions.

positive and significant in the household variables regression, indicating that off-farm income may be used to subsidize production and diversification in garden plots. All regression results control for variation in garden size, which is not significant in any of the estimated equations.

While marginal effects of household variables appear robust across different specifications of the econometric model, institutional variables, as they have been measured and specified here, have a much more limited influence on the fruit and nut diversity measured at the household level. Only one of the variables from the community group institutions, and none of the variables related to seed system institutions, has a statistically significant effect. An increase in the number of community groups to which a household belongs increases the probability that the household manages a more diverse orchard. A joint test of the significance of all of the institutional variables indicates that the community group variables are jointly significant in the two regressions (likelihood ratio tests demonstrate significance levels of 0.002 and 0.0017, respectively). Households can take advantage of community groups to acquire varieties and related information.

The joint test of the factors representing seed institutions does not support their statistical significance taken together, but separately from other institutional variables. While some relationship between institutions and the type of seed system is evident in Tables 12.8 and 12.9, clearly the effect is not big enough to influence the

Table 12.11. Poisson regression explaining total number of varieties in home garden, with and without institutions.

	No institutions		Indirect institutions		Direct institutions		Indirect and indirect institutions	
	Coefficient	t-ratio	Coefficient	t-ratio	Coefficient	t-ratio	Coefficient	t-ratio
Constant	1.956	18.088***	1.730	13.235***	1.883	5.519***	1.732	12.728***
Bulungur region	−0.228	−5.405***	−0.213	−4.454***	−0.210	−4.232***	−0.213	−3.832***
Household characteristics								
Age of household head	0.003	1.874*	0.004	2.444**	0.003	1.863*	0.004	2.395**
Number of adults	0.024	2.173**	0.026	2.347**	0.024	2.246**	0.026	2.351**
Cash income	1.04E−07	1.709*	9.16E−08	1.465	9.36E−08	1.523	8.57E−08	1.361
Number of off-farm workers	−0.019	−1.730*	−0.028	−2.546**	−0.017	−1.607	−0.028	−2.487**
Durable assets	0.010	1.595	0.011	1.718*	0.008	1.270	0.010	1.540
Farm characteristics								
Garden area	0.000	0.186	0.000	0.161	0.000	0.338	0.000	0.230
Community institutions								
Number of institutions			0.038	2.261**			0.038	2.183**
Number of sources of agricultural information			−0.002	−1.608			−0.002	−1.548
Number of meetings/month			0.002	1.627			0.002	1.506
Financial commitment			−0.037	−1.400			−0.037	−1.370
Index ranking of agricultural information			−0.001	−0.112			−0.001	−0.103
Seed institutions								
Use Shreder					0.062	1.337	0.035	0.738
Use bazaar					0.045	0.937	0.017	0.344
Use informal					0.047	1.038	−0.008	−0.156
N	368		368		368		368	
Pearson R^2	0.10		0.15		0.11		0.16	
			(LR test)	d.f.	(LR test)	d.f.	(LR test)	d.f.
Joint test of institutions variables			24	5***	4.24	3	24.7	8***

*10% statistical significance; ** 5% statistical significance; ***1% statistical significance.

overall level of fruit and nut diversity in household orchards.

Table 12.12 presents the results for regressions, testing the relationship between seed system institutions used by the household and participation in the community groups. Using the 'stated preference' variables on where a household 'would source new material', a set of three probit regressions was estimated with the same household and community institutions variables.

Findings are mixed, with conflicting signs apparent across seed system institutions for many of the independent variables. The coefficient on cash income is positive and significant for the use of bazaar, but negative and significant for the

Table 12.12. Probit regression of seed system institution used for access to foreign materials.

	Shreder		Bazaar		Other//informal	
Dependent variable	Coefficient	t-ratio	Coefficient	t-ratio	Coefficient	t-ratio
Constant	−1.938	−4.285***	−0.343	−0.916	−0.921	−2.414**
Household characteristics						
Age of household head	0.010	1.563	0.002	0.329	0.005	0.862
Number of adults	0.025	0.603	−0.028	−0.709	0.009	0.226
Cash income	0.000	−0.067	8.7E−07	4.079***	−8.6E−07	−3.646***
Number of off-farm workers	−0.022	−0.525	−0.064	−1.601	−0.066	−1.621
Durable assets	0.086	3.682***	0.006	0.246	0.001	0.030
Farm characteristics						
Garden area	−0.045	−3.328***	0.000	0.039	−0.002	−0.974
Community institutions						
Number of institutions	0.203	3.348***	−0.192	−3.072***	0.373	5.762***
Number of sources of agricultural information	−0.001	−0.341	−0.006	−0.695	0.065	5.554***
Number of meetings per month	0.013	2.421**	−0.001	−0.095	0.006	0.954
Financial commitment	0.044	0.447	−0.235	−2.288**	−0.002	−0.016
Index ranking of agricultural information	−0.053	−2.405**	0.103	4.124***	−0.134	−5.153***
N	368		368		368	
Log-likelihood	−193.22		−220.96		−200.27	
Chi-square (11 d.f.)	59.0***		46.9***		108.9***	

**5% statistical significance; ***1% statistical significance.

informal sources. Households with less cash might look to informal, exchange or barter-based sources of materials rather than the marketplace. The only other significant household variables are in the Shreder Institute regression, where the effects of wealth are positive and that of garden area is negative. Wealthier households are more likely to look to the formal, government source for new material.

The results for the community variables are compelling, and more variables are statistically significant than in the previous regression. The coefficient on the number of institutions is positive for the regression explaining a preference for the Shreder Institute and informal institutions as a source of genetic material, and negative in the regression predicting a preference for the bazaar. The more community institutions in which a household participates, the more likely that the household will look to the Shreder Institute or informal institutions as a source for material, and the less likely that it will seek materials at the bazaar. The use of institutions for agricultural information reduces the probability that the household would use either the Shreder Institute or information institutions to obtain seed, although it increases the chances it would use the bazaar. The interpretation of these findings is unclear, reinforcing the conclusion that numerous processes affect the search for planting material.

Conclusion

Building on the household model of on-farm diversity presented in Chapter 5, this chapter describes and explains the diversity of fruits and nuts in two districts in Samarqand, Uzbekistan, emphasizing the role of social institutions that convey planting material and agricultural information. This research had two motivations. First, the research has sought to understand better how rural households use home gardens to survive the process of economic transition. Second, it has sought to understand the interrelationships among social institutions, planting material systems and crop diversity. Home gardens in Uzbekistan are diverse with respect to species, with low numbers of varieties within each crop. Only a few of the crops, including apples and grapes, are planted by a majority of the households. Several other species are planted by roughly half of the households. The planting material system appears to be a robust combination of self-supply, informal local village networks, the bazaar and official sources. Household sources for most material are predominantly local, within the same village or the same district. Most of the diversity of fruits and nuts is contained in the home gardens, but does not appear to depend much on the size of the garden – at least in the range represented by the sample.

Statistical tests uncovered a statistically significant association between the extent of household participation in social groups and the level of fruit and nut diversity in the home gardens of Samarqand. Furthermore, a methodology to measure and describe the extent and intensity of community institutions in rural villages was presented. Community institutions range from a wide social sphere such as a wedding or meeting at a teahouse with limited financial or social obligation to work brigades and reciprocal exchange groups with small numbers of participants and more intense bonds of social commitment. Articulating the link between these social groups and crop biodiversity on household farms may require further, more nuanced analysis because these indirect effects appear in communities in multiple stages. Establishing causality is a challenge; community institutional participation may influence the availability of information or materials, but may be secondary to other processes determining which households plant diverse materials.

Institutions that convey planting material encompass formal, market and informal channels. In this study, they are envisioned as sources of both planting material and agricultural information. The majority of households acquiring information from a single institution depend on the bazaar, while a smaller number of households using multiple institutions rely on a combination of informal sources. This suggests that within the same village, different households follow distinct strategies for obtaining access to planting material and agricultural information. So far, data have revealed no direct linkage between the system for obtaining planting material and the diversity of fruits and nuts in the home garden. Subsequent regressions did demonstrate that household participation in community groups affects where they look for seed.

Implications

Data support the existence of an association between social institutions and management of fruit tree and nut diversity in two districts of Samarqand, Uzbekistan. This study addresses two key elements to consider in designing and influencing policies for sustainable management of crop biodiversity on farms. So far, applied economists have addressed neither element in research about on-farm conservation. Any conservation programme in locations such as these would need to target not only the households or communities with more diversity in their home orchards, but also: (i) the community institutions (formal and informal) that are influential for household agricultural decisions and (ii) seed system institutions (direct and indirect) most used by households. Community institutions can serve as useful avenues for a programme to disseminate information and varieties. To provide incentives through the seed system, its heterogeneous forms and household participation in these forms must be well understood.

Acknowledgements

We wish to thank Muhabbat Turdieva and Pablo Eyzaguirre (IPGRI) for their contributions, as well as Ruth Meinzen-Dick and the organizers of the CAPRi (System-wide Program on Collective Action and Property Rights of the CGIAR) workshop on 'Property Rights, Collective Action, and Local Conservation of Genetic Resources' held in Rome, September 2003. CAPRi and the Swedish International Development Agency

(SIDA) also provided funding for the field research and data analysis. Special thanks to the local communities in the study who were gracious hosts and partners in the research.

References

Bloch, P. (2002) Agrarian reform in Uzbekistan and other Central Asian countries. Working Paper no. 49. Land Tenure Centre, University of Wisconsin at Madison, Wisconsin.

Expert Centre for Social Research (1999) Consultations with the poor, participatory poverty assessment in Uzbekistan for the World Development Report, Expert Centre, Tashkent.

Kandiyoti, D. (1999) *Poverty in Transition and Ethnographic Critique of Household Surveys in Soviet Central Asia. Development and Change.* Institute of Social Change, Oxford, UK.

Lerman, Z. (1998) Land reform in Uzbekistan. In: Wegren, S. (ed.) *Land Reform in the Former Soviet Union and Eastern Europe.* Routledge, London and New York, pp. 136–161.

Lerman, Z., Csaki, C. and Feder, G. (2004) *Agriculture in Transition: Land Policies and Evolving Farm Structures in Post-Soviet Countries.* Lexington Books, Lanham, Maryland.

North, D. (1990) *Institutions, Institutional Change, and Economic Performance.* Cambridge University Press, Cambridge, UK.

Ostrom, E. (1990) *Governing the Commons: The Evolution of Institutions for Collective Action.* Cambridge University Press, Cambridge, UK.

Seeth, H.T., Chachnov, S., Surinov, A. and Von-Braun, J. (2003) Russian poverty: muddling through economic transition with garden plots. *World Development* 26(9), 1611–1623.

Thurman, M. and Lundell, M. (2001) Agriculture in Uzbekistan: private, deqhan, and shirkat farms in the pilot districts of the Rural Enterprise Support Project. Environmentally and Socially Sustainable Development. World Bank, Washington, DC.

Uphoff, N. (1986) *Local Institutional Development: An Analytical Sourcebook with Cases.* Kumarian Press, West Harford, Connecticut.

Vavilov, N.I. (1930) *Five Continents.* Translated by Doris Love (1997). IPGRI/VIR, Rome.

13 Community Seed Systems and the Biodiversity of Millet Crops in Southern India

L. Nagarajan and M. Smale

Abstract

This chapter expands the approach of Chapter 11, adding seed system characteristics to the set of factors that influence variation in crop biodiversity levels measured at the geographical scale of the community. In the subsistence-oriented, semiarid production systems of Andhra Pradesh and Karnataka, India, the environment is marginal for crop growth and often there is no substitute for millet crops. Across communities, farmers grow 13 different combinations of pearl millet, sorghum, finger millet, little millet and foxtail millet varieties, but individual farmers grow an average of only two to three millet varieties per season. The notion of the seed system includes all channels through which farmers acquire genetic materials, outside or in interaction with the commercial seed industry. Data are compiled through household surveys and interviews with traders and dealers in village and district markets. Based on the concept of the seed lot, several characteristics of local seed markets are defined and measured by millet crop, including seed transfer rates for farmer-to-farmer transactions and seed replacement ratios. Most seed transactions appear to be based on money. Seed supply channels differ by improvement status of the genetic material. Econometric results indicate the significance of the seed replacement ratios and seed volumes traded in determining the levels of crop biodiversity managed by communities, in addition to the household, farm and other market-related factors identified by previous studies. These are interpreted as indicators of market strength.

Introduction

Understanding systems for planting material is crucial for maintaining crop biodiversity. Institutional arrangements, instruments and policies designed to convey incentives for local conservation of crop biodiversity function through the planting material systems that embody genetic resources and enable their exchange. Farmer and community access to the genetic resources embodied in seed is affected by the extent to which it is traded on markets or through other social institutions, as well as by related legal frameworks, national and international agreements. Seed systems convey incentives for farmers to grow one crop variety rather than another, or to grow a set of crops and varieties rather than one. Markets are a component of seed systems, transmitting value through consumers' willingness to pay, including both consumers of planting material (seed) and consumers of products. In semi-subsistence agriculture, farmers are both the consumers of seed and the consumers of products.

This chapter relates seed systems to variation in crop biodiversity measured at the scale of the community. Building on the analysis of determinants of crop biodiversity in communities (Chapter 11), seed system parameters are introduced as previously omitted, explanatory variables. An integrated definition of the seed system that encompasses formal and informal channels and a combination of survey techniques were used to generate the household- and community-level data for the analysis.

This research is part of a larger effort whose purpose is to identify possible entry points for policies to promote farmer welfare through supporting their management of millet biodiversity (Nagarajan, 2004). In harsh, risky production environments, farmers often depend directly on the crop biodiversity that is also of public value for future crop improvement or the resilience of the farming system. Such is the case for the sites studied here.

Although India is a major world producer of millet crops, yield progress attained through the adoption of modern varieties has not been so impressive as for rice and wheat (Evenson and Gollin, 2003). One explanation is related to demand for research products. In certain states of India located in relatively favourable production environments (Maharastra, Punjab and Haryana), production and consumption of millet crops has declined as farmers replace them with other crops when income rises and food preferences change. In the subsistence-oriented production systems of India where the environment is marginal for crop growth (such as semiarid regions of Karnataka and Andhra Pradesh), there is often no substitute crop for millets. Millet crops have attributes that are desirable for semi-subsistence farmers who both sell and consume their harvest. These include higher value of micronutrients compared with major cereals, greater tolerance to pests and diseases and ability to yield under extreme soil conditions (Seetharam et al., 1989).

The following section defines the notion of the seed system used in this chapter, which had implications for the research design. Millet biodiversity and characteristics of seed systems in study sites are defined and summarized, including seed channels for marketing millet crops and varieties. Variation among villages in the biodiversity of millet crops is then explained with a community-level econometric model that includes seed system parameters. Implications are drawn in the final section.

Methods

Definition of a seed system

Typically, the notion of a seed system in economics has been limited to the 'formal' seed industry for developing, multiplying and distributing finished varieties as certified seed, which can be publicly and privately funded and organized in different ways. For example, maize seed industries are thought to develop along a path from pre-industrial organization to the maturity stage, characterized by entirely commercial organization with plant variety protection, patents and various financing arrangements (Morris et al., 1998). The notion of a farmer-based seed system, termed 'informal', is documented extensively by other social scientists (Sperling et al., 1993; Thiele, 1999; Zimmerer, 2003), ethnobotanists and geographers, but is most often treated separately by economists as vestigial or marginal to the process of economic development.

In this chapter, the seed system is broadly defined, including all the channels through which farmers acquire the genetic materials they demand and information about those materials, outside of, or in interaction with, the commercial seed industry. These channels include various farmers' organizations, weekly markets and social networks. Farmers' seed management consists of variety demand, selection of seed to plant the next season, seed storage and seed transfers, exchanges or mixtures (Louette, 1994; Bellon et al., 1997; Smale and Bellon, 1999). Varieties demanded may include either those saved and selected for many generations on farms (traditional, ancestral or landrace types) or modern varieties (hybrids or improved open-pollinated varieties (IOPVs)). Seed selection may include mass selection practices or farmer breeding, as well as re-use of hybrids or other commercial varieties.

When product markets are less fully developed or are incomplete, the demand for planting material is derived from: (i) the agricultural household's demand for the attributes of the goods its members choose to consume and (ii) the agronomic traits the household members select to best fit the variety to physical features of the farm and available technology. In semi-subsistence agriculture, purchases of improved seed may be periodic, and most of the seed is reproduced from the harvests of the previous seasons or the stocks maintained by community members, who may or may not trade seed with other communities.

Implications for research design

Economic analyses of incentives for maintaining crop biodiversity on farms have been based largely on models of decision making by agricultural households, applied with econometrics to household survey data; seed market studies are often compiled from secondary data. Neither household surveys nor secondary data are sufficient for analysing seed systems as defined above.

On the one hand, data collected from farm households reveal how individual farmers exchange seed and products but lead to few conclusions about market channels and the role of other institutions that affect exchange. On the other, secondary data are not variety-specific, and even when they are, names are likely to be inconclusive regarding farmer-managed units of biodiversity. The timing of seed exchange is particularly seasonal for farmers' cultivars (FCs) (just before planting), and may also occur in limited geographical areas (a few farmers; a few villages). Often there is no recognition of volumes traded because they are so minimal. Some millet crops grown in India are a good example, classified in the official statistics with 'other coarse grains'.

Furthermore, in some cases, planting material and product are also indistinguishable, particularly after poor harvests, when farmers may purchase seed from food grain if they are unable to find quality seed through other sources. Finally, those who participate in informal systems may not generally describe themselves as 'traders' by occupation, or may not engage in trading full-time. The nature of the transaction may include barter or another form of exchange without cash.

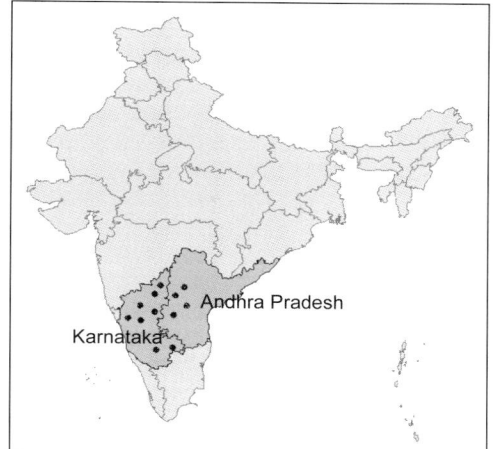

Fig. 13.1. Location of survey areas in Andhra Pradesh and Karnataka, India.

Data

Data were collected in two states (Andhra Pradesh and Karnataka, Fig. 13.1) of India between October 2002 and June 2003, spanning the cool, rainy season (*Kharif*, lasting mid-July to the end of October) and the post-rainy season (*Rabi*, from December to March). The domain was purposively selected to represent major areas of production for a number of millet crops in a semiarid environment, including some improved varieties, a range of FCs and diversity known from previous scientific research.[1] Within the states of Andhra Pradesh and Karnataka, 75 villages were selected in 61 *panchayats* of six districts: Mahabubnagar (Andhra Pradesh); Bijapur, Bellary, Chitradurga, Belgaum, Dharwad (Karnataka). *Panchayat* is the local term for 'village community', an administrative unit composed of four to five villages, depending on population size and geographical limits.

A self-weighting sample of 432 households was selected in villages using a random number table and a constant sampling fraction of 9–10%. Of the set of structured survey instruments developed and pre-tested, those used in this chapter elicited general household information about size and composition, income sources, assets and expenditures, area and plot characteristics for millet crops grown in each season and seed management information to

---

[1] Personal communication with Dr A. Seetharam, All India Coordinated research on Small Millets, UAS Bangalore; Dr K.N. Roy and Mr Gopal Reddy, Scientists, ICRISAT and Prof Naik, Millet Breeder, UAS Dharwad; Rabi Sorghum Germplasm Collection in Northern Karnataka and Adjoining Areas of Andhra Pradesh, Genetic Resources Progress Report-74 &85, ICRISAT; Rainy Sorghum Germplasm Collections in Karnataka and Adjoining Areas, Genetic Resources Progress Report-29, ICRISAT.

quantify the direction, frequency and nature of farmer transactions in formal and informal seed supply channels.

Local seed suppliers ('experts') were identified from each of the 60 *panchayats* included in the domain through key informants. A semi-structured questionnaire elicited information about the nature of their involvement in seed channels during good and bad cropping seasons, why they are considered to be experts and their social and economic characteristics. Twenty-nine input dealers, representing 10–12% of all millet seed dealers in each of the six district headquarters, were interviewed using a brief structured questionnaire about volume of sales, prices and varieties handled.

As the field research was implemented, it became clear that 'shandies' (*panchayat* weekly markets) cater specifically to local seed demand. Shandies operate weekly at various places, typically with a group of five to six villages, covering a radius of 10–15 km. For logistical reasons, it was not possible to cover all shandies simultaneously and a total of 25 were selected arbitrarily. When data were collected (which was not immediately before planting), seed flows were thin in shandies, although grain flows were not. Often those who engaged in transactions did not differentiate between seed and food grain, or between seed types. Participants were interviewed in groups with checklists, and asked to estimate the frequency of transactions and seed volumes, prices and the quality of material transacted.

Both farmer and scientist taxonomies were employed to assess the extent of biodiversity in millet crops. Farmers were asked to identify each cultivar grown by name for each millet crop and then describe its distinguishing characteristics (grain colour, shape and size; plant height; maturity and shape of spikelets). Representative seed samples were then collected from a mature crop stand or threshing floor, seed storage structures or seed stocks of farmers, and compared with descriptors used by the ICRISAT gene bank experts or seed companies, or those found in research reports (Prasada Rao, 1980; Gopal Reddy 1993, 1996).[2]

Description

Millet biodiversity

Millets refers to a group of annual grasses mainly found in the arid and semiarid regions of the world. Millets belong to five genera: Pennisettum, Eleucine, Setaria, Panicum and Paspalum. Sorghum is not classified under millets by genus but belongs to the same family classification as other millet crops (Monocotyledonae and the subfamily of Poaceae), and is often referred to in India as 'great millet'. These grasses produce small-seeded grains and are often cultivated as cereals.

Finger millet is grown widely in the southern part of Karnataka and in Tamil Nadu. Nearly two-thirds of the national output is produced in this region. Grown as an irrigated crop during the dry season in south India, finger millet is also intersown between rows of maize and other crops. Foxtail millet is grown as a food crop in Andhra Pradesh and Karnataka. Although it grows in dry weather, foxtail millet requires good soil. Although not extensively grown, it is of significance in certain sections of the lower Deccan plains and the highlands of Andhra Pradesh, Karnataka and Tamil Nadu. The food prepared from foxtail millet is considered to be good for pregnant women and invalids. Little millet is grown mostly in southern India, parts of Andhra Pradesh and Karnataka, apart from the central and hilly tracts of the north of India. Grown mainly as a rain-fed crop, on poor, infertile soils, little millet is often used as dry fodder for ruminants and the grains fed to poultry.

The counts of distinct cultivars grown in the rainy and post-rainy seasons are shown in Table 13.1, including their improvement status. The terms 'major' and 'minor' are often used to refer to the extent of research investment and commercial importance of the crop. Table 13.1 shows, for example, that the number of hybrids and IOPVs bred by professional plant breeders is much higher for pearl millet and sorghum, the 'major millets'. Minor millets refer to finger millet, foxtail millet and little millet. Some research

---

[2] Personal communication, 2003, Professor A. Seetharaman, Professor (Emeritus), ICAR Center for Small Millets, University of Agricultural Sciences, Bangalore.

Table 13.1. Total number of distinct cultivars grown by all farmers surveyed, by millet crop and improvement status.

Millet crop	Improvement status	Rainy	Post-rainy	Total number of distinct varieties
Pearl millet	Hybrid	5	0	5
	IOPVs	2	0	2
	FC	3	0	3
Sorghum	Hybrid	10	4	10
	IOPVs	7	3	7
	FC	10	16	19
Finger millet	IPLS	7	1	8
	FC	3	0	3
Little millet	FC	4	0	4
Foxtail millet	FC	2	0	2
All millet crops		53	24	63

Note: Little millet (samai); foxtail millet (Navane); pearl millet (bajra); sorghum (jowar); finger millet (ragi). FC = farmer cultivar; IOPV = improved open-pollinated variety; IPLS = improved pure-line selection.

effort by professional plant breeders is evident for finger millet, in the form of pure-line selections from FCs. All foxtail and little millet varieties are FCs, and they appear to be largely managed as a pool – that is, with fewer distinguishing characteristics (grain colour and texture).

By far the highest richness levels are found in sorghum followed by finger millet and pearl millet, considering all categories of crop improvement. Of all five millet crops and farmers surveyed, a total of 53 distinct varieties were grown in the rainy season and 24 in the post-rainy season. Since a number of these were grown in both seasons, there were 63 total distinct varieties of millet crops cultivated in the areas studied (Table 13.2).

These totals provide one perspective on the overall structure of millet biodiversity in the communities studied, although they conceal the complexity of millet cropping choices observable from farm to farm. On one hand, individual farmers plant an average of only one to two millet crops in one to three varieties in the rainy season, and even fewer in the post-rainy season. The average at the *panchayat* level is three millet crops in the rainy season, with seven total varieties. On the other hand, combinations of crops and varieties proliferate. During the rainy season in the survey period, 45% grew one of 13 different combinations of major and minor millets. Only 5% grew major millets only (pearl millet and sorghum). Forty-two per cent grew a millet monocrop (sorghum, pearl, finger or little millet), and of that group, farmers growing only sorghum were the most common. Hybrids and improved varieties dominated the seed planted of pearl millet and sorghum crops in the rainy season; in the post-rainy season, FCs of sorghum were far more frequently planted than were improved types.

A sizeable percentage of farmers classified their pearl millet or sorghum hybrids as ancestral or mixed (24 and 36, respectively), suggesting that they save seed and replant it. Farmers may deliberately mix the seed or the materials may become genetically mixed through pollen flows among varieties that are planted contiguously, as has been reported for pearl millet in Rajasthan (vom Brocke, 2001). In the rainy season, only 7% of the seed lot varieties of IOPV pearl millet planted, as compared with 63% of those of IOPV sorghum, were classified as ancestral or mixed. Possibly, farmers recycle the seed of sorghum IOPVs more because the crop outcrosses less and because yield advantages are more easily maintained for successive seasons than is the case for pearl millet. It may also be the case that pearl millet IOPVs are more recently released and so the seed has been more recently purchased. Some respondents also described FCs of finger millet as 'improved', perhaps because they consider those released by the State Department of Agriculture (pure-line selections from FCs) as their own.

Table 13.2. Varieties grown by the households in the survey areas. (Field surveys conducted during October 2002–June 2003, ICRISAT gene bank, and ICAR Center for sorghum and finger millet, UAS, Dharwad and Bangalore (2003–04).)

Number	Variety name	Improvement status	Number	Variety name	Improvement status
Sorghum			Pearl millet		
1	Allina jola	FC	1	Local dwarf bajra	FC
2	Bijapur jola	FC	2	Advante hybrid	HYB
3	Bili jola	FC	3	Bajra kaveri	HYB
4	Csh-1	HYB	4	Bajra paras	IOPV
5	Csh-11	HYB	5	Bajra agro	IOPV
6	Csh-14	HYB	6	Bajra seedtec hyb.	HYB
7	Csh-15	HYB	7	Hybrid bajra mahyco	HYB
8	Csh-16	HYB	8	ICMV-221	HYB
9	Csh-5	HYB	9	ICTP series(5 lines)	IOPV
10	Csh-9	HYB	10	Jawari bajra	FC
11	Dodda jola	FC	11	Jawari sajji	FC
12	Gangavati sorghum	FC	12	Kaveri	IOPV
			13	Paras bajra	HYB
13	Gidda maldandi	FC	Finger millet		
14	Gunduteni	FC	1	Annapoorna ragi	IPLS
15	Hala jola	FC	2	Black ragi	FC
16	Hombale jowar	FC	3	Dwarf ragi	FC
17	Itc jowar	HYB	4	Farm ragi	FC
18	Jawari jowar	FC	5	Godavari	IPLS
19	Jk-5	HYB	6	Gpu-22	IPLS
20	Jk-22	HYB	7	Gpu-28	IPLS
21	Kenjola	FC	8	Indof-5	IPLS
22	Kesari	FC	9	Kalyani	IPLS
23	M-35-1	FC	10	Pr-202	IPLS
24	Maldandi	FC	11	Short ragi	FC
25	Mugutheni	FC	12	V-20	IPLS
26	Muguti maldandi	FC	13	White ragi	FC
27	Msh-51	HYB	Little millet		
28	Nandiyal white	FC	1	Black samai	FC
29	Pac-501	IOPV	2	Hali samai	FC
30	Paras jowar	IOPV	3	Jawari samai	FC
31	Pioneer jowar	IOPV	4	Mallige samai	FC
32	Proagro-296	IOPV	5	Local samai	FC
33	Sorghum agro	IOPV	6	White samai	FC
34	Tella jola	FC	Foxtail millet		
35	Vikarbad local	IOPV	1	Hala Navane	FC
36	Yaniger	FC	2	Local Navane	FC

OPV = open-pollinated varieties, IPLS = improved pure-line selection.
ICAR – Indian Council of Agricultural Research and ICRISAT – International Crops Research Institute for semiarid Tropics, others include state agricultural universities and private sector companies.

Seed sources, transfer rates and replacement ratios

To analyse seed, we use the 'seed lot' as the unit of observation. As defined by Louette (1994), a seed lot is the physical unit of seed the farmer uses to reproduce a given cultivar in each season. Regardless of improvement status, seed lots are often mixed or replaced, partially or completely, since the time the original seed for the cultivar was acquired (Aguirre Gómez, 1999). The variety may have been grown for many years, but each season, a new seed lot is planted for that same variety.

In this study, farmers were asked to report how many years they had grown each cultivar planted in each survey season (rainy and post-rainy). The number of years grown by an individual farmer represents the age or longevity of the cultivar under that farmer's management. Farmers were then asked, for each cultivar, the number of times they have replaced the seed from any source other than their own harvest. Thus, as measured here, seed replacement does not include variety change. The seed replacement ratio was calculated as the frequency of seed replacement for a cultivar divided by its age on that farm. Farmers were also asked the number of times they had supplied seed for each cultivar to other farmers, by source and mode of transaction. The seed transfer rate is the number of times a seed lot for a given cultivar has been transferred from the farmer interviewed to another farmer divided by age of the cultivar on that farm.

The seed replacement rate or ratio is normally calculated by commercial seed organizations to forecast the demand for their varieties. A higher velocity (frequency) of seed replacement is thought to be desirable, especially for modern varieties. Seed replacement for the same variety protects against genetic deterioration; replacing seed for the purposes of changing varieties can promote yield enhancement (Heisey and Brennan, 1991). Seed replacement buffers pest and disease problems through maintaining genetic resistance or the diversity in sources of resistance over time (Apple, 1977). In landrace systems for cross-pollinating crops, some genetic studies indicate that mixture and replacement serves the purpose of protecting the genetic viability of the seed. Berthaud et al. (2002) argue that maize farmers in Mexico use introductions from other farmers as a tool to 'rescue' their cultivars from high rates of deleterious mutations. In a number of empirical studies, farmers have reported the need to replace their 'tired' cultivar or 'renew' seed (Almekinders et al., 1994; Li and Wu, 1996; Sperling et al., 1996; Louette and Smale, 2000). Heisey and Brennan (1991) developed a model to analyse farmers' demand for replacement seed, but in their simulations, they found that a wide range of seed replacement times were consistent with economically optimal behaviour.

Base yield and the seed-to-grain price ratio had almost no impact on optimal replacement time, although increasing the rate of yield improvement reduces the time to replacement.

During the rainy season, the 432 farmers surveyed planted five types of millet and a total of 165 seed lots of pearl millet, 381 of sorghum, 192 of finger millet, 77 of little millet and 25 of foxtail millet. FCs have clearly been grown for longer than any improved types (25–32 years). Little millet cultivars are the oldest, although farmers also appear to have grown their local sorghum cultivars for a long time. The average age of sorghum or pearl millet hybrids is 5–7 years, similar to that of improved selections of finger millet. IOPV sorghum are older, with a mean age of 10 years.

The frequency of seed replacements for varieties grown during the rainy season is lower for FCs relative to modern varieties of either pearl millet or sorghum, although it is the same (only twice on average since the original seed for the variety was obtained) across the minor millets (finger, little and foxtail). The frequency of seed replacements is higher for hybrids of pearl millet than for sorghum, perhaps because a greater range of these hybrids is available to farmers. By contrast, the number of seed transfers from farmer to farmer is greater for little and foxtail millets than for major millets, and for FCs of major millets as compared with improved types.

Here, seed replacement ratios are highest for pearl millet hybrids and IOPVs (these are replaced nearly annually), considerably higher for these than for sorghum hybrids and higher for sorghum hybrids relative to improved selections of finger millet. They are extremely low for FCs. In the case of finger millet, most of the improved cultivars available are publicly bred and the replacement rates are not as high as for sorghum and pearl millet. However, they are higher than for other minor millets since the government subsidizes and supplies seed of finger millet as a form of assistance. Farmers reported that they replace the seed of their FCs more often during drought years (which occurs once in 5–7 years in the semiarid regions), when local seed supplies dwindle. In general, mean seed replacement ratios demonstrate the expected positive relationship to improvement status. The rates at which farmers replace seed for FCs are much lower than for improved types, and are higher for hybrids than for IOPVs and higher for heavily outcrossing crops like pearl millet.

Farmers transfer seed to other farmers less frequently than they replace it. That is, when controlling for the number of years the cultivar is actually grown, it is more common for farmers to demand replacement seed from any source (farmers, traders, dealers) than for them to supply it to other farmers – suppliers are few relative to those who demand seed, as expected. Only for little millet and foxtail millet does this not appear to be the case, since seed for these crops is not supplied through formal channels at all. Furthermore, the informal seed market for these millet crops is 'generalized', suggesting little farmer-recognized differentiation of genetic material within these crops. (Table 13.3).

During the post-rainy season only sorghum and finger millet are grown, and farmers planted a total or 318 seed lots of sorghum and only 36 of finger millet, although FCs of both crops dominated. Little formal research has been devoted to sorghum varieties suited to post-rainy production, and the FC of *Maldandi* and its derivatives are the most popular post-rainy sorghum varieties among the farmers. Comparing the patterns for the two seasons reveals that the sorghum cultivars grown during post-rainy season are older, reflecting a higher proportion of FCs. This is not the case for finger millet cultivars. Seed replacement and transfer rates, as well as replacement and transfer ratios, are higher for the IOPV, improved pure-line selection (IPLS) and FCs of sorghum and finger millet grown in post-rainy seasons than for those grown in the rainy season (Table 13.4). One hypothesis is that high rates reflect local seed supply shortages during the post-rainy season for some individual farmers, but not for the community as a whole. The post-rainy season cultivars are those for which production risk is greater, seed quality tends to be poor, seed is saved from the harvest for modern varieties as well as FCs and most seed is farmer-supplied. Cultivars well suited to this season are relatively few, and for some farmers, a combination of poor harvests from which to save seed in the last planting period and rainfall uncertainty in the current planting period can mean a last-minute scramble for seed.

Historical seed transactions for varieties planted in the main rainy season reveal that although family and friends are important sources of original seed and replacement seed as well as recipients of transfers, the frequency of cash exchanges in all three categories is substantial. Even transactions with family and friends, referred to as 'gifts', involve 'token money'. Farmers acquired seed for the original seed for cultivars grown during the rainy season primarily through purchase, although less so, as expected, for FCs than for improved varieties. Even so, 33% of FCs of pearl millet, 52% of FCs of sorghum, 31% of FCs of finger millet, 61% of FCs of little millet and 48% of those of foxtail millet were originally obtained through purchases. Seed replacement transactions for these cultivars also occurred primarily in cash exchanges, typically through dealers for improved varieties and hybrids and through village traders for FCs. Farmers also supplied their own seed of these cultivars to others for 'token money', or through shandies.

Some original and replacement seed was also provided through the government as aid. From time to time, the Department of Agriculture purchases seed from farmers, especially for popular varieties such as *Maldandi*, and especially during drought cycles. Farmers supply seed at a nominal rate. Government purchase rates are always less than the market rates (Table 13.5).

Seasonal differences are again pronounced. The original source of seed as well as replacement source for sorghum and finger millet cultivars is even more heavily dominated by purchases during the post-rainy season. Compared with the varieties planted in the main season, virtually all had been replaced and transferred. When farmers supplied the seed of these cultivars to others, they did so to an even greater extent as a 'gift'. Almost all historical supply transactions for sorghum and finger millet cultivars grown in this season were among friends and family, for 'token money' (Table 13.6). Again, this supports the hypothesis of local supply shortages for some farmers met by local supply from other farmers in this season, as compared with the rainy season, where transactions involve formal seed supply channels. Local supply shortages for some farmers result from the unpredictable and uneven moisture in the post-rainy season – both spatially and temporally – which leads to poor harvests and insufficient seed lots to carry over for some, but not all, farmers.

Table 13.3. Seed replacement, transfer rates and age of cultivars grown in the rainy season, by millet crop.

	Pearl millet				Sorghum			Finger millet		Little millet		Foxtail millet	
	Total	Hybrid	IOPV	FC	Total	Hybrid	IOPV	FC	Total	IPLS	FC	FC	FC
Number of seed lots	165	95	46	24	381	201	38	142	192	131	59	77	25
Mean													
Number of years of growing the same named cultivar	7.6	4.5	4.7	25.3	15.6	6.8	10.4	29.6	12.5	6.9	24.8	32.7	29.1
Number of seed replacements per cultivar	3.9	4.2	4.3	1.9	2.3	3.0	1.4	1.6	2.0	2.0	2.0	2.0	2.1
Number of seed transfers per cultivar	2.0	0.0	0.0	2.0	1.5	1.1	1.8	1.7	1.0	1.0	1.0	2.7	3.7
Seed replacement ratio	0.79	0.9	0.9	0.10	0.35	0.58	0.16	0.07	0.25	0.32	0.09	0.02	0.08
Transfer rate	0.01	0.0	0.0	0.06	0.05	0.05	0.08	0.05	0.01	0.13	0.05	0.05	0.09

Note: See text for definition of seed lot. Seed replacement ratio = number of times the farmer has replaced the seed for the same named cultivar divided by the number of years the cultivar has been grown; transfer rate = number of times the farmer has transferred seed of a named cultivar to another farmer divided by the number of years grown. FC = farmer cultivar; IOPV = improved open-pollinated variety; IPLS = improved pure-line selection.

Table 13.4. Seed replacement, transfer rates and age of cultivars grown in post-rainy season, by millet crop.

Cultivar characteristic	Sorghum				Finger millet		
	Total	Hybrid	IOPV	FC	Total	IPLS	FC
Number of seed lots	318	41	14	263	36	13	23
Mean							
Number of years of growing the same named cultivar	24.5	10.8	7.6	27.6	11.7	20.9	6.5
Number of seed replacements per cultivar	2.7	3.1	4.6	2.6	2.9	2.9	3.0
Number of seed transfers per cultivar	3.0	2.6	1.2	3.2	1.6	2.0	1.3
Seed replacement ratio	0.2	0.3	0.8	0.1	0.4	0.5	0.2
Transfer rate	0.2	0.3	0.1	0.2	0.2	0.2	0.1

Note: See text for definition of seed lot. Seed replacement ratio = number of times the farmer has replaced the seed for the same named cultivar divided by the number of years the cultivar has been grown; transfer rate = number of times the farmer has transferred seed of a named cultivar to another farmer divided by the number of years grown. FC = farmer cultivar; IOPV = improved open-pollinated variety; IPLS = improved pure-line selection.

Seed supply channels

Seed supply channels appear to be differentiated by improvement status, as expected (Figs 13.2–13.4). Public hybrids are those bred by publicly funded research institutions; private hybrids are those produced by companies. Both privately and publicly bred modern varieties (hybrids and IOPVs) are supplied for pearl millet and sorghum, the major millets. Privately bred varieties are distributed at the state and district level through seed distributors, and at the villages level, through traders and dealers. Publicly bred varieties may be distributed through the same channel, but also through state seed corporations, seed farms and depots.

IPLSs of finger millet are exclusively publicly bred, although they too may be distributed locally by private seed dealers and village traders, as well as through seed depots and occasional government assistance programmes for farmers. Seed supply channels for FCs of finger, little and foxtail millet are 'autarkic' in the sense that they have no interface with private companies or public seed corporations. However, these seed types are traded, like all others, by village traders and shandies. Government programmes sometimes purchase leading FCs of minor millets for redistribution to farmers elsewhere (Figs 13.1–13.3).

In 60 *panchayats*, a total of 61 farmer seed suppliers were identified by respondents and key informants. They were roughly equally distributed among those with expertise in modern varieties, FCs or both. Although most experts were more likely to be men, some women experts were found among those with special knowledge about FCs. Most experts are farmers who own their land and have irrigation. The remainder either belong to the village but work outside the farm or are traders from that particular village who bring information or knowledge about seeds into the village. Experts in FCs were older on average, with fewer years of formal schooling, than experts in modern varieties. They were more likely to be farmers and owned more land than experts in modern varieties. Experts in both are intermediate between the other two groups with respect to the same characteristics.

Responses to open-ended questions provide some additional interpretation. Recognition as an expert in modern varieties appears most related to the exposure individuals have to information from 'the outside world'. For example, most of the experts dealing with modern varieties are village headmen or have a recognized official position in the village. Some experts in modern varieties have regular access to communication facilities such as the radio and newspaper; some have regular contacts with the extension agency officials. Many had attended farm schools conducted by agricultural departments (six to seven of them) and they update their knowledge periodically. One of the

Table 13.5. Mode of seed transactions for cultivars grown in rainy season, by millet crop.

Historical transactions	Pearl millet				Sorghum				Finger millet				Little millet	Foxtail millet
	Total	Hybrid	IOPV	FC	Total	Hybrid	IOPV	FC	Total	IPLS	FC			
Number of seed lots for cultivars planted	165	95	46	24	381	201	38	142	192	131	59		77	25
Source (%)														
Gift	21	0	41	67	27	9	45	48	40	26	69		39	52
Aid	24	36	11	0	19	34	16	0	20	29	0		0	0
Purchase	55	64	48	33	54	57	39	52	40	45	31		61	48
Number of past seed replacements for cultivars planted	165	95	46	24	339	201	28	110	183	123	60		24	25
Replacement (%)														
Gift	18	13	26	21	21	14	25	33	23	14	42		29	60
Aid	22	31	15	0	10	14	14	3	12	18	0		0	0
Purchase	61	57	59	79	69	72	61	65	65	68	58		71	40
Number of past transfers for cultivars planted	18	0	0	18	189	59	18	113	125	67	57		36	25
Farmer supply (%)														
Gift	78	0	0	78	75	76	61	77	78	94	60		64	56
Aid	0	0	0	0	4	0	28	3	0	0	0		0	0
Sales	22	0	0	22	21	24	11	20	22	6	40		36	44

FC = farmer cultivar; IOPV= improved open-pollinated variety; IPLS = improved pure-line selection.
Gift denotes that seeds are exchanged among family and friends for money, but at less than the market price (termed 'token money'). Seeds supplied through government programmes as a part of agri-input subsidies are 'aid'. Purchase and sales are exchanges through community markets or dealers. See text for definition of seed lot.

Table 13.6. Mode of seed transactions for cultivars grown in post-rainy season, by millet crop.

Historical transactions	Sorghum				Finger millet		
	Total	Hybrid	IOPV	FC	Total	IPLS	FC
Number of seed lots for cultivars planted	318	14	41	263	36	23	13
Source (%)							
Gift	41	7	24	46	8	9	8
Purchase	59	93	76	54	92	91	92
Number of past seed replacements for cultivars planted	318	14	41	263	36	23	13
Replacement (%)							
Gift	43	21	49	43	33	30	38
Purchase	57	79	51	57	67	70	62
Number of past seed transfers for cultivars planted	311	10	38	263	31	19	12
Farmer supply (%)							
Gift	87	10	74	89	94	100	83
Purchase	13	0	26	11	6	0	17

FC = farmer cultivar; IOPV = improved open-pollinated variety; IPLS = improved pure-line selection.
Seeds supplied through government programmes as a part of agri-input subsidies are 'aid'.
Purchase and sales are exchanges through community markets or dealers.

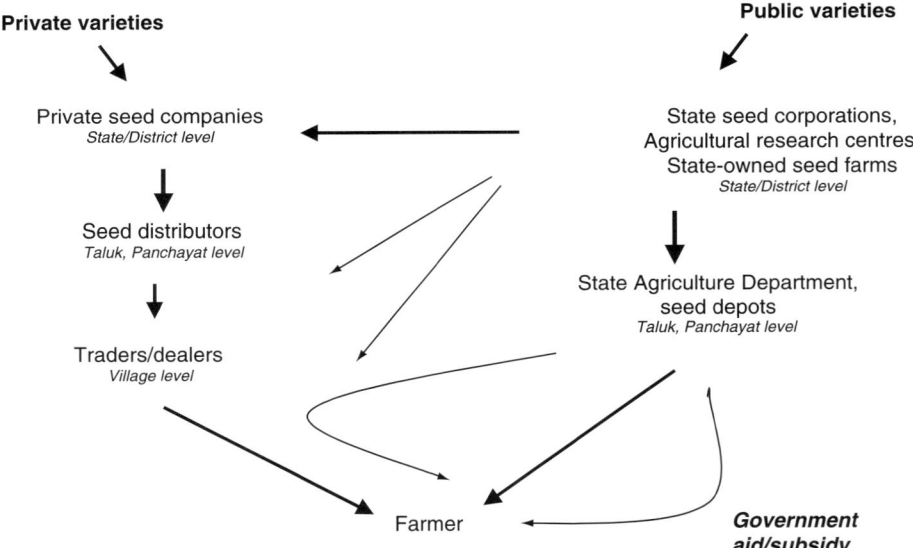

Fig. 13.2. Seed supply channels for hybrids and improved varieties of pearl millet and sorghum in Karnataka and Andhra Pradesh, India.

experts has a son who is an agricultural officer who provides him with information.

Recognition as an expert in FCs refers more to the depth of 'inside' knowledge. Most of the seed experts for FCs explained that they gained their skills through many years of farming experience, learning from their parents and grandparents and having grown FCs for as long as 40–50 years. They explain that they produce the best-quality seeds in their fields and they store it

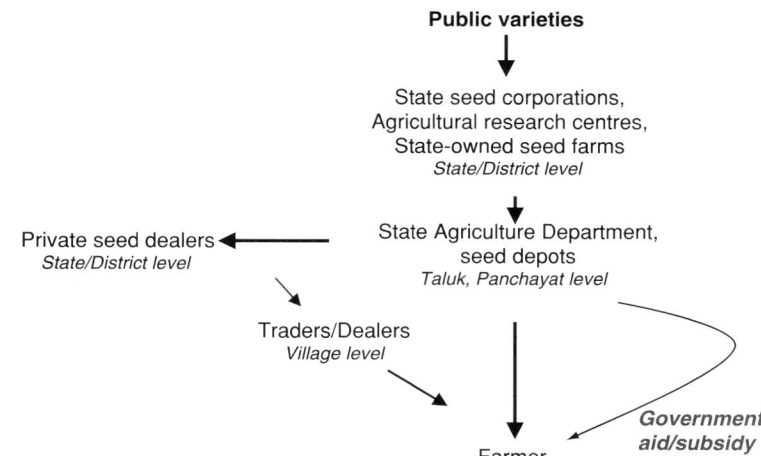

Fig. 13.3. Seed supply channels for improved pure-line selections (IPLSs) of finger millet.

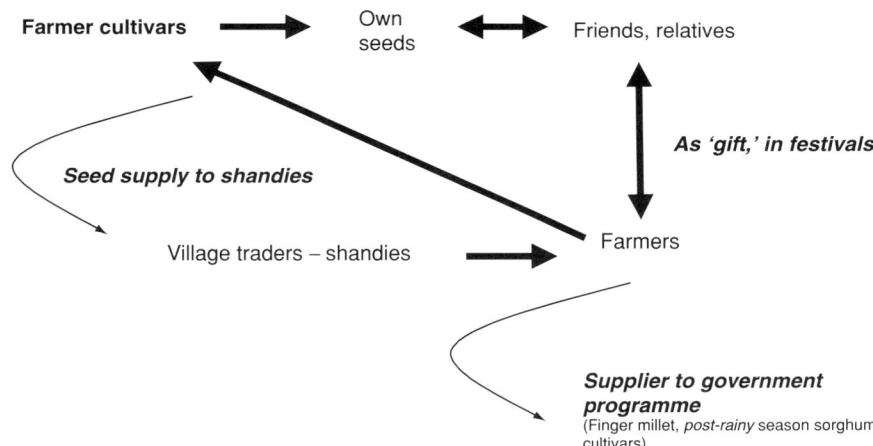

Fig. 13.4. Seed supply channels for farmers' cultivars (FCs) of finger, little and foxtail millet.

more carefully than do other farmers. They share their genetic materials (they sell their seeds) with other farmers from the same village and farmers from nearby areas because the quality of their seed is known to be good. In other words, they have 'credibility' and are trusted (Table 13.7).

Seed dealers are a vital link between farmers and the seed supply from the public seed corporations and private companies. They are the retailers in communities and are able to provision relatively large crop areas, given their knowledge of both formal and informal seed networks. Areas of operation usually extend to a radius of 50 km, and dealers may appoint other retailers to handle the small amounts demanded in remote villages. There is no credit provided to farmers for purchasing seeds, because it is risky and they cannot be accountable for poor germination. Seed dealers also sell other agricultural inputs such as fertilizers and pesticides, and rent farm equipment. Millet seed comprise less than 20% of seeds handled for any of the 29 dealers interviewed. The majority of dealers sold only pearl millet and sorghum seed, and only one dealer in Chitradurga sold finger millet seeds.

Table 13.7. Social and economic profile of farmer seed experts.

	Modern variety expert	Farmer variety expert	Expert in both
Number of observations	19	22	20
Mean			
Age (years)	47.8	62.5	54.2
Education (years in school)	5	2	3.5
Land owned (ha)	2.4	3.6	2.7
In percentage			
Men	100	87	100
Farming with irrigation	50	40	50
Primary occupation category			
Agriculture	84	95	89
Trade	2	3	5
Other[a]	14	2	6

[a]Other category includes teachers, government workers and factory workers.

All seed dealers purchase their seeds from seed companies or a dealer who represents a particular seed company at the district/state level. Depending on the volume of their business operations, they work directly either with a seed firm or through a seed distributor at the district level on a commission basis. The distributor handles the product on a wholesale basis. Generally the commission ranges from 10% to 12% of the distributor margin,[3] exclusive of their marketing cost. Dealers sell all kinds of proprietary hybrids and varieties (released by private firms) and in some cases, on demand, public varieties. Since the profit margin (15–30%) is much higher for improved types, dealers typically prefer to sell these. In some cases, dealers do sell truthfully labelled seed (TFL) materials procured from a well-known seed farmer or farm to cater to the local demand. This is more prevalent in the case of finger millet and sorghum cultivars grown in the post-rainy season, such as *Maldandi* (Table 13.8).

The flow of seeds and grains through shandies is thin but the product turnover high, especially before planting. These serve as 'exchange markets' where farmers – especially women – bring their produce and transact in order to meet immediate cash needs. Grain and seed cannot be differentiated, and specific varieties are difficult to recognize, although most of the millet grain is from local villages (in and around 10–15 km) and some distinct characters are distinguishable (Table 13.9).

Indicators of the extent to which villages and communities (*panchayats*) are autarkic in their supply and demand for replacement seed are shown in Table 13.9 by crop, improvement status and seasons. The three categories considered are within the village (a radius of less than 5 km), within the community (from 6- to 25-km radius), and outside the community (more than 25 km). Virtually no replacements occur at more than a 25-km radius. Most seed is traded within a 25-km radius of the farm household, in either the rainy or the post-rainy season. Within-village exchange is dominated by sorghum and finger millet, although in the rainy season many of these seed lots include sorghum hybrids. In the post-rainy season, more than 80% of all seed exchange within the village and 70% within the community is for FCs of sorghum. The single cultivar, *Maldandi*, represents roughly 30% of these exchanges, among the 30 or so cultivars exchanged at very low frequencies. The figures suggest strongly that the locus of the seed trade among the farmers surveyed is at the community level. This trade is autarkic, meaning that it is dominated by local FCs (of sorghum) in the post-rainy season, without introductions. In the rainy season, about 70% of all transactions were for improved seed.

---

[3] Distributor margin=((Farmer purchase price–distributor sales price)/farmer purchase price).

Table 13.8. Market profile for dealers of seed of modern varieties.

	Dharwad	Bellary	Belgaum	Chitradurga	Bijapur	Mahbubnagar	All districts
Total							
Total dealers	6	5	4	3	5	6	29
Crops dealt	2	2	2	2	2	2	3
Selling OPV seed	2	3	2	3	2	3	15
Selling hybrid seed	5	4	5	3	5	3	25
Clients	300	350	250	300	300	350	2150
Mean							
Distance covered	25	40	25	50	40	45	37.5
3-year average volume sold of sorghum seed (mt)	3.9	3.95	3.4	2.2	1.2	3.2	3.0
3-year average volume sold of pearl millet seed (mt)	1.15	1.8	1.0	0.6	4.4	0.6	1.6
3-year average volume sold of finger millet seed (mt)	0	0.7	0	2	0	1.7	0.8

Econometric Analysis

As in most of the chapters of this book, spatial indices adapted from the ecological literature (Margalef, Berger–Parker, Shannon) were used as the dependent variables in a regression model explaining the determinants of millet crop biodiversity in villages (see Chapter 1 for additional explanation). The conceptual framework for analysing the biodiversity of millet crops at the level of the *panchayat* is the household model of on-farm diversity (Chapter 5), measured at a higher level of aggregation (Chapter 11). Hypothesized determinants of millet biodiversity include market infrastructure, household and farm characteristics. Variable definitions and hypothesized effects are listed in Table 13.11. The number of explanatory variables is necessarily few, given the relatively small number of *panchayats* (58) in the sample.

Market infrastructure is measured as the total kilometres of paved roads in the community, drawn from secondary data. Household and farm variables are measured by calculating ratios and totals over sample data collected from households, by *panchayat*. Household characteristics include adult literacy rates, the gender structure of the adult population involved in farming, the extent of off-farm employment, asset and income levels. The farm characteristic that is hypothesized to affect millet biodiversity most is the proportion of rain-fed land in the total area cultivated. The significance of district-level fixed effects is tested in each regression.

Two seed system variables that could not be measured in the analysis presented in Chapter 11 and are the focus of the analysis in this chapter are included in these regressions: the seed replacement ratio and the quantities of millet seed traded through dealers or shandies, independent of seed type or identity. Both are interpreted as indicators of the 'strength' of the local market, registering both the local demand and the supply of seed, whether formal or informal. In the cases of sorghum and pearl millet, the amount traded through dealers is composed primarily of modern varieties. On some occasions, dealers do sell TFL procured from a well-known seed farmer or farm to cater to the local demand. This is more prevalent in the case of finger millet and sorghum cultivars grown in the post-rainy season, however. Only one dealer surveyed in Chitradurga sold finger millet seed. Most seed of minor millets is traded in shandies. The total seed quantities traded through market nodes could be related to greater farmer specialization,

Table 13.9. Profile of shandy traders.

	Dharwad	Bellary	Belgaum	Chitradurga	Bijapur	Mahbubnagar
Number of shandies	4	5	4	3	3	6
Number of millet crops dealt	3	4	3	3	2	4
Number of varieties handled	8	10	6	8	5	11
Mean						
Number of traders per shandy	6	8	6	5	5	6
Volume of sales/trader/transaction in peak season (kg)	10–12	15–20	10	15–20	5	15–20
Volume of sales/trader/transaction in lean season (kg)	3–5	2–5	5	5	5	5
Estimated total						
Volume of sales, peak season (mt)	0.160–0.200	0.240–0.320	0.160–0.200	0.250–0.320	0.150–0.160	0.240–0.320
Volume of sales, lean season (mt)	0.095–0.165	0.065–0.160	0.160–0.170	0.150–0.160	0.150–0.160	0.150–0.160

The peak seasons are before-rainy planting (May–June) and post-rainy (December–January) and the lean season is the remainder of the year. Traders confirm most grain or seed sold in shandies is of farmers' cultivars.

Table 13.10. Autarky and trade in millet seed replacements.

Percentage of past seed replacements for cultivars planted in survey season

Historical transactions	Pearl millet				Sorghum				Finger millet				Little millet	Foxtail millet	All	
	Total	Hybrid	IOPV	FV	Total	Hybrid	IOPV	FC	Total	IPLS	FC				Per cent	n
Rainy season																
Within village (0–5 km)	17.8	7.4	3.3	7.0	42.6	16.9	4.6	21.1	28.1	17.8	10.3		7.4	4.1	100	242
Within community (6–25 km)	23.8	15.4	7.0	1.4	48.6	32.9	3.5	12.3	23.2	16.2	7.0		1.2	3.1	100	487
Outside community (> 25 km)	85.7	28.6	57.1	0.0	0.0	0.0	0.0	0.0	14.3	14.3	0.0		0.0	0.0	100	7
Post-rainy season																
Within village (0–5 km)	0.0	0.0	0.0	0.0	90.9	1.4	7.8	81.7	8.5	4.9	3.5		0.0	0.7	100	142
Within community (6–25 km)	0.0	0.0	0.0	0.0	88.7	5.7	14.2	68.9	11.4	7.6	3.8		0.0	0.0	100	212
Outside community (> 25 km)	0.0	0.0	0.0	0.0	100.0	0.0	0.0	100.0	0.0	0.0	0.0		0.0	0.0	100	1

Table 13.11. Definition of explanatory variables.

Variable	Definition	Hypothesized effects
Household characteristics		
Per cent adult literacy	Proportion of adult literates in the community	(+,−)
Men as proportion of all adults in farming	Ratio of adult men to total adults engaged in farming in the community	(+,−)
Total value of livestock	Value of the livestock units owned in the community in 2002 (in Indian rupees)	(+,−)
Total land value	Value of both irrigated and rainfed lands in the community in 2002 (in Indian rupees)	(+,−)
Total expenditures (income)	Cash spent in the community during past season (in Indian rupees)	(−)
Farm characteristics		
Estimated total rainfed area	Proportion of rain-fed area in total area cultivated in the community	(−) (+)
Markets		
Kilometres of paved road	Length of structured (all weather) road in community	(−)
Labour market	Number of months worked off-farm by all adults in the community	(+,−)
Seed replacement ratio	Number of times the seed of a cultivar planted in the survey season has been replaced divided by the years grown, averaged over all varieties in the community	(+,−)
Seed quantities traded	Three-year average amount (kg) of millet seed sold by the dealers sampled in the market closest to the community or traded in shandies during peak season (depending on the millet crop)	(+,−)
District		
Location in Bijapur	Dummy variable = 1 if community located in Bijapur, else 0	(+,−)
Location in Bellary	Dummy variable = 1 if community located in Bellary, else 0	(+,−)
Location in Chitradurga	Dummy variable = 1 if community located in Chitradurga, else 0	(+,−)
Location in Dharwar	Dummy variable = 1 if community located in Dharwar, else 0	(+,−)
Location in Mahabubnagar	Dummy variable = 1 if community located in Mahabubnagar, else 0	(+,−)

in particular modern varieties; on the other hand, seed trade could contribute to the overall richness and evenness of seed types grown.

For each millet crop, the seed replacement ratio was calculated as the historical frequency of seed lot replacement divided by the number of years the cultivar was grown, averaged over all cultivars planted by farmers surveyed in the *panchayat*. Data from the main rainy season only were used because of the higher number of crops and varieties grown. The ratio is a proxy for the temporal rate of seed turnover in a community or the velocity of seed flows in past planting periods. The seed replacement ratio was not correlated with the amount of millet seed traded. The amount of millet seed traded was measured as a 3-year average.

Determinants of variation in *panchayat*-level millet diversity are shown in Table 13.12, for sorghum, pearl millet and minor millets. Results for household characteristics are mixed. As in the household-level findings for the highlands of Ethiopia, the greater the proportion of women involved in farming in a given community, the greater is the diversity of sorghum and pearl millet varieties. In contrast with many of the household-level findings reported in Parts III and IV, communities that are wealthier in livestock assets

Table 13.12. Determinants of community-level variation in intracrop millet diversity.

	Marginal effects								
	Sorghum			Minor millets			Pearl millet		
Explanatory variables	Shannon	Margalef	Berger–Parker	Shannon	Margalef	Berger–Parker	Shannon	Margalef	Berger–Parker
Household characteristics									
Per cent adult literacy	−0.0001	0.0043	0.7063**	0.0059		0.2161		0.0046	−0.4529*
Total value of livestock	−2.8E−07*	−1.5E−07	−2E−05*	−2.0E−07**		−2.8E−07		−1.7E−07	5.9E−07
Total land value	1.2E−07	−7.1E−07*	7.6E−07	2.4E−07*		9.7E−06*		−1.6E−05**	−1.6E−05
Farm characteristics									
Estimated total rain-fed area	−0.0005	0.0154*	0.5447*	−0.0010		>−0.1248		0.0186*	−0.1670
Market characteristics									
Kilometres of paved road	−0.0187	−0.0134	−16.1622**	−0.0176		−0.7739		−0.1244	8.8309*
Labour market	−0.0014	−0.0212	1.0903*	0.0148*		0.2709**		0.0087	−0.3192
Seed replacement ratio	0.1827*		8.1395**	0.2719*		12.0978*		0.1399**	17.7390*
Seed quantities traded	0.0001		−0.0067	0.0001**		0.0038**		0.0008*	0.0080*
District fixed effects									
Location in Bijapur		−8.1506							
Location in Bellary	0.1097	−2.0223*	10.9286*	1.0223*		18.8026*		1.4359*	7.1979
Location in Chitradurga	−0.7316*	−4.1228*	−42.7790*	0.6508*		6.9709		0.0391	13.2278
Location in Dharwar	0.8107*	−0.4016	182.7133**	0.5711**		8.4634		−4.2290	−142.5212
Location in Mahabubnagar	0.4054**	−1.9933*	−36.6735	1.1099*		23.5519*		−3.0541	−71.7447
								−1.3923	19.8935
	(Test statistics for log-likelihood ratio tests of joint hypotheses)								
Equation	93.05	60.1	101.99	48.94		84.69		73.61	106.57
District fixed effects λ (5, 0.05)	24.69	24.38	28.35	22.47		13.47		25.55	45.21
Seed system factors λ (2, 0.05)	10.89	5.32	12.04	10.3		18.75		9.81	53.14

Note: $n = 58$. Tobit regressions. Marginal effects are partial derivatives of expected value, computed at the means of variables; (*) 5% significance (**) 10% significance (for number of tails, see Table 13.11). In Bijapur, pearl millet is primarily grown.

generally have less evenness in sorghum or minor millet varieties. Some varieties may be more suitable for feed or fodder. Communities with lower land values, or the less wealthy in land assets, have more richness in sorghum and pearl millet varieties, perhaps because farmers seek to match varieties to soil types and conditions. The only positive association between wealth and crop biodiversity is for minor millets. Possibly, different minor millet cultivars are distributed in small pockets among the households in communities. Communities with lower average cash income levels also have more diverse minor millets. Higher cash income levels are associated with less dominance of the most popular pearl millet variety.

Physical factors and location of the farm have strong and consistent effects on millet biodiversity in this dry, challenging production environment. Larger rain-fed areas, as hypothesized, imply greater richness in pearl millet and sorghum varieties and less dominance by any single variety, although there is no effect among varieties of minor millets. District fixed effects are both jointly and individually significant across diversity indices for sorghum and minor millets. The effect of location in Bijapur is evident only for pearl millet, the dominant millet crop in that district.

Different types of market-related variables have contrasting effects on intracrop diversity of millets in these *panchayats*. Road density within the community lessens the dominance of the most widely grown variety of pearl millet, suggesting that market infrastructure enables competing modern types to be introduced. The opposite is the case for sorghum varieties, and by a very large magnitude (findings refer to the rainy season). An active off-farm labour market is positively associated with less dominance and more evenness of sorghum and minor millets grown in the community.

Seed system parameters, and in particular the seed replacement ratio, are related significantly to the level of intracrop diversity in almost all regressions. Historical rates of seed replacement in a community are positively correlated with the spatial richness and equitability among varieties of both major and minor millets grown in the rainy season. Larger average quantities traded through shandies enhance the diversity of minor millet varieties. Dealer trade contributes to the richness of pearl millet varieties and reduces the dominance of the most widely grown variety, but has no perceptible effect on sorghum diversity.

Conclusions

One indicator of the value farmers attach to biologically diverse crops is a revealed preference for growing them, in the presence of economic and social change. The villages studied in this research are found in one such location. Farmers in these villages grow a range of millet crop and cultivar combinations in a harsh, semiarid environment. Millet biodiversity is distributed across and within crops. Pearl millet, sorghum, finger millet, foxtail and little millet have varying rates of outcrossing. In some cases they occupy unique environmental niches and in others they compete for the same environmental niche.

Two methodological questions motivated this research. First, seed systems convey incentives for farmers to grow one or another crop variety, or a set of them as compared with only one. Too often, seed systems for improved materials and FCs have been treated as disjoint and addressed from the separate vantage points of economics and anthropology. A more comprehensive definition of seed system is necessary to advance the understanding of how these systems can deliver incentives to support farmer management of crop biodiversity.

Second, as evidenced by the chapters in this book, most applied economics research about managing biodiversity during economic change has so far been conducted using the household as the unit of observation. The dynamics of seed systems and the population genetics of crops and cultivars must be observed at a higher level of observation and analysis than the individual farm, although they must be grounded in the microeconomics of farmer decision making. This research represents an initial exploration into the relationship between seed system parameters and the levels of crop biodiversity maintained by communities, while controlling for other determinants identified in past studies. Two seed system parameters are proposed and measured at the level of the *panchayat* (literally, 'village community'). One is the seed replacement ratio, measuring the average historical rate of seed replacement for all varieties grown by farmers, through both formal and informal seed supply channels in a community. The other is the 3-year average seed quantity traded through formal and informal channels of seed supply, an indicator of market size. Both are an expression of the strength of the local seed market, including both supply and demand.

The descriptive data reveal that in the 60 communities studied, farmers grow a total of 63 distinct varieties of five millet crops, including hybrids, IOPVs, IPLS and FCs. About half of the 432 farmers in these communities grow some combination of both major and minor millets, and many combinations are observed. There is evidence of seed saving for pearl millet and for sorghum hybrids as well as for other materials and of farmers not necessarily recognizing the pure-line selections of finger millet as 'improved'.

Seed is replaced for the same cultivars and for the purposes of variety change. Replacement and transfers of seed are one measure of diversity in time, which can be important for buffering yields against biotic stresses (although these are few in this dry environment) and protecting against genetic deterioration. The age of varieties is negatively related to and the rates of seed replacement positively related to improvement status and whether the millet crop is major or minor; however, the higher the number of seed transfers, the less improved is the material, and the number is highest for the minor millets.

Surprisingly, most seed transactions (original, replacement, transfer) appear to be based on money, even when they are described as 'gifts', and occur between family and friends for 'token money'. Even for FCs, a larger proportion of transactions are between individuals that are neither family nor friend.

Seed supply channels differ by improvement status, although *all* categories of millet genetic resources (by crop and improvement status) exchange hands at the level of the village trader and shandy. Although distinguishing characteristics are observable for some of the materials traded in shandies, in general it is not possible to trace these to particular varieties. Varieties can be detected in seed transfers only when they are branded materials sold by designated dealers or, to a lesser extent, by local seed experts. Lack of transparency and credibility problems are well-known problems associated with seed markets (Morris et al., 1998; Tripp, 2001). Knowledge is asymmetric, held by few and not without cost to obtain. Implications for farmer-owned brands or proprietary rights as incentives for conserving millet biodiversity are therefore unclear.

Implications

This study demonstrates that seed system factors, defined in the broadest sense, are significant determinants of crop biodiversity levels on farms. The velocity of seed flows, as indicated by the average seed replacement ratio in a community, is positively correlated with the spatial richness and relative abundance of varieties of major and minor millets. Larger seed quantities traded through informal community markets (shandies) are associated with greater diversity in minor millet varieties. Although dealers sell primarily the seed of modern varieties, larger quantities traded positively influence the richness of pearl millet varieties and do not contribute to the dominance of any single pearl millet or sorghum variety grown during the rainy season. In other words, in this setting, the strength of the seed market does not appear to induce variety specialization. Further research should consider the role of seed systems in the sustainable management of crop biodiversity on farms through a more fully developed analytical framework.

Acknowledgements

This chapter was developed from part of the PhD research undertaken by the first author at University of Minnesota, USA, based on the research collaboration of IFPRI, FAO and the International Crops Research Institute for Semi-Arid Tropics (ICRISAT), Hyderabad, India. We gratefully acknowledge financial support from the European Union and the Department of Applied and Agricultural Economics, University of Minnesota. The authors thank Eric Van Dusen, Phil Pardey, Leslie Lipper, Richard Jones, Dr A. Seetharam, Dr Prem Mathur and scientists at the Gene Bank Facility at ICRISAT, India, for useful comments and suggestions.

References

Aguirre Gómez, J.A. (1999) Análisis regional de la diversidad del maíz en el sureste de Guanajuato. (A regional analysis of maize diversity in southeast Guanajuato). PhD thesis. Universidad Nacional Autónoma de México, Facultad de Ciencias, Mexico City.

Almekinders, C.J.M., Louwaars, N.P. and de Bruijn, G.H. (1994) Local seed systems and their importance for an improved seed supply in developing countries. *Euphytica* 78, 207–216.

Apple, J.L. (1977) The theory of disease management. In: Horsfall, J.G. (ed.) *Plant Disease: An Advanced Treatise*. Academic Press, New York.

Benin, S., Gebremedhin, B., Smale, M., Pender, J. and Ehui, S. (2003) Determinants of cereal diversity in communities and on household farms of the northern Ethiopian highlands. Agriculture and Development Economics Division (ESA) Working Paper 03–14. FAO, Rome. Available at: http://www.fao.org/es/ESA/wp/ESAWP03_14.pdf

Berthaud, J., Pressoir, G., Ramirez, C.F., Bellon, M.R. (2002) Farmers' management of maize landrace diversity. A case study in Oaxaca and beyond. Pages 79–88. Proceedings of the 7th International Symposium on the Biosafety of Genetically Modified Organisms, Beijing, 10–16 October, 2002.

Brown, G.M. (1991) Valuation of genetic resources. In: Orians, G.H., Brown, G.M. Jr, Kunin, W.E. and Swierbinski, J.E. (eds) *The Preservation and Valuation of Biological Resources*. University of Washington Press, Seattle, Washington.

Cleveland, D.A. and Soleri, D. (eds) (2002) *Farmers, Scientists, and Plant Breeding: Integrating Knowledge and Practice*. CAB International, Wallingford, UK.

Evenson, R.E. and Gollin, D. (2003) Review: assessing the impact of the green revolution, 1960 to 2000. *Science* 300(May), 758–762.

Gopal Reddy, V. and Sharma, V.D. (1996) Rabi sorghum germplasm collection in Maharashtra and adjoining areas of Karnataka. Genetic Resources Progress Report-85, A Joint Collection Mission by ICRISAT Asia Center and NBPGR (ICAR). January–February 1996, ICRISAT, Patancheru, India.

Gopal Reddy, V., Mathur, P.N., Prasada Rao, K.E. and Mengesha, M.H. (1993) Rabi sorghum germplasm collection in northern Karnataka and adjoining areas of Andhra Pradesh. Genetic Resources Report-74, ICRISAT, Patancheru, India.

Heisey, P.W. and Brennan, J.P. (1991) An analytical model of farmers' demand for replacement seed. *American Journal of Agricultural Economics* 73(4), 1044–1052.

International Crops Research Institute for the Semi-Arid Tropics (ICRISAT) (1980, 1985) Rabi sorghum germplasm collection in northern Karnataka and adjoining areas of Andhra Pradesh, Genetic Resources Progress Reports-No. 74 and 85, ICRISAT, Patancheru, India.

International Crops Research Institute for the Semi-arid Tropics (ICRISAT) (1985) Descriptors for finger millet: AGPG: IBPGR/86/106 June. ICRISAT, Patancheru, India.

International Crops Research Institute for the Semi-Arid Tropics (ICRISAT). (1989) Kharif sorghum germplasm collections in Karnataka and adjoining areas, Genetic Resources Progress Report-No. 29. ICRISAT, Patancheru, India.

Krutilla, J.V. (1967) Conservation reconsidered. *American Economic Review* 57(3), 777–786.

Li, Y. and Yu, S. (1996) Traditional maintenance and multiplication of foxtail millet. *Euphytica* 87, 38–48.

Louette, D. (1994) *Gestion Traditionnelle de Variétés de Maïs dans la Réserve de la Biosphère Sierra de Manantlán (RBSM, états de Jalisco et Colima, Mexique) et Conservation In Situ des Ressources Génétiques de Plantes Cultivées*. Thèse de doctorat, Ecole Nationale Supérieure Agronomique de Montpellier, Montpellier, France.

Louette, D. and Smale, M. (2000) Farmers' seed selection practices and maize variety characteristics in a traditional Mexican community. *Euphytica* 113, 25–41.

Magurran, A. (1988) *Ecological Diversity and Its Measurement*. Princeton University Press, Princeton, New Jersey.

Morris, M.L., Rusike, J. and Smale, M. (1998) Maize seed industries: a conceptual framework. In: Morris, M. (ed.) *Maize Seed Industries in Developing Countries*. Lynne Rienner and CIMMYT, Boulder, Colorado and Mexico, DF.

Nagarajan, L. (2004) Managing millet diversity: farmer's choices, seed systems and genetic resource policy in India. PhD thesis. Department of Applied Economics, University of Minnesota, Minneapolis, Minnesota.

National Bureau of Plant Genetic Resources (NBPGR) (1985) Descriptors for small millets in India. National Bureau of Plant Genetic Resources (NBPGR). New Delhi.

Prasada Rao, K.E. and Gopal Reddy, V. (1980) Kharif sorghum germplasm collection in Karnataka and adjoining areas. Genetic Resources Progress Report 29. ICRISAT, Patancheru, India.

Seetharam, A., Riley, K.W. and Harinarayana, G. (1989) *Small Millets in Global Agriculture*. Oxford and IBH, New Delhi.

Sperling, L. and Loevinsohn, M.E. (1993) The dynamics of adoption: distribution and mortality of bean varieties among small farmers in Rwanda. *Agricultural Systems* 41(4), 441–453.

Sperling, L., Scheidegger, U. and Buruchara, R. (1996) Designing seed systems with small farmers: principles derived from bean research in the Great Lakes region of Africa. Agricultural Research and Extension Network Paper No. 60. Overseas Development Agency, London.

Thiele, G. (1999) Informal potato seed systems in the Andes: why are they important and what should we do with them? *World Development* 27(1), 83–99.

Tripp, R. (2001) Seed Provision and Agricultural Development: The Institutions of Rural Change. Overseas Development Institute, London.

vom Brocke, K. (2001) Effects of farmers' seed management on performance, adaptation, and genetic diversity of pearl millet (*Pennisettum glaucum* [L.] R.Br.) populations in Rajasthan, India. PhD dissertation. University of Hohenheim, Germany.

Wood, D. and Lenné, J. (1999) Why agro biodiversity? In: Wood, D. and Lenné, J. (eds) *Agro Biodiversity: Characterization, Utilization, and Management*. CAB International, Wallingford, UK.

Zimmerer, K.S. (2003) Geographies of seed networks for food plants (potato, Ulluco) and approaches to agro biodiversity conservation in the Andean countries. *Society & Natural Resources* 16(3), 583–601.

14 Seed Supply and the On-farm Demand for Diversity: A Case Study from Eastern Ethiopia

L. Lipper, R. Cavatassi and P. Winters

Abstract

This chapter uses a household model similar to those used in Part III to analyse the relationship between seed supply and intercrop diversity (between crop species) on household farms, based on an assessment of the impact of a seed supply intervention at the household level in Ethiopia. A detailed description of the seed system for two major crops (sorghum and wheat) is used to support and explain the results. Econometric analysis shows that participation in a programme implemented by a non-governmental organization (the Hararghe Catholic Secretariat (HCS)) was positively related to one index of intercrop diversity and did not negatively affect other indices. Other information about the characteristics of the wheat seed system, particularly in relation to the sorghum seed system, reinforces this finding. Key characteristics are the relative availability of local versus improved genetic materials and the types of services they provide to farmers, given farmers' demand for genetic services. Farmers' demand for genetic services in turn reflects their specific consumption and production situations. It is argued that the nature of the seed supply intervention, such as the crop selected and the type of intervention, is an important determinant of programme impact on intercrop diversity.

Introduction

The International Treaty on Plant Genetic Resources (ITPGR) and the Convention on Biological Diversity (CBD) require signatories to adopt policies that will promote the sustainable utilization of plant genetic resources and its associated diversity. Sustainable utilization incorporates both environmental and development concerns. It involves improving the accessibility to, and productivity of, genetic resources, as well as the conservation of socially valuable genetic diversity. However, there is considerable uncertainty on how to attain sustainable utilization in practice. That is, there is insufficient understanding of how government interventions influence sustainable utilization and how interventions can be used to promote this objective. What is clear is that the system of seed supply, which includes traditional, modern and genetically modified varieties of crops, will certainly be a major determinant of sustainable utilization. In developing countries in particular, the local system of seed supply will have a major impact on the pattern of crop genetic resource utilization. It is towards an improved understanding of seed supply and how this relates to sustainable utilization that this chapter is concerned.

While seeds represent the vehicle and repository of crop genetic resources, they are also a crucial input in the production of crops for farmers. Thus seed utilization patterns generate both public and private goods (Chapter 1). The performance of a particular crop and variety depends on the genetic content of the material, farmer management and local environmental conditions. Through selection of crops and varieties, farmers satisfy their own needs and preferences, given a

© CAB International 2006. *Valuing Crop Biodiversity: On-farm Genetic Resources and Economic Change* (ed. M. Smale)

set of constraints. The set of crops and varieties farmers select for production results in a utilization pattern of crop genetic resources that may generate public goods by conserving important resources and their diversity, and by reducing vulnerability to pests and diseases. Utilization patterns that generate both public and private goods are also those that are most likely to be sustainable. Thus, identifying where these occur and the factors that drive them is critical to designing policies to promote sustainable utilization.

One of the most important constraints farmers face in choosing crops and varieties to plant is the availability and accessibility of seed. Availability refers to whether a sufficient quantity of seed of appropriate crops is present physically within reasonable proximity, and in time for planting. Accessibility refers to whether people have adequate information, income or other resources to acquire the seed that is available (Sperling and Cooper, 2003). Availability is similar to a supply constraint, accessibility to a demand constraint (Bellon, 2004). Seeds of desired crops and varieties might not exist if they have not been the focus of breeding programmes. Even if a crop or variety does exist farmers may face numerous barriers in accessing seeds of that variety. These barriers, including asymmetry of information, transport costs and other transaction costs, and uncertainty of crop and variety performance, will affect farmers' choice of seeds and therefore utilization patterns in a given region.

The seed system, which comprises all different channels through which farmers may access the crop genetic resources embodied in seeds, is then critical to the understanding of crop genetic resource utilization patterns. In this chapter, the relationship between seed supply and one measure of on-farm genetic diversity is analysed: the intercrop (between crop species) diversity that results from the diversification of crops chosen for planting (see Chapter 1 for definitions). The analysis is based on an assessment of the impact of a seed supply intervention on degrees of intercrop diversity at the household level, using empirical data from Ethiopia. A detailed description of the seed system for two key crops is used to support and explain the results obtained in the statistical analysis of the determinants of intercrop diversity.

The chapter is organized as follows. After a discussion on the motivation for this study, a discussion of how seed supply is incorporated in a household model of on-farm demand for crop species diversity is presented. The framework is situated within the economic literature on household-level determinants of diversity conservation (chapters presented in Part III). The empirical setting in eastern Ethiopia is then described. Statistics describing the seed system and seed supply interventions are presented. Next, a series of econometric models identifying the determinants of intercrop diversity on household farms and testing the impacts of a seed supply intervention are presented. Conclusions are summarized in the final section.

Motivation

An empirical study was designed to test the relationship between seed systems and crop utilization patterns in the eastern part of Ethiopia. Ethiopia was selected for several reasons: (i) it is a centre of origin and diversity for several agricultural crops; (ii) the population is highly dependent on low-productivity agriculture and food insecurity rates are high; and (iii) several studies on seed systems have been done in the country providing a rich base of data to work from (McGuire, 1999; Mulatu, 2000; Worede et al., 2000; Teshome, 2001). Maintaining crop diversity has been found to be a strategy adopted by farmers in order to exploit the highly heterogeneous agroecological conditions, as well as to efficiently utilize other factors of production such as labour and animal power (Chapters 5 and 11, Worede et al., 2000).

This study site is located in the Hararghe zone, an area in the eastern part of Ethiopia that has been a repeated recipient of both food and seed emergency relief supplies because of chronic food deficits and problems of seed insecurity. The main criterion used in selecting the site was the presence of readily identifiable variation in seed systems. A seed supply intervention conducted by a non-governmental organization, the HCS, provided such a variation. HCS has been active in the Hararghe region since the early 1990s with a range of seed system interventions, including seed selection, multiplication and distribution for both landrace and improved varieties of wheat, sorghum and haricot bean.

A major part of the HCS seed programme involved the distribution of wheat, sorghum and haricot bean seed within a selected set of communities, and to a selected set of participant households. This programme involved the provision of improved varieties of wheat and haricot bean and selected landrace varieties of sorghum seeds to households. Seeds were provided under a credit arrangement that required repayment in the form of seed with a 15% interest charge. In addition, agricultural tools were provided as part of the intervention package. Participation in this programme in 2000 was used as a basis for selecting the study sample as discussed in further detail in the section on sampling.

Sorghum and wheat seed systems were selected for detailed analysis, due to their importance to food security, to the conservation of crop genetic diversity, and in the HCS programme. The two crops provide some interesting contrasts: sorghum is a long-season crop, grown mainly for subsistence purposes, while wheat is more likely to be marketed and has a short growing season (Mulatu, 2000). Sorghum is the most important crop in the drought-prone areas of Hararghe. Hararghe is considered a primary centre of origin for sorghum and most varieties planted in the region are landraces, although formal sector breeding has been undertaken for almost 25 years (McGuire, 1999). Having a long tradition with sorghum, farmers have developed good storage systems and technology to save seed, which is not the case for wheat seed. Sorghum is a multi-purpose crop used for many different applications (food, housing materials, livestock feed, etc.) and according to local experts landraces are preferred to modern varieties because modern varieties generally provide one rather than several traits.

In Ethiopia, wheat is grown primarily by subsistence farmers under rain-fed conditions, with a growing segment of modern variety adopters, particularly in high-productivity zones near Addis Ababa. Ethiopia is a centre of origin for durum wheat, and much of the production in the country relies upon durum landraces. However, most of the modern varieties released from the formal system in Ethiopia are bread wheat (Beyene *et al.*, 1998; Mulatu, 2000). The Hararghe region of Ethiopia is neither a centre of origin for durum wheat nor a major wheat production area of the country, although wheat is an important crop in terms of area planted. Since wheat is a short-season crop it is planted to capture the benefits of early rains or as a relay crop in sorghum plots to exploit residual moisture or late rains (Mulatu, 2000). Most of the wheat varieties planted in the Hararghe region are improved varieties introduced through the extension system and in most cases were substituted for other short-season grain crops such as barley (Dr T. Tesema, Ethiopian Agricultural Research Organization, Ethiopia, personal communication).

In his recent study, Mulatu (2000) provides data and historical information on the formal sector breeding policies for sorghum and wheat as well as for the quantity of modern varieties distributed over recent years. He notes that wheat has been bred in Ethiopia for some time, with durum wheat breeding based largely on local materials, while bread wheat has used imported lines and there has been greater emphasis on release and distribution in the latter category. The primary objectives of Ethiopian wheat breeding programmes have been to promote both high yield and high disease resistance. A striga (wheat disease)-resistant variety of wheat has been introduced from the USA in collaboration with Purdue University, but it was rejected by farmers, primarily because it did not provide traits that the farmers found desirable (Zegeye *et al.*, 2001).

Multiplication and distribution of improved wheat varieties has been considerable in Ethiopia, in comparison with sorghum. In the late 1990s, the High Input Extension Package (HIEP) programme was initiated, which boosted the distribution of improved seeds throughout the country. The programme focused on seeds of improved varieties of wheat, maize and teff. Under this programme farmers had to satisfy a set of criteria on land, labour and animal power availability, as well as the capacity to pay 25% of the costs of the input package (including fertilizer, seeds and pesticides) in order to participate. The HIEP programme was active in the Hararghe region with wheat distribution and a survey of households in the area from 1998 to 1999 revealed that over 40% of the wheat producers had obtained their wheat seeds from the programme (Mulatu, 2000).

In the case of sorghum there has been much less formal sector breeding activity and seed dissemination. The breeding programme has focused on yield characteristics, ignoring key characteristics desirable to farmers such as disease resistance. Less attention has been given to multiplication and distribution of modern sorghum varieties, and this is likely to be one reason for limited use of improved sorghum seed. Formal sector sorghum breeding in Ethiopia is mostly based on pure-line selection of local materials. Breeders think it is unlikely that an introduced variety will replace local varieties, due to the multiple attributes farmers demand from sorghum, which local varieties are more likely to provide (Dr A. Ketema, Alemaya University, Ethiopia, personal communication).

This chapter investigates the impact of the HCS seed supply intervention on intercrop diversity at the household level. The primary question of interest is whether participation in the HCS seed programme encouraged farmers to specialize or diversify in terms of the portfolio of crops they planted. The main assumption of the study is that the impact of a seed supply intervention on intercrop diversity will depend on: (i) the type of intervention (increasing the crop or variety choice set or reducing costs of access); (ii) the features of the local seed system where the intervention is taking place (the set of crop genetic resources available to the farmer from both the formal and the informal seed system and the attributes they embody); and (iii) farmers' demand for a set of services from their crop genetic resources derived from a farm-level-constrained utility maximization.

The working hypothesis of the analysis presented here is that these features of the seed system will filter the impact of a supply side intervention and determine the direction of the relationship between seed supply intervention and diversity. These hypotheses are tested through the use of descriptive statistics and econometric analysis using data from the case study as well as from secondary data sources.

Conceptual Approach

Previous literature (Chapter 1), and particularly the chapters in Part III of this book, has emphasized farmers' motivations for growing diverse crops and crop varieties, building largely on the model of the agricultural household and models of variety choice in the technology adoption literature. Other applied economics research analysed optimal rates for seed replacement (Heisey and Brennan, 1991), seed supply and demand in farming systems dominated by modern varieties (Morris, 1998), and their effects on regional-level diversity (Morris and Heisey, 1998). Applied research in other disciplines investigated the genetic diversity effects of farmer exchange of seed lots and mixtures in traditional systems (Sperling and Loevinsohn, 1993; Louette, 1994; Rice et al., 1998; Bellon and Risopoulos, 2001). A village-level constraint for planting material was introduced in the household model for banana diversity in Chapter 7, and maize diversity in Smale et al. (2001). Chapters 12 and 13 have tested the effects of seed system factors on crop biodiversity at the household and village levels, and Chapter 15 examines institutional aspects. However, the manner in which seed supply fits within the theoretical framework of the household farm model has not been fully articulated.

Variants of the household model of on-farm diversity have been presented in the chapters of Parts III and IV. The household is the basic unit of decision making and is both a consumer and a producer of goods in these approaches. The household chooses over a set of consumption goods, including crops that it may produce, to maximize utility subject to a number of constraints. Under the assumption of perfect markets, the household's consumption and production decisions are separable and thus can be treated in a recursive fashion where the household maximizes profit and then maximizes consumption over consumption goods. However, in the presence of market imperfections, such as missing markets or markets with high transactions costs, the household's consumption and production decisions are non-separable, making the household decision dependent on production consideration and vice versa.

Findings reported in Parts III and IV generally support the working hypothesis that household production and consumption decisions are non-separable for agricultural households in the empirical contexts studied. They signal the importance of agroecological factors, market integration and labour market imperfections as determinants of on-farm crop diversity. Results presented in Chapters 12 and 13 reveal associations between seed system factors and crop biodiversity measured at farm and village levels.

The household model of on-farm diversity recognizes that there may be missing markets for credit, labour, consumption goods and attributes of goods. These constraints alter household decision making, changing the derived demand for seed so that it is a function of both production traits and consumption attributes of goods. As noted, the functioning of the seed system can have a significant impact on the crops and varieties that are planted by households. In this chapter, as in Chapter 9, the effects of a specific seed market intervention on crop biodiversity are assessed.

Consider the standard non-separable household model where markets are missing for credit, labour and certain consumer goods but where the seed system functions as a normal 'perfect' market. In such a case, all crops and varieties would be available to farmers for a given price (without any additional transaction costs). In this model, the household consumption decision over consumer goods would only be influenced by the price of seeds that produce the goods.

Alternatively, consider a more complex seed system where farmers retain their own seeds, obtain seed from informal channels, such as through gifts, exchange or other institutional arrangement, or through formal channels, including through market transactions, as part of a free technological package from the government, or through non-governmental organizations. Under such conditions, seed availability for production depends on a number of factors such as participation in informal seed exchange networks or participation in government or the seed programmes of non-governmental organizations. The household decision may depend not only on the price of seed but also on access to networks and organizations. Crop choices will be constrained by the amount of seed available from the household's own sources and these other sources. The derived demand for a particular crop becomes a function of not just price, household characteristics, farm characteristics, labour and expenditure constraints but also a function of the characteristics of the seed system.

Correspondingly, crop diversity becomes a function of the characteristics of the seed system. Thus, in evaluating on-farm diversity the characteristics of the seed system should be explicitly incorporated. Failure to incorporate the seed system constraints in econometric analysis of on-farm diversity may lead to omitted variable bias in results when such constraints are binding. In the analysis presented, seed system variables are explicitly incorporated.

Data Design

The sampling strategy was designed to evaluate the effects of the HCS intervention and to minimize sources of variation not related to seed systems. The sample was limited to *woredas* (counties) where HCS had been active. Three *woredas* (Dawa, Chiro and Meta) were selected for sampling. Within these *woredas* three major agroecological zones defined by elevation (lowlands, midlands and highlands) are present. With varying rainfall, harvesting and planting dates, each zone has its own set of crops and varieties. The sample included peasant associations (PAs) only within the mid- and highland areas, which have similar agroecological zones and fairly uniform cropping patterns. In selecting the sample, the degree of market integration (high, medium, low) was considered to evaluate the importance of market access on diversity. Finally, PAs that participated with the HCS programme and those that did not were included in the sample. In the three *woredas*, a total of 30 PAs were selected: 15 PAs in which the HCS project had been implemented and 15 similar PAs in which HCS did not distribute seeds. A map of the survey area is shown in Fig. 14.1.

Within PAs, the household sample was divided into three groups: (i) households that participated in the HCS seed programme (HCS); (ii) households that did not participate, but lived within communities where the programme was implemented (non-HCS I); and (iii) households that did not participate and lived in communities where no programme was implemented (non-HCS II). The rationale for this design was to provide information about the direct as well as indirect impacts of the project.

The total number of households that had participated in the HCS seed programme up until, and including 2001, was used as the sampling frame for project participants. Approximately 24 households from each of the 15 HCS

Fig. 14.1. Map of the survey area.

PAs were randomly selected from a list of names of HCS participants. The sample represented about 5% of the households that participated in the HCS programme in 2001 (362 households of a total of 7257). The remainder of the total sample of households was equally divided between the two types of non-participant groups.

The principle governing the selection of non-participant households in the project sites (i.e. the control group) was to identify PAs and households as similar as possible to the HCS project areas and households. According to the HCS, the programme targeted farmers who were known to be good farmers and with good farming conditions (in terms of land owned, type of soils, etc.), but who had fallen into debt due to crop failures beyond their control. Within the communities that HCS selected for their project, the PA committee nominated candidates for project participation based on HCS criteria. Nominated households could refuse to participate, but households not selected by the PA committee could not participate.

Non-participants in project areas were selected for the sample with the assistance of the PA committees. PA committees were asked to identify farmers within the community that fit the criteria but who had not (yet) participated in the HCS project. Since the demand for project participation was greater than what the HCS could meet, there were ample numbers of households on the waiting list for HCS participation. This list was used as the non-HCS I sample frame.

To sample non-participants in non-project areas (non-HCS II), 15 PAs located in highland or midlands agroecological zones were identified. In these 15 PAs sorghum and wheat are the primary crops, and seed insecurity had been identified as a problem. Households within these areas were selected for inclusion in the PA sample frame through a process of consultation with PA committees.

A number of different survey instruments were used to collect data on various aspects of the sorghum and wheat systems. Household, community and market questionnaires were used for data about seed supply and use and the impacts of seed systems on farmer welfare. Focus group interviews were used to elicit farmer descriptors for sorghum and wheat agromorphological characteristics in order to classify varieties. This chapter is based primarily on the household and community data, with some reference to other data. Of the 720 households in the sample, data for 699 were complete enough for this analysis. The scope of the survey is the cropping season of 2002. The household survey instrument was implemented in two rounds. The first round was conducted towards the end of the Meher (main crop) planting season in August 2002. The second round was done after the harvest of the Meher crop in early 2003. In each of the 30 PAs surveyed, data on community characteristics were gathered through the use of a community-level survey instrument administered to key informants, usually PA leaders.

Descriptive Statistics

In this section, descriptive statistics of key variables are presented as part of the analysis of the impact of the HCS intervention on household-level intercrop diversity. Starting with a description of the key outcome of interest, e.g. the dependent variables in the econometric analysis, comparisons of crop utilization patterns and three measures of intercrop diversity among participants and non-participants are presented. Descriptive statistics on the explanatory variables or the characteristics of the surveyed households and communities likely to influence levels of intercrop diversity follows, again differentiating between HCS participants and non-participants. The section concludes with a set of descriptive statistics on the wheat and sorghum seed systems and the impact of the HCS intervention in these systems, which is used to interpret the regression results.

Crop utilization and intercrop diversity

Table 14.1 displays the area share planted by crop over the entire sample. Sorghum is the most popular crop in the area, occupying almost 40%

Table 14.1. Percentage area share by crop. (From FAO–Netherlands Partnership Programme: Seed System Impact on Household Welfare and Agricultural Biodiversity – Household Survey.)

Crop	Percentage area share
Sorghum	39
Maize	23
Wheat	12
Haricot bean	7
Chat	5
Barley	4
Vegetables	3
Faba bean	2
Teff	1
Other	4

of the total area under cultivation. Maize and wheat follow with 23% and 12%, respectively. Haricot bean is the fourth most planted crop and it is often intercropped with either sorghum or maize. Chat, which is a stimulant and mild narcotic as well as a profitable cash crop, is the fifth most important crop in terms of area. Figure 14.2 shows the distribution over the sample of the number of crops planted. The distribution appears like a log-normal distribution, with very few farmers growing more than four crops. The mean is about 2.7 crops, with a minimum of 1 and a maximum of 7 crops.

Among farmers that grow just one crop, the most commonly grown crop is sorghum (74%). Among those that grow two crops, the most frequent crop combination is sorghum and maize, followed by wheat and maize, sorghum and chat and wheat and sorghum. Similarly, for those that grow more than two crops, either sorghum and/or wheat is always chosen.

Three indices of spatial diversity are used to measure intercrop diversity at the household level, adapted from the ecological literature, as discussed in Chapter 1. The richness index is a count of the total number of crops that the household reports planting over the season of interest. The Shannon index expresses proportional abundance or evenness, accounting for the land shares allocated to each crop as well as to the number of crops. The index gives less weight to rare species than to common ones, but is more sensitive to differences in small degrees of relative abundance than the Simpson index, another

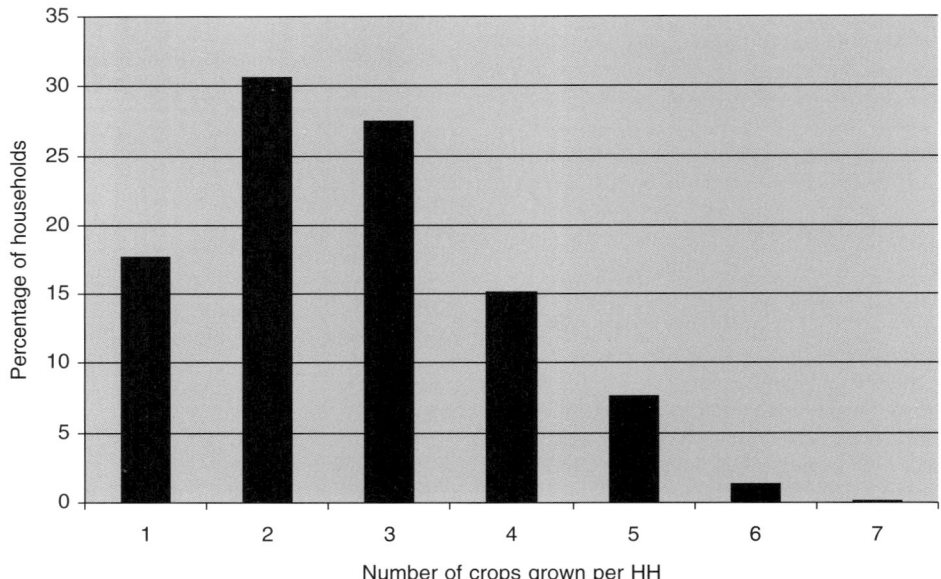

Fig. 14.2. Frequency distribution of households by number of crops grown. (Adapted from FAO–Netherlands Partnership Programme: Seed System Impact on Household Welfare and Agricultural Biodiversity – Household Survey.)

widely used evenness index measure of diversity (Magurran, 1988; Baumgartner, 2002). The Berger–Parker index of inverse dominance reflects the relative abundance of the most common species (Magurran, 1988; Baumgartner, 2002), or in the case of this study, the most widely grown on each household farm. The formulas for the three diversity indices are given in Chapter 1. In Table 14.2, the mean values of the three indices have been summarized for the entire sample and for each of the three sample subgroups.[1]

The data indicate that there are significant differences in the intercrop diversity measures

Table 14.2. Summary statistics for diversity indices. (From FAO–Netherlands Partnership Programme: Seed System Impact on Household Welfare and Agricultural Biodiversity – Household Survey.)

	Mean		
	Count	Shannon	Berger–Parker
Entire sample (N = 699)	2.73	0.79	1.92
HCS participants (N = 361)	2.94 (0.00)*	0.88 (0.00)*	2.03 (0.00)*
Non-HCS I (N = 161)	2.60 (0.18)	0.74 (0.15)	1.85 (0.21)
Non-HCS II (N = 177)	2.41 (0.00)*	0.66 (0.00)*	1.75 (0.00)*

Note: Values in parentheses indicate results of *t*-tests for mean values. Each group is compared with the whole sample. See Chapter 1 for index definition.

---

[1] The sample sub-groups are: (i) HCS participants, (ii) Non-participants located in participant communities (Non-HCS I), and (iii) Non-participants in non-participant communities (Non-HCS II).

among the three sample groups, with HCS participants showing the highest levels on all three. The count index is relatively easy to interpret – HCS participants seem to grow more crops than either of the non-HCS sample groups. Non-HCS I households also grow more crops than non-HCS II households. Additionally, HCS households have higher levels of evenness in the area shares planted to each of the crops under production. Non-HCS I households also rank higher than non-HCS II on this diversity measure. The same pattern is found with the Berger–Parker index. Overall, these results indicate that HCS participants are more likely to have a richer and more evenly distributed pattern of crops, as compared with non-HCS participants. Furthermore, non-HCS participants located in HCS communities are more likely to have richer and more even cropping patterns than non-participants in control communities. This suggests that households in HCS communities, both participants and non-participants, have some observable differences with households in non-HCS communities.

Characteristics of HCS participants and non-participants

Table 14.3 compares the social and economic characteristics of the three groups of sample farmers. The sample population consists of large-size families, with fairly young heads of households, low levels of education and high dependency ratios. HCS households are significantly larger and with higher education levels than households in either of the non-participant groups. Despite their larger size, HCS households have significantly lower labour/land ratios than non-HCS households, due to the greater number of plots and larger land area they operate.

Table 14.3. Demographic and socio-economic characteristics of sample farmers comparing HCS and non-HCS participants. (From FAO–Netherlands Partnership Programme: Seed System Impact on Household Welfare and Agricultural Biodiversity – Household Survey.)

	Mean values			
Variable	Non-HCS (N = 338)	HCS (N = 361)	Non-HCS I (N = 161)	Non-HCS II (N = 177)
Age of household head	39.66	39.59	40.18	39.51
Years of education	1.01	1.27(*)	1.01(*)	1.00
Family size	6.73	7.07(*)	6.75	6.71
Dependency ratio	0.5	0.52	0.52	0.47(*)
Labour/land ratio	1.30	1.11(*)	1.19	1.41(*)
Altitude	1977	2129(*)	2096	1867(*)
Distance to market	9.65	9.15	9.35	9.93
Total area operated (in timmad)	3.67	4.38(*)	4.24	3.14(*)
Number of plots owned	1.64	1.98(*)	1.80(*)	1.47(*)
Total value of agricultural assets (in birr)	85.44	95.26	81.30(*)	89.15
Total value of non-agricultural assets (in birr)	46.22	59.25	48.56	44.34
Number of oxen owned	0.4	0.42	0.44	0.36
Number of rooms	1.57	1.6	1.6	1.54
Off-farm income	0.55	0.49(*)	0.53	0.57
Slope	0.56	0.73(*)	0.58(*)	0.54
Soil	0.58	0.57	0.57	0.59
Irrigated	0.15	0.51(*)	0.1(*)	0.19
Organization	0.44	0.48	0.50	0.39(*)
Extension	0.76	0.86(*)	0.86	0.66(*)
Seed exchange	0.63	0.67	0.66	0.61

Note: (*) indicates significant difference in means calculated through t-tests. Each group is compared with the one on its left. Non-HCS group includes both non-HCS I and non-HCS II. 8 Timmad correspond to 1 ha; US$1 corresponds to about 8.6 birr.

No significant difference is found between the means of variables measuring other aspects of wealth, including the value of agricultural and non-agricultural assets, oxen ownership and number of rooms in the house.

An examination of the difference in means between the two non-participant groups indicates that the differences between the groups may be even greater at the PA than at the household level. HCS communities have significantly higher average size land holdings than do non-HCS communities, as well as higher average number of plots cultivated. In addition, the HCS communities have a significantly higher mean altitude.

To further explore differences between HCS and non-HCS households, the results of a probit regression on the probability of HCS participation are presented in Table 14.5 (variables defined in Table 14.4). The choice of explanatory variables for the probit follows from the earlier discussion of factors influencing diversity and confirms the findings of the descriptive statistics. Essentially, the results indicate that HCS participants tend to farm more extensive and better-quality lands. Their plots are flatter, with good soil quality and irrigated land. Overall, HCS participants have a significantly higher amount of land under operation. As expressed in the descriptive statistics, HCS participants have fewer livestock assets and they operate a significantly higher amount of land than non-participants. HCS communities tend to be located at higher elevations, and closer to markets and major cities than non-participant communities.

Taken together, the results indicate a number of observable differences between HCS and non-HCS households and between HCS communities and non-HCS communities that must be controlled for in the regression analysis of diversity presented below. Furthermore, it is important to carefully consider the effects of sample selection bias due to unobservable differences between HCS and non-HCS households and communities.

Sorghum and wheat seed systems and the HCS intervention

In this section, information about the sorghum and wheat seed systems and the HCS intervention is presented with the intention of providing

Table 14.4. Variable definitions. (From FAO–Netherlands Partnership Programme: Seed System Impact on Household Welfare and Agricultural Biodiversity – Household Survey.)

Variable name	Definition
Family size	Total number of family members
Education	Average years of education among household adults
Age	Age of household head
Off-farm income	Dummy = 1 if household has a source of off-farm income
Operated area	Area under operation in 2002 production year in timmad
Slope	Dummy = 1 if household has gently sloped or terraced plots, else 0
Soil	Dummy = 1 if household operated black soil, else 0
Irrigated	Dummy = 1 if household has irrigated plots of land, else 0
PA altitude	PA altitude in metres
Livestock assets	Value of livestock assets in birr
Agricultural assets	Value of agricultural assets in birr
Non-agricultural assets	Value of non-agricultural assets in birr
Distance to market	Average number of km to nearest market
Distance to city	Distance to nearest city in kilometres
Organization	Dummy = 1 if any HH member participates in a non-HCS organization; else 0
Extension	Dummy = 1 if Extension is present in the PA; else 0
Seed exchange	Dummy = 1 if household engages in seed exchange for sorghum or wheat with other farmers
Woreda 2	Dummy = 1 for Chiro *Woreda*
Woreda 3	Dummy = 1 for Dire Dawa *Woreda*
HCS participant	Dummy = 1 if HH participates in HCS seed distribution programme, else 0

Table 14.5. Factors explaining the probability of participation in HCS. (From FAO–Netherlands Partnership Programme: Seed System Impact on Household Welfare and Agricultural Biodiversity – Household Survey.)

Variable name	Coefficient	$P>|z|$
Family size	0.0233	0.216
Education	0.0352	0.719
Age	0.0047	0.611
Off-farm income	0.1107	0.524
Operated area	0.0310	0.000
Slope	0.1179	0.027
Soil	0.1174	0.024
Irrigated area	0.1283	0.000
PA altitude	0.0003	0.000
Livestock assets	0.0001	0.004
Agricultural assets	0.0007	0.993
Non-agricultural assets	0.0006	0.472
Distance to market	0.0078	0.014
Distance to city	0.0021	0.008
Organization	0.1152	0.180
Extension	0.1621	0.456
Seed exchange	0.1138	0.534
Woreda 2	0.3662	0.000
Woreda 3	0.4666	0.588
Constant	0.8147	0.000

Number of obs = 679.
Log pseudo-likelihood = –365.15871.
Pseudo R^2 = 0.2225.
Note: Dependent variable: HCS participant.

insights that will assist in interpreting the results from an econometric analysis of the impact of the HCS programme on intercrop diversity. Ideally, this analysis would be presented for all the crops grown in the sample. However, the survey was limited to a detailed analysis of the seed systems of only two crops and data are not available for all crops. Since sorghum and wheat were the main crops of focus of the HCS programme, as well as two of the three most widely grown crops in the survey area, considerable insight can be obtained even from this more limited analysis.

Although the focus of the study is intercrop diversity, an understanding of utilization patterns at the variety level is useful in understanding crop-level decision making, and thus selected information on variety utilization patterns is included here. The average number of sorghum varieties grown per household is 1.3, with a maximum of 3. The average number of wheat varieties grown per household is only 1.1, with a maximum of 2. Considerable variation was found in the number of varieties reported over the entire sample when comparing sorghum and wheat. A much more diverse set of sorghum varieties were reported, with a total of 38 varieties planted by survey respondents, as compared with only 15 varieties of wheat. For both wheat and sorghum, one variety dominates in terms of both frequency grown and area planted across the sample and at the household level. However, the frequency and area share are much higher for the dominant wheat variety (80% growers and approximately 77% of total area planted to wheat) versus the dominant sorghum variety (30% of growers and about 26% of total area planted to sorghum).

Table 14.6 presents farmers' perceptions of the improvement status of the wheat and sorghum varieties they planted, and indication of genetic structure. According to farmers, 85.5% of wheat varieties grown are improved, while 88.5% of the sorghum varieties are local landraces. Several secondary sources of data were used to complement the survey data on the genetic content of the varieties, including interviews with sorghum and wheat breeders and

Table 14.6. Farmers' perceptions of genetic origin of sorghum and wheat varieties planted. (From FAO–Netherlands Partnership Programme: Seed System Impact on Household Welfare and Agricultural Biodiversity – Household Survey.)

Crop	Improved		Landraces		All varieties	
	Number	Per cent	Number	Per cent	Number	Per cent
Sorghum	63	11.1	505	88.9	568	100
Wheat	224	86.8	34	13.2	258	100

published studies (McGuire, 1999; Mulatu, 2000) and these support the conclusions found here. With the exception of one variety, all of the sorghum varieties grown by sample farmers are local landraces, while almost all the wheat varieties are improved varieties released over the past 10 years by the Ethiopian agricultural plant breeding system The improvement status of the various wheat varieties reported is shown in Table 14.7, as well as the breeding programme of origin and initial release year.

These results confirm that significant differences exist between sorghum and wheat in terms of the degree to which farmers access formal versus informal systems for their supply. The formal system for wheat is more highly developed with a higher number of variety releases over the past 10 years, and virtually no sources of local diversity for wheat are available. The opposite is true for sorghum, where the informal system is the main supply source, most likely due to the fact that a rich source of local genetic diversity is present, but also because sorghum has not received as much attention in the formal breeding sector as wheat. These data also indicate a higher diversity of sorghum varieties planted as compared with wheat, although this analysis is conducted using variety names, rather than morphological characteristics or molecular-level analyses, thus precluding any firm conclusions on the respective levels of genetic diversity in the two seed systems.

The next aspect of the wheat and sorghum seed systems examined is farmer motivation for growing a variety of a crop. Farmers were asked to list the most important advantage associated with each variety they had planted for sorghum and wheat. As seen in Table 14.8, a total of 19 different advantages were reported as the primary reason for choosing to grow a crop variety. These 19 advantages have been aggregated into three broad categories for the two crops of particular interest: sorghum and wheat. Variety advantages were also aggregated to the crop level to highlight the differences in farmer motivations for growing sorghum and wheat. The three categories or 'services' include high return (either high yield or good market value), production risk management and household consumption.

Table 14.9 shows the frequencies of farmer classifications of varieties by their most important advantage, aggregated by crop and genetic service category. Wheat varieties have a much higher score on high return than risk, whereas sorghum varieties are fairly closely split between the two. Data indicate that sorghum is more likely than wheat to be grown for multiple purposes. Consumption characteristics do not emerge as a primary advantage for either crop – although they were important as secondary advantages. These data suggest that at an individual crop level, sorghum provides a wider range of services to the farmers than wheat. Thus supply side interventions that increase the availability of wheat

Table 14.7. Improvement status of wheat varieties. DZARC refers to Debre Zeit Agricultural Research Centre. (From Zegeye et al., 2001; NSIA, 2003.)

Variety name	Improvement status	Origin	Release year
Har 1685 (kubsa)	Bread advanced/improved	CIMMYT cross release	1995
Har 710 (wabe)	Bread advanced/improved	CIMMYT cross release	1994
Pavon 76	Bread advanced/improved	CIMMYT cross release	1982
Har 1868 (shinna)	Bread advanced/improved	Possibly CIMMYT	1999
Enkoy	Bread advanced/improved	Kenya/Ethiopia	1974
Bohai	Durum advanced/improved	CIMMYT cross release	1982
Asassa	Durum advanced/improved	DZARC variety	1997
Kemedi dima (red wheat generic)	Not possible to distinguish but most likely improved bread wheat		
Kemedi adi (white wheat generic)	Not possible to distinguish but most likely improved bread wheat		
Shemame	Local landrace from Tigray area, most likely imported via immigration		

Table 14.8. Grouping of most important advantages of crop varieties into 'service' groups. (From FAO–Netherlands Partnership Programme: Seed System Impact on Household Welfare and Agricultural Biodiversity – Household Survey.)

Category I: high return	Category II: risk management	Category III: consumption
Good yield in grain	Resistance to drought	Taste of food
Good yield in residuals	Resistance to frost	Consumption quality
Good fodder quality	Resistance to pest	Cooking quality
Good grain quality	Resistance to disease	
Good market acceptance	Resistance to weevil	
Easy to thresh	Resistance to bird attack	
Late maturity	Good adaptability	
Uniform maturity	Early maturity	

Table 14.9. Frequencies of 'service' category selection by crop. (From FAO–Netherlands Partnership Programme: Seed System Impact on Household Welfare and Agricultural Biodiversity – Household Survey.)

	Sorghum		Wheat	
Service	Number of households ranking planted varieties per category	Per cent households ranking planted varieties per category	Number of households ranking planted varieties per category	Per cent households ranking planted varieties per category
High return	251	51.1	193	74.5
Risk management	212	43.2	36	13.9
Consumption	28	5.7	15	5.8
No service	0	0.0	15	5.8
Total	491	100.0	259	100.0

seed are unlikely to induce farmers to specialize in the crop and reduce their sorghum production, in the absence of a means of replacing the services they derived from sorghum. In contrast, we might expect that supply side interventions for sorghum are more likely to lead to specialization in this crop because it provides a wider range of services.

The next aspect of the wheat and sorghum seed systems explored is the nature of the HCS seed programme intervention in these systems. Among the sample of HCS participants, wheat was the primary emphasis of the HCS supply intervention, with over 70% of the participant households receiving wheat seeds from the programme. The impact of the HCS intervention can be seen in the renewal rates for seed by crop, as well as in the means by which seeds were acquired. Table 14.10 shows the percentage of farmers reporting seed renewal for wheat and sorghum. 'Seed renewal' means that the farmer obtained seed for planting during the survey season from an outside source, rather than saved it from own production, but replaced the seed of a variety already grown rather than changing the variety.

Almost 80% of the wheat seeds planted in the 2002 season had been renewed, while 83.3% of sorghum planted comes from farmer-saved seed from the previous harvest. This finding is not surprising as most of the wheat growers in the selected sample are HCS participants (79%) and the HCS intervention essentially consisted of wheat seed supply. Thus among wheat growers,

Table 14.10. Percentage of sorghum and wheat seeds renewed and retained 2002 planting season: HCS and non-HCS households. (From FAO–Netherlands Partnership Programme: Seed System Impact on Household Welfare and Agricultural Biodiversity – Household Survey.)

Source	Sorghum			Wheat		
	Total	HCS	Non-HCS	Total	HCS	Non-HCS
Renewed	13.85	36.76	63.24	79.46	80.98	19.02
Retained	83.3	46.34	53.55	13.95	63.89	36.11
Both sources	2.85	69.23	35.71	6.59	88.24	11.76
Total	100	45.62	54.58	100	79.07	20.93

HCS participants are much more likely to have renewed their wheat seeds in the survey year than non-HCS farmers (81% vs 19%).

Finally, the means by which seeds were acquired among households that renewed their seeds is examined. Table 14.11 reports the impact of the HCS intervention in the wheat seed system. Wheat is much more likely to be acquired under a loan – which is clearly linked to HCS participation. Out of the 68% of farmers that acquired wheat seeds through credit, more than 92% were HCS participants. For non-HCS participants, both sorghum and wheat are purchased with cash, with very little credit used by these groups. In the sample of HCS farmers, the nature of the intervention was to facilitate access primarily to wheat seed, through the provision of relatively low-cost credit to farmers, who are unlikely to have been able to afford such seeds without the intervention. HCS was not introducing a crop that was new to the area; the programme facilitated access to an existing, albeit minor, crop in the area.

Econometric Analysis

To further examine the factors influencing diversity, the approach proposed and applied in Part III was followed. Dependent variables are the three diversity indices. The explanatory variables, defined in Table 14.4, include indicators of the blocks of factors identified as exogenous determinants in the household model of on-farm diversity: (i) household characteristics including wealth and labour supply; (ii) farm physical characteristics or agroecological factors; and (iii) market-related factors. Additional measures of seed supply factors are included, such as access to extension services and seed exchange networks.

Table 14.11. Means of acquisition for renewed seeds. (From FAO–Netherlands Partnership Programme: Seed System Impact on Household Welfare and Agricultural Biodiversity – Household Survey.)

Means of seeds acquisition (%)	Sorghum (82 HHs)			Wheat (222 HHs)		
	Total	HCS	Non-HCS	Total	HCS	Non-HCS
Purchased paying cash	52.44	39.53	60.46	26.58	57.63	42.37
Purchased through loan	3.66	66.67	33.33	68.02	92.05	7.95
Exchange	14.63	33.33	66.67	3.60	75.00	25.00
Gift	29.27	45.83	54.17	1.35	33.33	66.67
Other	0	0	0	0.45	100.00	0.00
Total	100	41.46	58.53	100	81.53	18.47

The impact of HCS on diversity, the supply factor of obvious significance given the study design, is included through a dummy variable for participation. The potential problem with including an HCS dummy variable is that it assumes the variable only measures the impact of the programme and not some unobservable characteristics of HCS participants. For example, the coefficient on the variable may indicate that HCS positively affects diversity or it may indicate that HCS participants are more likely to have more diverse crops. To control for the observable difference between HCS and non-HCS households, variables found to significantly influence participation are included in the analysis. However, controlling for unobservable difference is not possible. The standard method for doing this is an instrumental variable approach in which HCS participation is predicted using a set of instruments that are correlated with participation but not diversity measures. Attempts to obtain such instruments have not been successful. This means that the coefficient on HCS variable may be biased. The direction of the potential bias is discussed with the results.

When using count data, such as the number of crops planted, an appropriate econometric approach is a Poisson regression, as has been applied in several other chapters of this book. Both the Shannon and Berger–Parker indices are censored (at zero and one, respectively), so a Tobit model was applied for these regressions.

Regression results are reported in Table 14.12. The results indicate that area operated and value of agricultural assets are statistically significant and positively related to all measures of diversity; i.e. producers with more land and agricultural assets are likely to be growing more crops, that are more evenly distributed in terms of land area. These findings are consistent with

Table 14.12. Factors explaining the intercrop diversity on household farms. (From FAO–Netherlands Partnership Programme: Seed System Impact on Household Welfare and Agricultural Biodiversity – Household Survey.)

Independent variables	Count Coefficient	$P>\|z\|$	Shannon Marginal effects	$P>\|t\|$	Berger–Parker Marginal effects	$P>\|t\|$
Family size	−0.0066	0.532	−0.0121	0.114	−0.0096	0.275
Education	0.0201	0.186	0.0083	0.397	0.0161	0.226
Age	0.0008	0.669	0.0017	0.147	0.0022	0.168
Off-farm income	0.0717	0.138	0.0683	0.021	0.0824	0.039
Operated area	0.0276	0.003	0.0245	0.000	0.0208	0.015
Slope	0.0783	0.149	0.0500	0.123	0.0577	0.190
Soil	0.0234	0.643	0.0345	0.260	0.0345	0.406
Irrigated area	0.0743	0.187	0.0130	0.709	−0.0551	0.245
PA altitude	0.0000	0.869	0.0000	0.652	0.0000	0.920
Livestock assets	0.0000	0.349	0.0000	0.180	0.0000	0.524
Agricultural assets	0.0006	0.048	0.0007	0.001	0.0008	0.004
Non-agricultural assets	0.0001	0.697	0.0001	0.621	0.0001	0.460
Distance to market	0.0010	0.766	0.0011	0.584	0.0007	0.795
Distance to city	−0.0012	0.143	−0.0018	0.001	−0.0030	0.000
Organization	0.0485	0.332	0.0909	0.003	0.0714	0.085
Extension	0.0299	0.662	−0.0093	0.820	−0.0145	0.793
Seed exchange	0.0180	0.722	0.0264	0.389	0.0562	0.176
Woreda 2	−0.1336	0.379	−0.2226	0.017	−0.3639	0.004
Woreda 3	−0.6526	0.001	−0.5818	0.000	−0.7671	0.000
HCS participant	0.0704	0.201	0.1167	0.000	0.1623	0.000
Constant	0.7773	0.028	0.5307	0.015	1.396	0.000
Number of obs	679		679		679	
Log-likelihood	−1087.9989		−472.5549		−883.9597	
Pseudo R^2	0.0506		0.2311		0.1121	

those reported for most of the developing country contexts studied in Part III. In the northern highlands of Ethiopia, Benin *et al.* (2003, Chapter 5) also found land area to be positively associated with intercrop diversity. However, results from this study (in eastern Ethiopia) indicate that labour and livestock assets are not important predictors of intercrop diversity levels, in contrast to their findings.

Several variables are found to be important predictors of the evenness of the crops planted, but not of the overall number grown. These include off-farm employment (positively associated with evenness and relative abundance), involvement with external organizations (positively associated with dominance) and distance to the nearest city (negatively associated). Access to off-farm employment may assist in obtaining income that can be used to purchase seeds that would have been difficult to obtain without cash income. Similarly, establishing a relationship with an external organization or intervention in the community may help farmers to procure seed that would otherwise be difficult to find. Proximity to a large city is associated with specialization, which may be related to greater participation in markets located in these cities, although the variable measuring proximity to the market was not significant in any of the regressions. Distance to the city reflects proximity to larger markets, while distance to the market includes smaller markets, and thus this result may indicate a differential impact of market participation depending on the nature of the market in question.

In terms of location, Dire Dawa and Chiro appear to have lower levels of diversity relative to Meta. Producers in Dire Dawa *woreda*, which lies at a lower elevation than the other two *woredas*, have significantly lower levels of intercrop diversity in all three regressions.

Participation in the HCS programme is positively associated with intercrop diversity in terms of the Shannon index. These results indicate that HCS participation does not increase the number of crops grown, but does increase the evenness in area distribution among crops, relative to non-participation. This finding is probably related to the influence of HCS participation on the likelihood of participating in wheat production. By promoting wheat production, HCS is promoting crop diversification among producers. The results from the descriptive statistics indicate that wheat is a crop that is grown to meet a fairly narrow range of attributes, as compared with sorghum, and is unlikely to meet the many production and consumption objectives of the producers. Since under the conditions present in the study site, these production and consumption objectives are unlikely to be met through market interactions, producers are thus constrained to growing crops that can meet this more diverse set of needs, in addition to wheat. In the study site, sorghum is the most common 'base' crop grown to meet a wide variety of needs, while wheat, as well as other crops such as maize and chat, is grown for a narrower range of purposes.

As noted however, the coefficient on HCS is potentially biased by self- and programme selection. In particular, it appears to be the case that wheat producers are more likely to participate in HCS. If so, the HCS coefficient could be upwardly biased, its statistical significance reflecting the fact that wheat producers have greater diversity rather than the fact that HCS leads to greater diversity. Whether HCS participation increases diversity thus remains unclear. Given the positive coefficient of the HCS variable and the reported P values, however, it seems reasonable to conclude that HCS participation does not reduce interspecific diversity. Therefore, at a minimum, HCS is an intervention that appears neutral with respect to interspecific diversity.

Neither of the other two characteristics of the seed supply system (seed exchange and the presence of extension in the community) are statistically significant in explaining levels of intercrop diversity. One explanation for the insignificance of the seed exchange variable, which is a dummy on whether or not the household exchanges seed for any crop with other farmers, may be that the supply impact occurs primarily at the intracrop level. In other words, seed exchange increases the availability of varieties available for one crop, but not the number of crops. Since informal seed systems depend on seeds derived from farmers' fields in the locality, they are less likely to be a source of seeds for crops not already grown in the area. The dummy variable on the presence of extension in the community was also insignificant. This result may reflect the fact that impacts occur at the household rather than at the community level, rather than that there is no effect of extension on intercrop diversity. According to Mulatu (2000) the HIEP programme was a key source of wheat and maize seeds to farmers in the region who would otherwise have been unable to grow a short-season crop in addition to sorghum, suggesting a positive relationship between extension and intercrop diversity. Further work on the impacts of seed supply factors on crop specialization and diversification is needed to adequately explain these findings.

Conclusions

This chapter argues that including supply side factors in the analysis of the farm-level determinants of crop genetic diversity is important. The study presented focuses on the impact of a seed supply intervention implemented by a non-governmental organization (NGO) on three different measures of intercrop diversity at the household farm level. Econometric results showed that participation in the NGO programme was significantly and positively related to two measures of intercrop diversity – the Shannon index of proportional abundance and the Berger–Parker index of relative abundance. Together with survey data and expert information, this finding indicates that programme participation increased the area under production for certain crops – particularly wheat – relative to other crops, although not the total number of crops in production.

This finding can be explained by considering the characteristics of the wheat seed system, particularly in relation to the sorghum system. The key characteristics of these systems are the relative availability of local versus improved genetic materials and the types of services they provide to farmers, as compared with farmers' demand for genetic services given their specific consumption and production situation. In addition, the nature of the seed supply intervention – the crop selected for the intervention and the nature of the intervention – are important determinants of the programme impact on intercrop diversity.

In this case study, the supply side intervention consisted of the provision of seeds under credit, focusing primarily on wheat. The purpose of the seed supply intervention was to reduce the costs of growing a crop that was already well established in the area, although a minor crop. Very little availability of local genetic diversity was found for wheat, even though there were several improved varieties of bread wheat available to farmers in the area, supplied by both the extension system and the NGO. In contrast, sorghum is the major crop grown in the area and considerable local genetic diversity is available to farmers, although relatively few improved varieties. Grouping these advantages into service categories shows that wheat varieties are selected primarily for their productivity advantages, while sorghum varieties have a much wider range of risk-management-related advantages such as drought and disease resistance.

Implications

The hypothesis tested here was that the characteristics of the wheat and sorghum seed systems, together with the nature of the HCS activities, determine the impact of the supply side intervention on farm levels of intercrop diversity. Due to possible problems of sample selection bias, it is not possible to conclude definitively that HCS participation increases intercrop diversity. It is reasonable to conclude, however, that HCS participation does not reduce it, and that at a minimum, participation has a neutral effect. Findings also imply that expected impacts on intercrop diversity of seed system intervention will vary depending on the crop selected for the intervention and its relation to the farming system.

Acknowledgements

We gratefully acknowledge the support of the FAO–Netherlands Partnership Programme Agro-biodiversity Component for the design and implementation of the project.

References

Baumgärtner, S. (2002) Measuring the diversity of what? And for what purpose? A conceptual comparison of ecological and economic measures of biodiversity. Paper presented at the Conference 'Healthy Ecosystems, Healthy People – Linkages between Biodiversity, Ecosystem Health and Human Health', Washington, DC, 6–11 June, 2002.

Bellon, M.R. (2004) Conceptualizing interventions to support on-farm genetic resource conservation. *World Development* 32(1), 159–172.

Bellon, M.R. and Risopoulos, J. (2001) Small-scale farmers expand the benefits of maize germplasm: a case study from Chiapas, Mexico. *World Development* 29(5), 799–812.

Benin, S., Gebremedhin, B., Smale, M., Pender, J. and Ehui, S. (2003) Determinants of cereal diversity in communities and on household farms of the northern Ethiopian highlands, ESA Working paper, No 03–14, Agricultural and Development Economics Division, FAO, Rome. Available at: http://www.fao.org/es/ESA/pdf/wp/ESAWP03_14.pdf

Beyene, H., Verkuijl, H. and Mwangi, W. (1998) Farmers' seed sources and management of bread wheat in Wolmera Woreda, Ethiopia. CIMMYT and IAR, Mexico DF.

Heisey, P.W. and Brennan, J.P. (1991) An analytical model of farmers' demand for replacement seeds. *American Journal of Agricultural Economics* 73(4), 1044–1052.

Louette, D. (1994) Gestion traditionnelle de variétés de maïs dans la Réserve de la Biosphère Sierra de Manantlàn (RBSM, états de Jalisco et Colima, Mexique) et conservation *in situ* des resources génétiques de plantes cultivées. Thèse de doctorat, l'École Supérieure Agronomique de Montpellier, Montpellier, France.

Magurran, A. (1988) Ecological diversity and its measurement. Princeton University Press, Princeton, New Jersey.

McGuire, S. (1999) Farmers' management of sorghum diversity in eastern Ethiopia. In: Almekinders, C. and de Boef, W. (eds) *Encouraging Diversity: the Conservation and Development of Plant Genetic Resources*. Intermediate Technology Publications, London, pp.43–48.

Morris, M.L. (ed.) (1998) *Maize Seed Industries in Developing Countries*. Lynne Rienner Publishers and CIMMYT, Boulder, Colorado and Mexico.

Morris, M.L. and Heisey, P.W. (1998) Achieving desirable levels of crop diversity in farmers' fields: factors affecting the production and use of commercial seed. In: Smale, M. (ed.) *Farmers, Gene Banks and Crop Breeding: Economic Analyses of Diversity in Wheat, Maize and Rice*. Kluwer Academic Press, Norwell, Massachusetts.

Mulatu, E. (2000) Seed systems and small-scale farmers: a case study of Ethiopia. PhD thesis, University of the Free State, South Africa.

NSIA: National Seed Industry Agency (2003) Crop Variety Register, Issue no. 5, NSIA, Ethiopia.

Rice, E., Smale, M. and Blanco, J.-L. (1998) Farmers' use of improved seed selection practices in Mexican maize: evidence and issues from the Sierra de Santa Marta. *World Development* 26(9), 1625–1640.

Smale, M., Bellon, M.R. and Aguirre Gomez, J. (2001) Maize diversity, variety attributes, and farmers' choices in southeastern Guanajuato, Mexico. *Economic Development and Cultural Change* 50(1), 201–225.

Sperling, L. and Cooper, D. (2003) Understanding seed systems and strengthening seed security. Internal document. FAO, Rome.

Sperling, L. and Loevinsohn, M.E. (1993) The dynamics of adoption: distribution and mortality of bean varieties among small farmers in Rwanda. *Agricultural Systems* 41(4), 441–453.

Teshome, A. (2001) Spatio-temporal dynamics of crop genetic diversity and farmers' selections *in-situ*, Ethiopia. PhD dissertation. University of Ottawa, Canada.

Worede, M., Tesfaye T. and Regassa F. (2000) Keeping diversity alive: an Ethiopian perspective. In: Brush, S. (ed.) *Genes in the Field*. International Development Research Center, Ottawa, Canada.

Zegeye, T., Taye, G., Tanner, D., Verkuijl, H., Agidie, A. and Mwangi, W. (2001) *Adoption of Improved Bread Wheat Varieties and Inorganic Fertilizer by Smallscale Farmers in Yelmana Densa & Farta Districts of Northwestern Ethiopia*. Agricultural Research Organization (EARO) and International Maize and Wheat Improvement Center (CIMMYT), Mexico DF, Ethiopia.

15 Institutions, Stakeholders and the Management of Crop Biodiversity on Hungarian Family Farms

G. Bela, B. Balázs and G. Pataki

Abstract

This chapter builds on Chapters 3 and 8, analysing the management of crop biodiversity on Hungarian family farms from the perspective of institutional environmental economics. Prospects for seed management by growers of maize and bean landraces are emphasized. Research findings presented are based on extensive qualitative interviews with representatives of the different stakeholder groups in the seed system. Special attention has been given to stakeholders with the most to lose and the least power to influence – farmers who save seed. The research revealed that policy makers face a number of constraints imposed by international agreements, and there are conflicts and discrepancies among the interests of stakeholders. Access to crop genetic resources is being shaped in a politically contested terrain where diverse and competing interests conflict. There are clear incentives for commercially oriented farmers to use varieties released by the formal seed industry, but these do not fully serve the needs of small-scale farmers who also grow crops for home consumption. Trade-offs between profitability and public attributes embodied in farmers' seed are less visible. Much work is needed to improve communications among stakeholders before any feasible policy can be formulated and put into practice.

Introduction

This chapter analyses the management of crop biodiversity on Hungarian family farms from an institutional environmental economics perspective. It depicts the current and potential institutional context of agrobiodiversity in Hungary, emphasizing seed management by growers of maize and bean landraces. Prospects for a seed system to sustain maize and bean diversity on small farms are also considered. Findings reported in this chapter are part of an interdisciplinary research project to value agrobiodiversity, using several methods including forms of revealed and stated preference approaches (Chapters 3 and 8), as well as institutional and social analysis. The concepts used here to characterize the seed system, and the notion of seed as a mixed good, are found in Chapters 1, 12, 13 and 14.

Research findings presented in this chapter are based on extensive qualitative interviewing about the management of crop biodiversity on family farms in Hungary, with representatives of the different stakeholder groups that constitute the seed system. Special regard has been taken of stakeholders with the most to lose and the least power to influence. In the context of this research, these stakeholders are Hungarian farm families who grow maize and bean landraces.

The contribution of institutional environmental economics to an understanding of the policy problem is summarized next. Stakeholder maps enable the identification of conflicting interests, identities and objectives of key actors in the seed system. The institutional context is described, including historical circumstances, policy regimes relevant to crop biodiversity issues and the field of organizational players. The structural features of the seed system are characterized, including the commercial seed industry and the informal, farmers' system. The chapter proceeds with a general discussion of the stakeholders' views, definitions and problem perception in managing crop biodiversity, as well as their disparate attitudes towards conservation. In conclusion, it is suggested that any sensible modelling of farmers' seed choice be preceded by a historical and institutional analysis of the dynamics of the seed system. Each seed system has a unique institutional context and a related stakeholder environment that keeps the system functioning and changing.

Theoretical and Methodological Perspective

Institutional economics

To the best of the authors' knowledge, no comprehensive studies about crop genetic diversity as it relates to economic aspects of farmer decision making making have been implemented previously in Hungary. This underscores the importance of comprehending the main parameters, or variables, involved in farmers' decision making before undertaking modelling exercises. Beforehand, it is necessary to understand the seed system, its institutional context and the related stakeholder environment. It is in this sense that the problem has been approached in this chapter: from the perspective of institutional economics.

Institutional approaches in social science research are quite diverse (Nielsen, 2001). From the perspective of the history of economic ideas, two main schools of institutional economics can be distinguished. The first, old American, is represented by the work of Thorsten Veblen, John Commons and Wesley Mitchell, and their contemporary followers; the second is typified by that of Oliver Williamson, who dubbed the approach 'the new institutional economics' (Williamson, 1985).

New institutionalists, like neoclassical economists, propose that individual economic actors, 'along with their assumed behavioural characteristics', be taken 'as the elemental building block in the theory of the social or economic system' (Hodgson, 1994, pp. 69–70). Institutions are treated like constraints or parameters that are exogenous to the optimizing behaviour of individuals. Although new institutional economics places special emphasis upon transactions as well as transaction costs related to certain institutional arrangements, it does not consider that institutions have their own momentum in explaining human behaviour.

By contrast, institutional economics in the vein of the older school posits that institutions affect individuals in fundamental ways. From this point of view, 'institutions possess causal power above that of individuals alone' (Hodgson, 2000, p. 324). Contemporary institutionalism, following Veblen in particular, does not hold a deterministic view of the individual. Rather, 'institutions are the outcome of individual behaviour and habituation, as well as institutions affecting individuals' (Hodgson, 2000, p. 326). Agents and institutions constitute each other in a dynamic way. An institutional analysis should reveal the interactions between the main institutional structures and the most significant groups of agents related to the problem under investigation.

Institutional environmental economics

Research that is primarily of an exploratory type (aims at understanding from within) usually involves the application of qualitative interviewing. Yet, most environmental valuation research applies quantitative methods and models in order to calculate monetary values attached to the different levels of biodiversity, from genetic diversity and species diversity to diversity at the habitat or ecosystem level (e.g. Drucker *et al.*, 2001; Scarpa *et al.*, 2003; Birol, 2004). Recently, researchers valuing environmental goods have applied methods grounded in the qualitative empirical tradition of scientific inquiry (e.g. Kaplowitz and Hoehn, 1998, 2001; De Marchi *et al.*, 2000;

Gregory and Wellman, 2001; Kontogianni *et al.*, 2001). Some of these studies (e.g. De Marchi *et al.*, 2000) have embraced an institutional perspective, thus exemplifying an approach that has been termed 'institutional environmental economics' (Jacobs, 1994). An institutional environmental economist asks: 'what is actually going on when people "value the environment"?' (Jacobs, 1994, p. 84). What is actually going on when farmers choose one seed variety, instead of another?

Instead of revealing or eliciting preferences (i.e. pre-given preferences are revealed or stated through application of a method), institutional environmental economics considers the valuation exercise as a social process of forming preferences. Therefore, empirical methods should be chosen in order to tease out information about how farmers think and behave towards different aspects of crop biodiversity conservation. Research methods should be applied to understand and make room for alternative types of valuation and consequent decisions.

Stakeholder analysis

A common thread in recent endeavours is the stakeholder approach, which is developed and utilized in business management and organizational studies (see Mitroff, 1983; Freeman, 1984). Stakeholder analysis can be a powerful tool for policy analysis and formulation in the field of natural resource management (see Grimble and Wellard, 1997; Lochner *et al.*, 2003).

Stakeholder analysis aims at identifying key actors or stakeholders of a system or a problem under examination. Here, a stakeholder is an agent that can influence or can be influenced by the operation of the seed system. Typically, the seed system has multiple stakeholders with numerous, and often conflicting, interests, identities and objectives. Stakeholders range from non-market actors, such as regulatory or state agencies and non-governmental organizations (NGOs), to market actors, including private, profit-making corporations and trade associations. Farmers themselves may be both market and non-market actors, depending on the context.

The stakeholders who might benefit or lose the most by decisions or actions within the system are called primary stakeholders; the others, with a much smaller stake, are secondary stakeholders. Stakeholders may also be categorized according to two important dimensions: importance (the strength of the one's stake) and influence (the power to act in one's interest). As shown in Fig. 15.1, stakeholders in area A have the largest stake but are also the most vulnerable, since their power to influence the course of actions is relatively weak. Typically, farmers who conserve crop biodiversity belong to this stakeholder group, cultivating marginal lands and belonging to the least advantageous and politically the least powerful class of society with relatively few economic resources at their disposal (see the contexts described in the chapters of Part III).

Interviewing stakeholders

In the research reported here, qualitative interviewing sought to gather evidence on the role of stakeholders in seed system, as well as on subjective perceptions and ways of processing the reality of seed system. The primary stakeholders of our research are the small-scale growers of maize and bean landraces with the most to lose and the least power to influence. The secondary stakeholders were identified by interviewing key informants and by reviewing relevant laws and regulations. Altogether, 25 semi-structured interviews were completed with representatives of various organizational stakeholders involved in the formal seed system, and 23 interviews were conducted with farmers in two of the environmentally sensitive areas (ESAs) in Szatmár-Bereg and

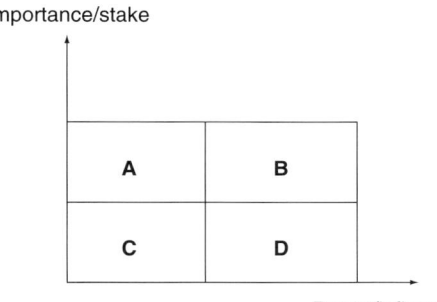

Fig. 15.1. Categories of stakeholders according to their importance and influence. (Adapted from Grimble and Wellard, 1997.)

Dévaványa környéke regions (Chapter 3). The individuals selected for the interview were the persons who were most likely to possess the necessary knowledge to answer questions or those who might be considered as decision makers.

To ensure that all stakeholders address common issues, a checklist of research questions was prepared for all interviews (shown in Box 15.1). Some questions depended on the characteristics of the stakeholder, and the degree and mode of his or her commitment or stake. The interviews were longer and more in-depth with those who were more affected. Depending on the degree of the respondent's stake, self-reflection and eloquence, the actual procedure extended from narration to structured questioning (see Strauss and Corbin, 1990).

The conversation techniques were applied flexibly in the light of the circumstances of the individual situation. The actual interview process started with establishing direct contacts with the stakeholders personally or through telephone. The discussion proceeded with questioning that generated storytelling (entering into a conversation, general explorations and ad hoc questions), as well as strategies that generate understanding (specific explorations with elements of references to previous answers, questions directed towards understanding and confrontation). Directly after the interview, scripts were written by the researchers to complement the tape recording. Scripts entailed an outline of the discussion topics, and comments about the interview situation, including nonverbal aspects and the interviewee's own focus. The stakeholder interviews were carried out in parallel with desk research that gathered secondary data about the seed market, and a household survey (Chapter 8).

Box 15.1. Interview guidelines.

Do the words biodiversity and genetic diversity mean anything to them?

Attitudes and perceptions: To assess the various attitudes among stakeholders and farmers, personal perceptions about seed saving, landraces and biodiversity were gathered from all interviewees. These were the questions: What importance do stakeholders and farmers ascribe to the conservation of agrobiodiversity and landraces/farm-saved seeds? Why is it important or unimportant? Are there any current benefits or expected future benefits from conserving agrobiodiversity?

Understanding decision making: To explore the PGR conservation issues in the case of maize and beans the following questions were posed: Are there any state/local/other incentives in the form of legal, economic or moral support for the conservation of landraces? What resources or power does the farmer have to mobilize for the preservation of landraces? What roles can the interviewee's organization fill?

Knowledge and experience: In order to evaluate the familiarity of the participants with the research topic, general questions were asked regarding the knowledge and experience of the participants on the topic: How do they understand the notion of landrace and farm-saved seed, and in what context do they use these terms? With which landraces are they familiar? Do they differentiate between landraces for fields, for garden plants and for fruit trees? Is there any cooperation or is it conceivable that there could be cooperation between the various stakeholders to conserve PGR? What kind of information/communication structure is needed to conserve landraces effectively? What changes in rules and incentive systems would be needed to conserve landraces?

Information strategies: To map the information sources stakeholders use in their decision making the following questions were asked: Are there any written rules, missions, guidelines or plans that influence the decisions and behaviour of the interviewee or interviewee's organization regarding the conservation of genetic diversity? What data are available to them?

Profile of interviewee and organization: To gather demographic information about the interviewee and describe the profile of the stakeholder organization, several characteristics were assessed: How long has the interviewee been working with agriculture, or with the issue of agrobiodiversity? What degree of competence does he or she have? What is his/her role in the organization? How is the work organized in the given stakeholder group?

Institutional Context

Historical patterns

The Carpathian Basin is characterized by heterogeneous ecological conditions. The geography is variegated, and three climatic zones are found within the borders of the country (Atlantic-alpine, continental, sub-Mediterranean). Hungary is a home to a great diversity of potentially valuable plant and animal species, whose preservation is of global value.

Plant cultivation activity is about 80,000 years old in the Carpathian Basin. Most of the cereals arrived in the basin during Neolithic times, and the major part of the leguminous plants were implemented by 'Tell cultures' in the Middle Bronze Age. Fruits and grapes became widespread during the Roman conquest. The cultural flora was enriched with additional species after the discovery of the New World. Thus, due to the particular agroecological conditions in the Carpathian Basin, the long duration of cultivation and the traditional selection practices, a great diversity in plant genetic resources has emerged (Surányi, 2002).

Landrace cultivation is thought to have flourished at the turn of the 19th century, when the highest levels of agrobiodiversity are believed to have existed in Hungary (Ángyán et al., 2002). As a consequence of the burst of plant breeding activity at the beginning of the last century and later hybridization programmes, crop landraces were displaced from large- and middle-scale farming. They continued to be cultivated mainly on small-scale, traditional farms and home gardens in marginal areas (Berkó, 1993).

After the First World War, the collectivization of agriculture 'forced' cooperative production. Despite these programmes, considerable scope was left for the persistence of traditional small-scale family farming in home gardens. Home garden production encouraged local saving of seed, and the passing of farmer-selected seeds and varieties from generation to generation. The process of societal transformation in the 1990s witnessed the bankruptcy and dissolution of most of the 'cooperatives'. Although the acute agricultural crisis that followed could not eliminate family gardens, rural areas experienced a drop in living standards and quality of life.

Today, the main source of income of Hungarian villagers is not derived from farming; only a few live entirely from the land. The average farm size in Hungary is 4.8 ha, and according to the census of the farmers in 2000, all 697,336 households have kitchen gardens (KSH, 2001). The area in gardens totals 41,193.66 ha, implying an average garden size of 591 m^2. The primary purpose of home garden cultivation has changed from recreation to subsistence farming, followed by supplementary income.

Yet, the small plots and gardens are insufficient to provide the necessities of life for most families, and with few rural employment opportunities, young people move to towns from village communities. Similar to international trends, the ageing of farmers is also considerable in Hungary: 59% of workers in the farm sector are middle-aged or older. The average age of men and women in private holdings is 53 and 60 years, respectively. The average wage in agriculture is 73% of the industrial sector and payment is usually uncertain (KSH, 2001). The ageing of the farm population and the migration of the young means that there is rapid decrease in the number of those who live from the land – whether as a supplementary source of income or for subsistence (see Harcsa et al., 1994; Juhász, 2001). Beyond the important role that home gardens and small plots play in supplying healthy food for local families and in rounding out household income, they are the most significant venue for crop biodiversity in Hungary.

Policy regime

The legal backbone of the Hungarian nature protection regime is the Nature Conservation Act (NCA) (Act of LIII 1996). The aim of NCA is the general protection of biodiversity. The nature protection regime plays a crucial role in maintaining the ecological conditions upon which the availability of wild relatives of crop plants depends.

Hungary has signed all the important international agreements relating to the protection of biodiversity in general and plant genetic resources in particular, such as the International Convention on Biological Diversity (CBD), the International Treaty on Plant Genetic Resources for Food and Agriculture (ITPGR), Patent Cooperation Treaty (Act of XXXIII 1995), the World Trade Organization's TRIPS agreement (Act of IX 1998) and UPOV (Act of LI 2002).

The Seed Act defines the conditions for variety certification and sets the rules for the formal seed system. The relevant regulation consists of two statutory rules issued by the Ministry for Agriculture and Rural Development: Production and Sales of Seeds (89/1997 XI.28. FM decree) and Preservation and Usage of Genetic Materials (95/2003. VIII. 14. FVM decree).

Most farmers' seeds do not satisfy the Distinct, Uniform and Stable (DUS) criteria for registration. Farmers' seeds are typically genetically heterogeneous and do not constitute a variety in the legal sense.[1] Inability to register farmers' seeds may contribute to the process of genetic erosion process in Hungary. During the process of variety registration and certification, only certain scientific views and commercial interests are taken into consideration. The limited publicity of the process also precludes the participation of local people. There is no institutional guarantee that public opinion and wider social interests will be involved in the decision making process.

During the accession process, Hungary formally adopted all European Union (EU) directives and adjusted the legal regime accordingly. Seed regulation changed considerably during 2003 and a new Patent Act (Act of XXXIX 2002) has been enacted to define special rules regarding plant varieties. According to the new regulation, Hungary must accept all the varieties that are certified by any members of the EU. The Common Variety List of the EU has been adopted. Furthermore, at least in the short term, farmers are not expected to change their cultivated varieties, in spite of the diversity in new types of seed supplied, due to their wariness about new unknown varieties. Farmers are typically considered risk-averse in this regard.

The 98/95/EC directive allows local varieties to be marketed as 'varieties for *in situ* conservation' or those for organic agriculture, and a special registration system is to be established. For these varieties, factors that are difficult to evaluate by conventional procedures have to be considered, and certification by authorities must be based on experience gained during production, propagation and use. However, the mechanism of this alternative registering system has yet to be identified nationally. Although Hungary has adopted this directive, the new registration system does not exist. Moreover, there is widespread uncertainty among decision makers about how this system might work, or which solution best fits the EU system.

Several national programmes exist, or are under construction, that will likely influence the function of the informal seed system, the laws and written rules that determine the power of actors. These programmes may have either favourable or adverse impacts on efforts to conserve agrobiodiversity on farms.

For instance, in the sixth article, the CBD (Act of LXXXI 1995) affirms the obligation of states to establish national strategies for biodiversity protection. The Ministry for Environment in Hungary has prepared a draft Action Plan for Agro-biodiversity Preservation (Ángyán, 2000), which outlines the important strategic steps to meet the CBD requirements, and identifies the organizations responsible for various actions. A number of paragraphs of the strategy stress the importance of agrobiodiversity protection and draft concrete provisions for enhancing the effective functioning of local, informal seed systems. The action plan is well prepared scientifically, but the practical implementation is at an early stage.

The realization of the plan is intended to occur through the National Agri-environmental Programme (NAEP), although it extends beyond the NAEP in scope. The NAEP is the primary national policy for supporting sustainable agriculture (104/2003 IX. 11. FVM decree and 290/2002 XII. 27. Gov. decree). The programme is based on an extensive agroecological analysis of Hungarian landscapes. The NAEP,

---

[1] Furthermore, the high level of heterogeneity (such as variable levels of quality) and the wide morphological and agronomic variations that characterize landraces may well cause problems with regard to their marketability. As some authors (Negri, 2003; Bardsley and Thomas, 2004) point out, this characteristic of landraces is not appreciated by modern marketing and consumers who are typically used to having standard products.

which started in 2002, aims to support the establishment of farming practices that are based on a sustainable utilization of natural resources, the preservation of natural values and biodiversity, the protection of landscape values and the production of healthy products. Although NAEP does not explicitly address the issues of plant (or, for that matter, animal) genetic diversity, the programme could have major indirect effects on the maintenance of crop biodiversity on farms. For example, NAEP intends to support – through a land-based subsidy scheme – those farmers who cultivate marginal lands with high ecological diversity. This group of farmers might be potential cultivators or users of landraces. NAEP also includes a horizontal programme for support to organic farming – another group of farmers with potential demand for landraces. NAEP promises the establishment of Regional Agro-environmental Centres, partly in order to explore and conserve traditional cultivation practices that are appropriate for specific regions. Subsidies might be available to support breeding for specific niches, such as organic production of unique environmental conditions.

The support of agroenvironmental goals from 2004 is a part of the Agriculture and Rural Development Operational Programme[2] (ARDOP). The NAEP objectives are in line with the National Rural Development Plan (NRDP) objectives, and its target programmes are integrated into the agri-environmental management measure of that plan.

An important funding mechanism within the policy regime for conserving agrobiodiversity in Hungary is the so-called Biological-base Tender. The tender has been operating for 10 years, and consists of two parts. One is a non-compensatory subsidy that is available for *ex situ* conservation to maintain specific varieties. The target group of this tender includes large institutions and gene bank collections, which means that it is not available for individual farmers or for farmers' associations. On the other hand, candidates can apply for a non-compensatory investment subsidy as well for covering costs of certification of new varieties. The tender also finances some countrywide research on the exploration of ecological factors that have a significant impact on cultivation.

Organizational field

There are several types of stakeholders that are connected to formal seed systems and local, informal seed systems in Hungary. A stakeholder map was developed to categorize institutions and organizations before planning interviews and secondary data collection (Fig. 15.2).

To implement the Seed Act in Hungary, the National Institute for Agricultural Quality Control oversees all activities in that respect: control of seed propagation, variety registration. The Hungarian Patent Office is responsible for enforcing the Patent Act. Additional relevant 'Regulatory Authority' connected to the seed system is the General Inspectorate for Consumer Protection. According to the Consumer Protection Act (Act of CLV 1997) the essential tasks of this inspectorate are the protection of consumers' interests and the provision of adequate information to consumers. Here, this actor is considered to be a secondary stakeholder of plant genetic resources protection.

The 'Legislative Institutions' connecting to the seed system are the Ministry for Agriculture and Rural Development and the Ministry for Environment and Water Management (MEWM). These two actors take part in the codification of laws and decrees without pre-defined shared roles and responsibilities in plant genetic resources conservation. Usually all subjects relating to the protection of wild relatives fall within the domain of the MEWM.

A number of national 'Education and Research Institutions' have a research project on

---

[2] According to the European Commission negotiations with the Hungarian government, Hungary prepared a Community Support Framework (CSF). The CSF represents the legal framework for financial support, and contains the financial commitments of the EU and the member state related to the development programmes for the member state launched in the given EU budget period. The CSF for Hungary 2004–2006 will be implemented by five operational programmes. One of these is the Agriculture and Rural Development Operational Programme.

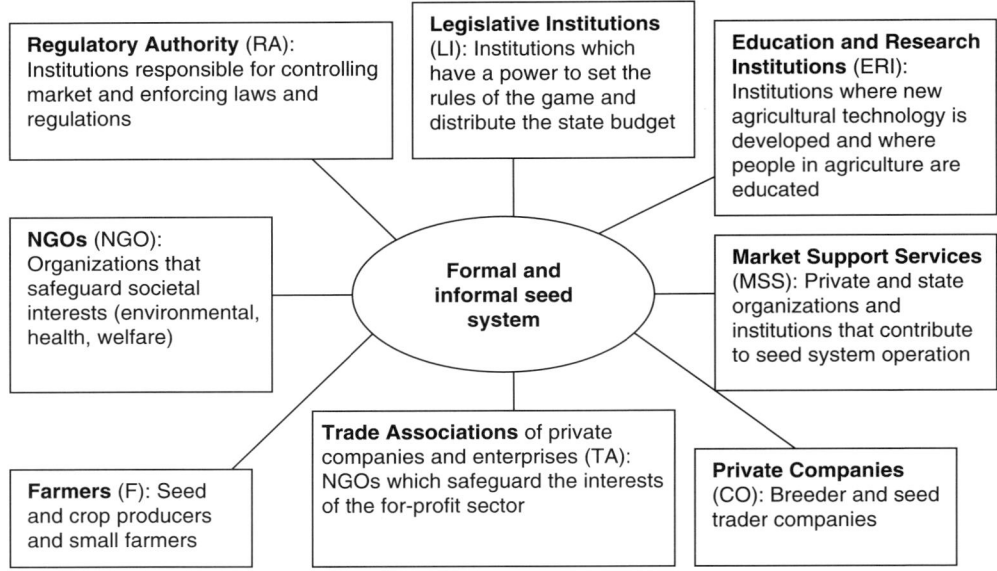

Fig. 15.2. Stakeholder map of the seed system in Hungary.

exploring and describing valuable genetic materials in Hungary. The important actors are St István University,[3] University of Debrecen,[4] Cereal Research Non-profit Company,[5] Institute for Agrobotany (Tápiószele),[6] Agricultural Research Institute of the Hungarian Academy of Sciences.[7]

The Market Support System (MSS) consists of the private and state organizations and institutions that contribute to seed system operation. Banks and financial institutions, the members of the farmers' notary network, local market operators, inspection bodies (such as the Biokontroll Hungária,[8] Hungária Ökogarancia), and the agrointegrators[9] are the elements of the MSS system. Locally adapted materials for breeding have several collections. The largest collection of field and vegetable crops can be found at the Institute for Agrobotany, but private and university breeding programmes also have their collections.

A special stakeholder group was identified as composed of lobbyists with for-profit interests. Termed 'Trade Associations', this group encompasses the Association of the Hungarian Breeders,[10] Crop Products Committee,[11] Association for Organic Agriculture, Chamber of Agriculture[12] and Association of the Hungarian Seed Distribution Companies.[13]

---

[3] http://www.szie.hu/
[4] http://www.klte.hu/
[5] http://www.gk-szeged.hu/
[6] http://www.rcat.hu/
[7] http://www.mgki.hu/
[8] http://www.biokontroll.hu/
[9] Agricultural integrator: actors who manage the agricultural production contracts between producers and commodity processors that detail an arrangement for raising agricultural commodities. Integrators have a decisive role in the success of maize growing on several hundred thousand hectares in Hungary.
[10] Magyar Növénynemesítők Egyesülete.
[11] http://www.vetomagtermektanacs.hu/
[12] http://www.agrarkamara.hu/
[13] Magyar Vetőmagkereskedők Szövetsége

The number of NGOs dealing with the issue of agrobiodiversity conservation is limited. Organic farmers are those who have shown the most interest in landraces (Biokultúra Association[14] and the Association of Conscious Consumers[15]). NGOs dedicated to environmental protection (Hungarian Environmental Partnership Foundation,[16] Öko-service[17] and the National Society of Conservationists[18]) could have a significant impact on information dissemination relating to landraces and conservation of agrobiodiversity. Small landowners do not have strong representative organizations. Some farmers' and gardeners' clubs exist in the countryside.

Roughly the same number of interviews was organized with representatives of each of the different stakeholder groups, constituting the seed system in Hungary (shown in Table 15.1). Before analysing stakeholder perceptions and attitudes, the most important actors and trends in the maize and bean seed system are presented. The following section presents the main characteristics of maize and bean seed systems in Hungary. Private companies, breeders and seed traders are treated in the subsection referring to the seed market. Small-scale farmers, being the primary stakeholders in this research and constituting the local, informal seed system, are dealt with in a separate subsection.

Seed System

The seed market

Fierce competition on the seed market after liberalization, combined with profound changes in social conditions as a result of economic transformation, had adverse impacts on the local, informal seed system. In the case of certain species (e.g. paprika seedlings and some bean varieties), local/informal seed exchange and trade is more extensive than in the tightly controlled, commercial species (maize, sunflower, wheat, etc.).

The seed market is an open market: provided that a seed has been certified by the National Institute for Agricultural Quality Control anyone is entitled to trade. At present there are 936 companies in the formal seed sector; quite a number of them trade in seeds. The size and functioning of the seed system of plants studied here differs. The maize seed industry is vertically integrated and concentrated, with a few multinational companies sharing the overwhelming part of total sales (Table 15.2).

In 2001 the harvested area of maize was 1,258,120 ha, from which 29,017 ha was for seed propagation. After quality control and certification, a major share of the planting material (seed) (59% – 32,471 t) was exported mainly to countries in Western Europe.

Large seed companies focus on large-scale farmers (with minimum 3–4 ha) and neglect home gardeners as a target group. The market of the so-called 'colour-packaged seeds'[19] is associated with Hungarian companies, such as ZKI and Hortseed, that are engaged in the breeding and use of domestic, locally adapted material in product development. There are several home delivery services with imported seed types.

The propagation area for bean (including green bean) was just 97 ha in 2001, and the total harvested production is not enough to satisfy domestic demand, so that bean imports are required. (KSH, 2002) The bean seed industry is not so concentrated and is relatively small. Bean is typically grown in home gardens. Only a few large-scale farmers grow beans, thus some of the seed companies do not distribute beans at all.

Plant breeding was strongly encouraged by the government especially from the 1960s until the 1980s. The state-established hybridization programmes diffused high-yielding varieties (HYVs), adopted first by large-scale farmers and cooperatives and later by smaller-scale farmers (Berkó, 1993). In parallel with the change in the agricultural support scheme, the direct funding for plant breeding was reduced. At present there is competition between multinational breeding

---

[14] http://www.biokultura.org/
[15] http://www.tudatosvasarlo.hu/
[16] http://www.okotars.hu/
[17] http://www.okoszerviz.hu/
[18] http://www.mtvsz.hu/
[19] Colour-packaged seeds are those destined for gardening.

Table 15.1. List of interviewees, by organization and stakeholder.

Institutions and organizations	Stakeholder grouping[a]	Primary (P)/secondary (S) stakeholders	Number of interviews
1. Agrobotany Institutions	MSS, LI	P	2
2. Ministry of Agriculture and Rural Planning Department of Sector Relations	LI, RA	P	2
3. Ministry of Agriculture and Rural Planning Department of Agro-environment	LI, RA	P	1
4. Ministry of Environment and Water Management	LI, RA	P	1
5. Cereal Research Non-profit Company	ERI	P	
6. Agricultural Research Institute of the Hungarian Academy of Sciences	ERI	P	
7. St István University Institute for Environmental and Landscape Management	ERI	P	2
8. St István University Department of Plant Production	ERI	S	
9. Debrecen University	ERI	S	
10. Association of the Hungarian Breeders	TA	S	1
11. Breeding companies	CO	P	3
12. Companies dealing with seed production	CO	P	2
13. Seed trading companies	CO	P	1
14. Banks and other financial institutions	MSS	S	
15. National Institute for Agricultural Quality Control	RA	P	3
16. National Agricultural and Breeding Committee	LI	S	
17. Hungarian Patent Office	RA	S	2
18. Crop Products Committee	TA	P	2
19. Biokontroll Hungary	MSS	S	1
20. Association for Organic Agriculture	TA	S	
21. Environmental Partnership Foundation (environmental NGO)	NGO	P	1
22. Chamber of Agriculture	TA	S	
23. Consumer Protection Office	RA	S	
24. Association of the Hungarian Seed Distribution Companies	TA	S	
25. Local market	MSS	P	
26. Farmers' notary (adviser for farmers)	MSS	S	
27. Bethlen Gábor Technical School for Agriculture	ERI	S	1
28. Small- and large-scale farmers	F	P	23

[a]See Fig. 15.2.

companies and the publicly financed, poorly funded national breeders. Heszky et al. (2002) analysed the pedigrees of the varieties of major crops that were developed by national researchers and certified from 1998 to 2000. The vast majority (85.2%) of inbred lines for hybrids originated from domestic gene stocks. It is regrettable for local conservation that crop area in maize varieties bred in Hungary has decreased continuously since the introduction of imported varieties from abroad, so that there are no apparent incentives to use local genetic materials in research.

Companies with the biggest market share do not have breeding programmes for maize and beans. The market potential of domestic seed

Table 15.2. Estimated market shares of seed companies in Hungary in 2003. (From Kleffmann, 2003.)

Company	Market share (%)
Pioneer Hi-Bred International	38
Monsanto Commercial Ltd	25
Syngenta Seed Ltd	13
Agricultural Research Institute of the Hungarian Academy of Sciences	8
Limagrain Hungary Ltd	4
Cereal Research Non-profit Company	3
Kiskun Research Center Ltd	3
KWS-RAGT HYBRID Ltd	2
Others	4

markets in the cases of maize and bean is small, so that large seed companies are not interested in developing varieties for particular environmental agroecological conditions using Hungarian genetic materials.

These companies test material developed somewhere else. The best adapting varieties are screened according to how well they fit domestic agri-ecological circumstances. For some varieties (sunflower, tomato, paprika) company breeding programmes have started to distribute seed to other East European markets. Data are not available on the companies experimenting with domestic crop resources, because neither the gene bank nor the farmers were informed about what happened with the seed taken from them. Interviews with breeding companies indicate that domestic breeding material is more interesting for breeding programmes that set their field testing within the limits of the country.

To summarize, the maize seed industry is vertically integrated, with a few multinational companies sharing the overwhelming part of total sales, while the bean seed industry is much less concentrated, and relatively small. The maize seed market generates incentives for imported varieties and commercially oriented farmers. Bean is typically grown in home gardens, with only a few large-scale farmers and seed companies interested in growing, breeding and distribution. Varieties released by the formal seed industries do not fully serve the needs of small-scale farmers who grow crops for home consumption.

Farmer seed systems

The seed market is usually identified with the formal seed system rather than the informal, or local, seed system (see Fig. 15.3). Although there may be farmers who exclusively participate in the formal seed system, here only those farmers are characterized who are involved, at least to some extent, in the informal, local seed system.

Since trade with local varieties is prohibited, there are no precise market data about the frequency of exchange and the market size. An estimate of the frequency of usage of local varieties in the study sites is provided by the household survey. Of the 323 farm families interviewed for the farm household survey, 142 of them stated that they cultivated landraces of beans or maize. By region, 26.9%, 52.3% and 52.7% of all households with garden and small plots in Dévaványa, Őrség-Vend and Szatmár-Bereg regions, respectively, have at least one landrace of maize or bean in their small farms (Chapter 8).

Farmers' interviews demonstrated that there are two markedly different groups of home gardeners or small-scale farmers maintaining landraces in the local seed system. The first category consists primarily of elderly farmers, with limited labour capacity, who manage kitchen gardens and/or very small plots. According to our interviews, a lot of these farmers had gained experience in intensive farming methods while working in state-owned 'cooperatives' during the totalitarian socialist regime. At the same time, they had usually acquired traditional knowledge about plant cultivation from their parents. They frequently experiment with mixing local and HYVs and intercropping beans as an understory of the maize plants. Although younger generations also know some locally adapted varieties by name, it is typically the elderly who maintain the informal seed system of beans and maize.

This category is composed of the most committed landrace conservationists. Saving seed is highly valued among these elderly farmers for a number of reasons. First, seed saved and reproduced on the farm originated with the ancestors, and is thus regarded as a cultural heritage or product, and a patrimony of the local community. Second, landraces furnish some of the essential ingredients of regional cooking recipes, or lend an idiosyncratic taste to a regular,

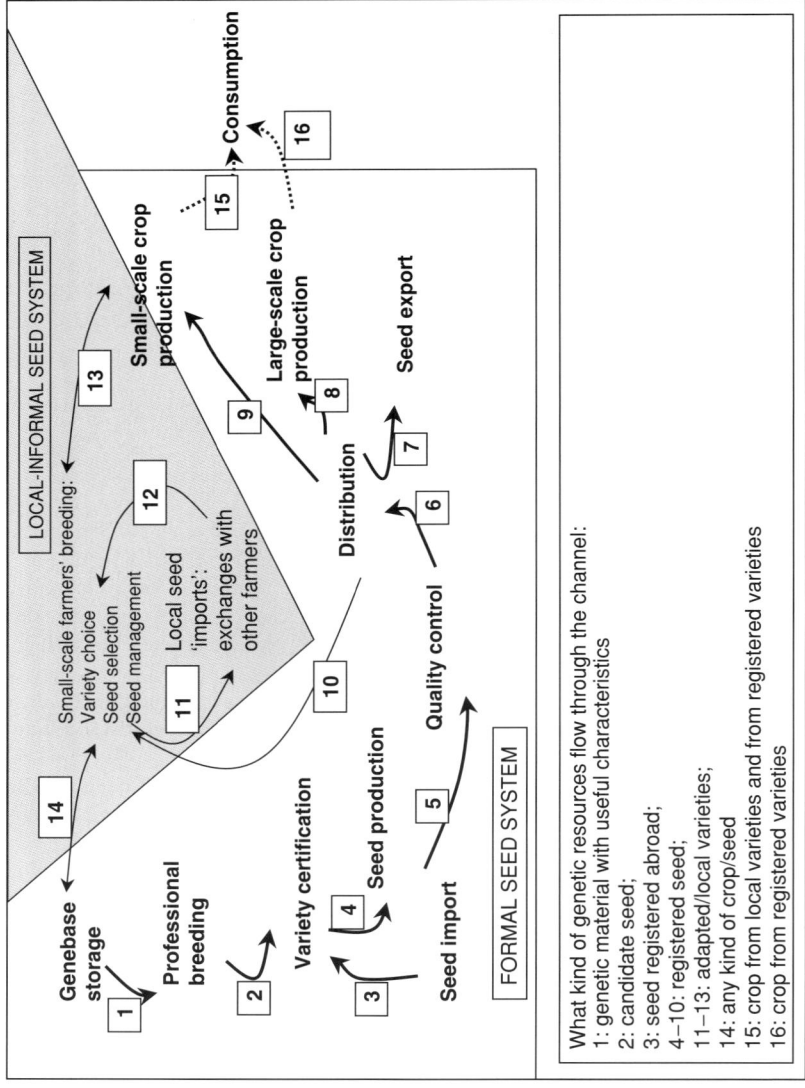

Fig. 15.3. Flow of genetic resources in the seed system.

traditional meal valued by family members. Livestock prefer the grain of old maize varieties and the meat of animals fed with this grain tastes better. Growing gardens and saving seed is an important recreational rural activity. Finally, this category of farmers considers that saving seed of the special, locally adapted variety is considered to be conservation, which also implies a certain responsibility.[20]

The second category of farmers includes middle-aged farmers who have gained less farming knowledge from their parents. In specialized agricultural schools and socialist ('forced') cooperatives, these farmers became familiar with the application of fertilizers, herbicides and pesticides, as well as the cultivation practices and attributes of HYVs. Today, mainly because of missing incentives for home garden production and seed saving, typically these farmers are contracted by a seed company that provides fertilizer and chemicals through the local integrator. They prefer market-oriented cultivation methods, use their own machinery, take advice frequently from expert consultants and read professional journals and books. They used or tried landraces for various motives. Some growers are only doing what others, relatives or friends, do, thus they grow and use others' landraces too. Some farmers are more interested in experimenting with interesting plants, and trying out new crops. Several growers believe that landrace seed is the key to seed supply problems, since it can be planted instead of hybrid seed when it is out of stock. Most of these farmers inherited landrace seed from generation to generation. For others interested in animal husbandry the grain sometimes supplements forage. Other farmers explained that landraces are more suitable for organic farming purposes. This younger group of farmers includes a core subset that grows landraces each year and intends to continue growing them.

In sum, farmers' interviews enabled the identification of two markedly different groups of landrace growers. The elderly, isolated farmers are the most committed conservationists of maize and bean landraces, because of their cultural heritage, taste and recreation value of landraces as well as sustainability of seed saving. The middle-aged farmers are usually contracted to the formal seed system, and prefer market-oriented cultivation methods while growing landraces for a range of reasons. Birol *et al.* (2004) also analysed the profiles of farmers most likely to grow maize and bean landraces in study regions using econometrics and descriptive statistics. The conclusions of the qualitative analysis and quantitative analyses are similar.

Stakeholder Perspectives

Definition and problem perception

In Hungarian scientific literature the most comprehensive scientific definition of landrace[21] is provided by Ferenc Gyulai (Ángyán *et al.*, 2002):

> They are currently being produced, are more primeval types in comparison with foreign varieties and are phenologically different from those, yet they are to be understood rather as genotype hybrids. In other words, former landraces were rather variety hybrids in the modern sense. They reached production primarily through mass selection and they formed a steady, so-called balanced population in that particular area. Landraces are resistant as a result of their genetic features and they fulfil the needs of extensive production too. As for yield, they generally lag behind modern varieties; although, landraces many times surpass modern varieties in quality.

Concerning the legal framework of plant genetic resources, neither the Seed Act nor the Decree on the Conservation of Plant Genetic Resources, 95/2003. (VIII.14.), of the Ministry for Agriculture and Rural Development deals explicitly with landraces. The term 'genetic resources' is defined generally, consisting of all materials of vegetable origin that represent a

---

[20] The finding that elderly – typically resource-poor – farmers are the most committed on-farm conservationists of landraces is supported by other research undertaken in the European context (see Virchow, 1999; Negri, 2003).
[21] For a detailed overview of landrace definitions found in the international literature, see Zeven (1998).

current or potential value both in nutrition and in agriculture, including propagation materials that contain a functioning unit of reproduction. This definition highlights the private-use value of genetic resources, understating its option value for biodiversity as well as public value. The two laws mentioned above apply a number of related terms: gene-source, variety of Hungarian origin, ecotype, landrace, traditional variety, genetic resources of industrial crops, genetic resources registered as national genetic resources. However, the laws do not define or describe these terms precisely.

Stakeholder interviews present a nebulous picture about which plant genetic resources ought to be conserved. The interviewees representing the formal seed system, when discussing genetic resources that should be conserved, used the following terms to describe them: landrace, old variety, traditional variety, straggling variety, primeval variety, Hungaricum. In the majority of cases, the usage of these terms was not coherent: some stakeholders meant the same by all the listed terms, while others differentiated the terms from each other.

For formal seed system stakeholders, the term 'landraces' usually meant the varieties that had been created as a result of some earlier domestic breeding programme and their usage is characteristic of a larger region (e.g. the Hungarian apricot of Gönc, the wheat of Bánkút). Some interviewees emphasized in their definition that landraces are not hybrids. According to an interviewee from the Regulatory Authority, a landrace can be regarded as a variety, if it is officially registered.

Varieties that had been popular during the lifetime of their grandparents were defined as old, traditional or primeval varieties. Interviewees from formal seed system usually thought of these with nostalgia and saw little chance that they have survived or would survive in current farming systems. They could cite primarily fruit and grape varieties and could not name old varieties of maize or bean. One person among the interviewees from the Regulatory Authorities denied the existence of such varieties in the case of maize, but others also doubted that such varieties would still exist in the case of any crop and fodder.

Apparently, neither the specialized scientific literature on landraces nor the legal regulation of plant genetic resources has managed to forge a consensual terminology in Hungary. Yet, such terminology is essential in order to define a common policy problem for various stakeholders with contesting interests.

Based on farmers' interviews, nine notions of landraces were distinguished: (i) old variety; (ii) named after the farmer who reproduces the seed (e.g. Gerö's bean); (iii) named after characteristics of the plant (e.g. colour or shape of the grain); (iv) named after the place of origin (e.g. specific landscape, village); (v) has no specific name, as compared with HYVs; (vi) other, indefinable name, such a 'baktipaszuly'; (vii) parents' varieties; and (viii) primeval varieties.

Several beliefs and convictions were formulated during the farmers' interviews about the utility and quality of old varieties. Discussants generally believed it is the older farmers that bond strongly to their varieties or to traditional cultivation practices that grow old varieties. Although farmers could list the good attributes they associate with old varieties, the say they usually choose to grow an HYV because 'everybody chooses these'. Considering this process, many farmers expressed fear that 'these (old) varieties will be lost in the future'.

Farmers' opinions varied on the actual quality of farm-saved seed. Many farmers believe that the seed of old varieties is inferior to commercial seeds. Several arguments against their cultivation were advanced during interviews. According to some, old varieties fail in drought years. One interpretation of this statement is that old varieties have longer maturity periods, and are less adapted to tolerate drought through 'drought escape'. Their stalks break easily, particularly in combine-harvesting, since they are adapted to non-mechanized agriculture. Addressing either of these problems is typically essential to a scientific plant-breeding programme.

On the other hand, many farmers insisted on sowing the old varieties because of their long-run sustainability. For example, they stated that old varieties produce healthier plants than the commercial varieties, usually because they require less chemical vaporization. Older varieties of maize produce two to three ears per plant and are resistant to smut (ustilaginales). Farmers reported that they are usually able to sow the crop again year after year, because these varieties retained high rates of germination for a long time.

Organic farmers, who constituted a specific target group of the interviewing process, thought that crops from farm-saved seed are not marketable and had a poor market price because there is no demand for them on the part of urban consumers. Additionally, organic farmers are reluctant to cultivate them because of yield uncertainty. However, most of the organic farmers associated old varieties with good quality in on-farm consumption and resistance to plant disease and pests with old varieties. Access to these plant genetic resources is difficult: 'the landraces are got frayed (no enough seed), and the reproduction of sufficient amount seed is time-consuming'.

Attitudes towards the utility of conservation and the informal seed trade

Attitudes towards conservation, its utility and the role of the informal seed trade were fragmented. According to most interviewees: 'Diversity ought to be retained!' If primeval varieties are still found here but are then lost, 'then we have lost something, which is a pity'. Interviewees who are part of the formal seed system were sceptical about the utility and value of landraces, mainly based on the perspective that the use of landraces is economically inefficient and unprofitable. An individual from the Regulatory Authority asserted:

> It is a little romantic to think that we can do big things with primeval varieties.... Just because a landrace is very valuable and an old lady call Kate is playing around with it on 5 spots in her small garden, it will not represent a national value.[22]

As far as the informal seed trade is concerned, the majority of the interviewees from the categories of Regulatory Authority, Trade Association and Private Companies said that in the case of species that cannot be hybridized, the 'black market' is still operating despite the strict regulation. The informal trade of landraces 'collides seriously with the interests of breeders and ought to be put under more severe control'.

Among the players of the formal sector there was a general consensus (except for the Gene Bank) that primeval varieties do not meet consumers' needs or the requirements of agrotechnology, and thus the materials circulating in trade are much better than the old varieties. As a pro-conservation interviewee asserted: 'The question of agro-biodiversity is underestimated (in Hungary).'

Almost without exception all interviewees were quite perplexed about who should be responsible for the establishment of a protection strategy. Naturally, they designated the Institute for Agrobotany responsible, while their own role remained unclear. The decree on the conservation and use of plant genetic resources issued by the Ministry of Agriculture and Regional Development (95/2003. (VIII.14.)) also identifies the state, and within that, the Gene Bank Council, as the chief participant of protection. On the other hand, the Institute for Agrobotany of Tápiószele is defined as a base institute, whose responsibility consists of the maintenance of the genetic resource collection, as well as the operation of the gene bank database. According to the experience of several interviewees from the categories of Regulatory Authority and Legislative Institution, farmers, NGOs and other market players demonstrate no need for landraces to enter commercial circulation: '... nobody is knocking on the door to tell you that he has got a good variety he wants to sell'.

Stakeholders' discourses of plant genetic resources conservation are dominated by the needs of consumers, the benefits of breeders and the requirements of agrotechnology. Altogether this implies that the landraces are inferior to market-oriented varieties. The strife for gaining public (economic, social, political and ecological) benefits through responsible protection strategy has little chance of being commercially successful.

---

[22] 'Egy kicsit romantikus elképzelés, hogy az ös fajtákkal tudunk nagy dolgot csinálni.' ...'Attól hogy egy tájfajta nagyon értékes és 5 helyen Kati néni eljátszik vele a kiskertjébe azzal nem fog nemzeti értéket képviselni.'

Strategies proposed by stakeholders

Two strategies to protect valuable genetic materials were explicitly proposed by stakeholders during the interviews.

One is to utilize landraces in organic farming because the use of varieties well adapted to local agroecological conditions is essential particularly for these farmers. The organic seed market is characterized by excess demand, which is mainly the consequence of domestic legal regulations. Under directive 2091/92/EC as well as the national legislation, organic farmers are obliged to use organically produced seed and propagating material. The law does not involve any requirement with regard to variety choice. Many companies offer organically grown seeds (in the case of maize, these are primarily intermediate varieties) of varieties that are popular in the domestic market anyway.

Landraces may be important primary products for organic seed breeding. At present there is no organic seed breeding in Hungary. According to the representatives of multinational companies interviewed, the market potential of organic seed is not large enough to justify launching a distinct breeding programme. At the same time, many stakeholders from the formal sector emphasized that the future of landraces should lie in their further breeding in order to produce seeds adequate for organic farming. The use of local seeds in the organic sector is not possible as long as the alternative registration system for valuable farm-saved seed does not function.

The other strategy proposed was to establish rules for trade and exchange of landraces. Some stakeholders believe that landraces do not need to enter commercial trade, arguing that it would be more sensible to provide the possibility of making them usable in a closed system, while not excluding farmers who plant landrace seed from government subsidization programmes. A farmer producing a landrace in larger amounts as a commodity would be required to register it. Some of the interviewees suggested a registration system similar to the French 'amateur list' of varieties, in combination with restricted trade of landraces. In this system, 'softer' rules regarding genetic uniformity rules could be devised and an alternative list could summarize all relevant materials. Variety maintenance could be resolved through these means. A list of varieties would definitely support their usage through recognition: this list would specify where a certain seed can be produced and circulated and what features it has. At the same time, seed with good quality should be allowed to circulate in small bags by the name of its species alone rather than variety. Recognizing it as quality landrace seed, buyers and consumers would hold different expectations than in the case of varieties developed by professional breeders. If this 'alternative registration system' of landraces is elaborated, its usage can be facilitated and simplified also by the farmers, doing the seed certification on their own.

Conclusions

Elaborating a policy for on-farm conservation of plant genetic resources that is politically feasible and in harmony with the national legal system poses a great challenge. Policy makers face a number of constraints imposed by international agreements, and there are conflicts and discrepancies among the interests of stakeholders. Identifying the actors with whom policy makers are able to work on plant genetic resource conservation is a first step in elaborating this policy. Then, analysing the situation systematically is essential for identifying good policy options, economic instruments and legal measures. Especially when dealing with social, economic and ecological issues involving multiple stakeholders, institutional economics, as compared with other approaches, offers a research perspective that enables the discovery of policy solutions as part of a process.

An institutional economics perspective favours research methodologies that recognize the importance of process in farmer valuation of landraces. Therefore, qualitative methods of social inquiry (primarily different types of interviewing techniques) were applied, enabling the research to paint a more comprehensive picture and understand better the complexity of the problem at hand.

One of the main results of the research was the identification and description of different groups of farmers with regard to their attitudes, values perception and landrace cultivation practices. In line with previous research in other cultural contexts as well as the quantitative analysis conducted in Chapters 3 and 8 of this book, elderly farmers managing small, marginal landholdings were found to be the main actors involved in conserving Hungarian crop genetic resources *in situ*.

Our other major finding relates to the legal and policy context of plant genetic resources conservation in Hungary. The research revealed that the institutional arrangement to support *ex situ* conservation of plant genetic resources is relatively effective and well managed in Hungary. However, *in situ* conservation efforts face an unsupportive and adversarial legal and policy context. The Hungarian legal and policy setting provides no incentives for farmers' *in situ* conservation of plant genetic resources, encouraging them to use commercial HYVs offered by the formal seed market. There are no actors operating in the formal, market seed system who are financially interested or in any way promote the conservation of plant genetic resources. The problem is not only that the informal, local seed system of farmers is not operating efficiently any more but that it is de-legitimized and even de-legalized by the current legal and policy framework. Consequently, there is no cooperation among stakeholders to form an effective lobby or a joint policy platform for the preservation of plant genetic resources.

Moreover, the general demographic, social and economic trends prevailing in Hungary contribute to the erosion of plant genetic resources. The social status of farming is low and probably still declining; the cultural cohesion of rural communities is quickly disappearing; the economic opportunities of people living in the countryside are very restricted compared with those of the urban population. These trends, to a great extent, demonstrate the lack of a comprehensive, integrated and effective policy for rural development in Hungary.

Implications

In sum, the findings paint a very bleak and unpromising picture of the possibilities of conserving plant genetic resources and the related cultural traditions of plant cultivation, including traditional ecological knowledge, in the future and for future generations in Hungary. The process of genetic and cultural erosion and the consequent loss of biological and sociocultural diversity might only be halted if an effective and comprehensive public policy and programme for financially encouraging *in situ* conservation by farmers is designed and carefully implemented.

Acknowledgements

We are grateful for the financial support of the European Union, IPGRI and IFPRI, and the comments of colleagues in the international research community during various workshops held by IPGRI, with special thanks to Melinda Smale. We would also like to express thanks to József Ángyán and László Podmaniczky who supported this project on the part of IELM. We are grateful to all interviewees in Hungary who willingly shared their opinions, and to the graduate students at IELM who were involved in the fieldwork. We thank Mariann Hajdú and Ildikó Kókai, who explored the legal background of the problem. Ekin Birol, Ágnes Gyovai, László Holly and István Már have also contributed at various stages of the project. András Takács-Sánta provided insightful comments.

References

Ángyán, J. (ed.) (2000) *Mezőgazdasági biodiverzitás megőrzési stratégia.* (*Action Plan for Preservation of Biodiversity: Strategy for the Hungarian Agriculture Sector*). Hungarian National Academy of Sciences (MTA), Budapest.

Ángyán, J., Tardy, J. and Vajnáné-Madarassy, A. (eds) (2002) *Védett és érzékeny természeti területek mezőgazdálkodásának alapjai.* (*Agriculture for Environmentally Protected and Sensitive Areas*). Mezőgazda Kiadó, Budapest.

Bardsley, D. and Thomas, I. (2004) In situ agrobiodiversity conservation in the Swiss inner Alpine zone. *GeoJournal* 60, 99–109.

Berkó, J. (1993) *A hibridkukorica magyarországi elterjedésének és a kukorica vetőmagipar kialakulásának története.* (*The Spread of Maize Hybrids and the History of Maize Seed Sector in Hungary*) Mimeo, M. Agrártud. Egy. Kiadó.

Birol, E. (2004) Valuing agricultural biodiversity on home gardens in Hungary: an application of stated and revealed preference methods. PhD dissertation. University College London, University of London, London.

Birol, E., Smale, M. and Gyovai, Á. (2004) Sustainable use and management of crop genetic resources: landraces on Hungarian small farms. Paper presented at the Thirteenth Annual Conference of the European Association of Environmental and Resource Economics, 25–28 June, 2004, Budapest.

De Marchi, B., Funtowicz, S.O., Lo Cascio, S. and Munda, G. (2000) Combining participative and institutional approaches with multicriteria evaluation. An empirical study for water issues in Troina, Sicily. *Ecological Economics* 34, 267–282.

Drucker, A.G., Gomez, V. and Anderson, S. (2001) The economic valuation of farm animal genetic resources: a survey of available methods. *Ecological Economics* 36, 1–18.

Freeman, R.E. (1984) *Strategic Management: A Stakeholder Approach*. Pitman, Boston, Massachusetts.

Gregory, R. and Wellman, K. (2001) Bringing stakeholder values into environmental policy choices: a community-based estuary case study. *Ecological Economics* 39, 37–52.

Grimble, R. and Wellard, K. (1997) Stakeholder methodologies in natural resource management: a review of principles, contexts, experiences and opportunities. *Agricultural Systems* 55(2), 173–193.

Harcsa, I., Kovách, I. and Szelényi, I. (1994) A posztszocialista átalakulási válság a mezögazdaságban és a falusi társadalomban. (Postsocialist transitional crisis in agriculture and rural society) *Szociológiai Szemle* 3, 15–43.

Heszky, L., Bódis, L. and Holly, L. (2002) A magyar növényi génkészlet jelentösége Magyarországon. (The significance of Hungarian crop gene stock in Hungary) *Növénytermelés* 51(1), 133–140; (2), 247–252; (3), 353–358.

Hodgson, G.M. (1994) The return of institutional economics. In: Smelser, N.J. and Swedberg, R. (eds) *The Handbook of Economic Sociology*. Princeton University Press, Princeton, New Jersey, pp. 58–75.

Hodgson, G.M. (2000) What is the essence of institutional economics? *Journal of Economic Issues* XXXIV(2), 317–329.

Jacobs, M. (1994) The limits to neoclassicism: towards an institutional environmental economics. In: Redclift, M. and Benton, T. (eds) *Social Theory and the Global Environment*. Routledge, London and New York, pp. 67–91.

Juhász, P. (2001) Mezögazdaságunk és az uniós kihivás (Hungarian Agriculture and the EU Challenge). Beszélö April.

Kaplowitz, M.D. and Hoehn, J.P. (1998) Using focus groups and individual interviews to improve natural resource valuation: lessons from the mangrove wetlands of Yucatán, Mexico. Paper presented at the World Congress of Environmental and Resource Economists, Venezia, Italy, 25–27 June, 1998.

Kaplowitz, M.D. and Hoehn, J.P. (2001) Do focus groups and individual interviews reveal the same information for natural resource valuation? *Ecological Economics* 36, 237–247.

Kleffmann (2003) *Maize Market Research in Hungary*. Kleffmann & Partner Co., Hungary.

Kontogianni, A., Skourtos, M.S., Langford, I.H., Bateman, I.J. and Georgiou, S. (2001) Integrating stakeholder analysis in non-market valuation of environmental assets. *Ecological Economics* 37, 123–138.

KSH (2001) Általános Mezögazdasági Összeírás (ÁMÖ) – (Agricultural Census in 2000). Központi Statisztikai Hivatal (Hungarian Statistical Office), Budapest.

KSH (2002) Fontosabb növények vetésterülete 2001 (Propagation Area of Important Agricultural Plants, 2001). Központi Statisztikai Hivatal (Hungarian Statistical Office), Budapest.

Lochner, P., Weaver, A., Gelderblom, C., Peart, R., Sandwith, T. and Fowkes, S. (2003) Aligning the diverse: the development of a biodiversity conservation strategy for the Cape Floristic Region. *Biological Conservation* 112, 29–43.

Mitroff, I. (1983) *Stakeholders of the Organizational Mind*. Jossey-Bass, San Francisco, California.

Negri, V. (2003) Landraces in central Italy: where and why they are conserved and perspectives for their on-farm conservation. *Genetic Resources and Crop Evolution* 50, 871–885.

Nielsen, K. (2001) Institutionalist approaches in the social sciences: typology, dialogue, and future challenges. *Journal of Economic Issues* XXXV(2), 505–516.

Scarpa, R., Ruto, E.S.K., Kristjanson, P., Radeny, M., Drucker, A.G. and Rege, J.E.O. (2003) Valuing indigenous cattle breeds in Kenya: an empirical comparison of stated and revealed preference value estimates. *Ecological Economics* 45, 409–426.

Smale, M. and Bellon, M.R. (1999) A conceptual framework for valuing on-farm genetic resources. In: Wood, D. and Lenné, J. (eds) *Agrobiodiversity: Characterization, Utilization, and Management*. CAB International, Wallingford, UK, pp. 387–488.

Strauss, A. and Corbin, J.M. (1990) *Basics of Qualitative Research: Grounded Theory, Procedures and Techniques*. Sage, Newbury Park, London, New Delhi.

Surányi, D. (2002) Tájfajták a Kárpát-medencében (XVIII. sz. – 1950). (Landraces in the Carpathian Basin) Agrártörténeti Szemle, 321–406.

Virchow, D. (1999) Conservation of plant genetic resources for food and agriculture: main actors and the costs to bear. *International Journal of Social Economics* 26(7,8,9), 1144–1161.

Williamson, O.E. (1985) *The Economic Institutions of Capitalism*. Free Press, New York.

Zeven, A.C. (1998) Landraces: a review of definitions and classifications. *Euphytica* 104, 127–139.

16 Cooperatives, Wheat Diversity and the Crop Productivity in Southern Italy

S. Di Falco and C. Perrings

Abstract

This chapter presents an empirical study testing the relationship of cooperatives to crop diversity and productivity, using regional, time-series data. The potential role of agricultural institutions has been neglected in previous literature that has investigated the productivity effects of crop biodiversity. However, institutional structures, such as agricultural cooperatives, can influence aggregate level of crop biodiversity through their food processing and marketing role. Data are drawn from southern Italy, a megadiversity area for wheat where a broad range of old and new varieties, including landraces, are grown. Different wheat varieties have differentiated characteristics that can create farmer demand for diverse types, derived from processor and consumer demand for differentiated products. A Cobb–Douglas production function is estimated with a first-difference model and a dynamic panel model. Cooperative concentration is associated with a higher level of wheat diversity. Furthermore, intracrop diversity is beneficial to wheat productivity at the regional level, over the decade considered.

Introduction

In Chapter 1, agricultural biodiversity was defined as a component of biodiversity, referring to all diversity within and among species found in crop and domesticated livestock systems, including wild relatives, interacting species of pollinators, pests, parasites and other organisms (Qualset et al., 1995; Wood and Lenne, 1997). Genetic diversity is the sum of genetic information that is contained in the genes of individuals of plants, animals and microorganisms. Species diversity is the diversity of species within which gene flow occurs under natural conditions.

Different components and levels of agricultural biodiversity are interlinked. For instance, the narrow genetic base of major crops and the concentration on a small number of crops can increase vulnerability to pests and pathogens, contributing to yield variability (National Research Council, 1972). This is because the greater the diversity between or within species and functional groups, the greater is the tolerance or resistance to pests. Pests have more ability to spread through crops with the same genetic base or genetic sources of resistance (Priestley and Bayles, 1980; Sumner, 1981; Altieri and Lieberman, 1986).

In the applied economics literature, few studies have investigated the relationship between district-level crop yields and genetic diversity using standard production or cost function analyses. In a study of the Punjab of Pakistan, Smale et al. (1998) related wheat productivity with several indicators of genetic diversity using a Just and Pope (1978) stochastic production function. The authors found that the production environment determines the sign of the relationship between diversity and productivity. For instance, among rain-fed districts, genealogical distance and number of varieties grown were associated with higher mean yields and lower yield variability. In the irrigated areas, instead, a high concentration of wheat area among fewer varieties, or greater genetic uniformity, had an important, positive effect on expected yields. A similar approach was employed by Widawsky and Rozelle (1998). They used a more generalized function form and an area-weighted, Solow–Polasky index to test the impact of rice variety diversity on the mean and the variance of yields using township data. Widawsky and Rozelle found that the number of planted varieties reduced both the mean and the variance of yields, although the effect on variance was not statistically significant.

Meng *et al.* (2003) modelled the productivity–diversity relationship as endogenously determined for modern wheat varieties in China, using a cost share system. Although the econometric results indicated that evenness in morphological groups was a positive factor in per hectare costs of wheat produced, potentially important cost savings were implied by some of the input share equations. For example, diversity may have contributed to a more efficient use of pesticides, which otherwise would have been required to maintain a similar level of production stability.

Given the emphasis of these studies on formal econometric modelling of diversity–productivity relationships, and the preoccupation with defining appropriate diversity indices, the role of institutions was neglected, other than a brief mention of markets, plant breeding and extension programmes. As argued by the authors of studies included in Part IV of this book, institutions are probably one of the driving forces of diversity loss or conservation.

Institutions are humanly devised constraints that structure political, economic and social interactions (North, 1991). They consist not only of formal constraints, such as laws and property rights, formal agreements, but also of informal constraints, such as more general customs or code of conduct. Institutional structures such as cooperatives or producer associations might play an instrumental role in influencing crop biodiversity levels within a region, affecting in turn the long-run productivity of that region.

The purpose of this chapter is twofold: (i) to test the relationship of cooperative organization of production and marketing with levels of crop biodiversity maintained at the regional level and (ii) to test the relationship of crop biodiversity with productivity. Data are drawn from an important area for cereal production in Europe – southern Italy. This region is a Vavilovian megadiversity spot and a centre of diversity for durum wheat (Vavilov, 1951; Harlan, 1992). Agriculture in southern Italy is also characterized by a large number of cooperatives involved in producing, processing and marketing crops.

The following section presents the conceptual background of this study, explaining the research interest in cooperative behaviour. This is followed by a site description, presenting the historical context of cooperatives and cereals production in southern Italy. The empirical approach used to investigate the relationship of cooperatives to cereal biodiversity levels, and the relationship of diversity to productivity, is summarized. Analysis is based on regional, time-series data. Findings are then discussed, and conclusions are drawn in the final section.

Conceptual background

Heisey *et al.* (1997) used the theory of impure public goods to relate the variety choices of farmers, the rusts of wheat and the genetic diversity in the Punjab of Pakistan, using district-level data. They showed that in the aggregate, farmers often chose to grow varieties that were higher yielding, but not necessarily less susceptible to rust. In some years, both aggregate wheat yields and latent, genetic resistance to rust might have been increased by growing a different combination of varieties; in others, a more genetically diverse mix would have incurred private costs in terms of yield forgone, with possible social consequences in that lower-income nation.

The analysis by Heisey *et al.* (1997) assumed that each farmer acted without the knowledge of other farmers' actions or the interests of other farmers in mind. In contrast, this chapter asks the question: does the institutional structure that conditions farmers' choices affect the genetic diversity of the crop they cultivate? Genetic diversity is club good (a type of impure public good), whose level is determined by the actions of the members of agricultural cooperative.

The agricultural cooperative is a voluntary group of individuals who derive mutual benefit from the coordination of production decisions, shared access to inputs, enhanced market power and more effective lobbying capacity.

The cooperative can process the crop and market the product in order to create added value and to distribute the revenues to members. From the cooperative perspective, having crop biodiversity implies capability to produce different products. Crops are the raw material for food processing. Varieties have different qualitative characteristics, such as protein content, colour and grain moisture or humidity. These qualitative differences can create a demand at the farm level for differentiated grain, derived from processor and consumer demand for differentiated products.

At the same time, maintaining genetic diversity in crops in the long run can affect the sustainability of a cropping system. Individual varieties respond differently to adverse biotic pressures and environmental conditions, which can contribute to stability over time (Tilman and Dowling, 1994; Tilman *et al.*, 1996). The performance of individual species also varies with climatic and other environmental conditions; greater species diversity in a geographical location enables the system to maintain productivity over a wider range of conditions (Naeem *et al*, 1994). This reasoning leads to the hypothesis that, under certain market conditions, agricultural cooperatives can represent a viable way to internalize some of the economic and ecological benefits of biodiversity.

In terms of diversity measurement, the analysis conducted by Heisey *et al.* (1997) and Smale *et al.* (1998) focused on modern varieties, since most (but not all) varieties grown in the higher-potential growing environment of the Punjab are modern varieties. Diversity indices were constructed from the coefficients of parentage, based on pedigree data. Pedigree data are available only for modern varieties. The analysis of Meng *et al.* (2003) also focused on modern wheat. The authors constructed diversity indices from a combination of ecological concepts, trial data, genealogies and a statistical model. Here, genetic diversity is captured by the Simpson index, measuring the proportional abundance of durum wheat varieties grown in southern Italy. The index of genetic diversity encompasses both landraces and modern varieties, spanning improvement status. The Simpson index is also used across cereal crops to represent intercrop diversity, although results were inconclusive.

Site Description

Economic, cultural and climatic characteristics make agriculture an important sector in southern Italy. Agriculture accounts for 8% of overall European Union (EU) agricultural land (in the eight southern Italian regions) and the average ratio of added value in agriculture against added value in industry was 0.4 from 1960 to 1993. During the time span considered in this study, the regions were all designated 'Objective 1' for development by the EU. The areas under 'Objective 1' are given a high priority for development, supported by substantial levels of financial assistance and ad hoc policy interventions by decision makers.

The production of cereals is particularly favoured by the dry, warm weather in southern Italy. Yields are negatively affected, instead, by cold, frosty winters or sudden changes in temperature. These weather conditions also reduce the spread and proliferation of pests, which spread more when humidity is high. In some areas the soil is sandy, reducing the ability of plant roots to absorb fertilizers and, hence, the benefit in using the nutrient. For this reason, application of pesticides and fertilizers appears to be relatively unimportant for the growth of cereals in southern Italy.

Southern Italy is roughly composed of eight regions. These regions differ somewhat in climate and topography, but the agricultural sectors, and particularly the cereals production sectors, are reasonably homogeneous. Table 16.1 compares average yield levels for the past three decades. Data represent 3-year averages around

Table 16.1. Durum wheat yields, three decades' comparison, southern Italy. (From authors' calculation on ISTAT (n.d.) data.)

	1969–1971	1979–1981	1989–1991
Abruzzi	2.22	2.43	2.74
Molise	1.95	2.21	2.67
Campania	1.79	2.16	2.8
Puglia	2.20	2.30	2.3
Basilicata	1.71	1.84	1.78
Calabria	1.50	1.9	1.8
Sicilia	1.60	1.79	1.8
Sardegna	1.35	1.74	1.38

Yield is in metres per hectare, 3-year averages.

1970, 1980 and 1990. Productivity ranges from 1.3 up to 2.7 mt/ha. In the major regions for durum wheat production (Sicilia and Puglia), average yield levels appear not to have changed much over the three decades. In Abruzzi, Molise and Campania, average yield levels appear to have increased.

A large proportion of land in each of the eight regions is sown to cereals. Table 16.2 shows the average share of all agricultural land allocated to cereals, and the share of cereals area sown to durum wheat, from 1970 to 1990. Cereals occupied 28% to 55% of all agricultural land at the regional level during these decades. In Puglia and Basilicata, cereals account for over half of all agricultural land, and in Sicilia, they represented 45%. Among cereals, durum wheat is the most widely grown, with more than 38% of the land share for all regions taken together. Aside from rice, which is grown in a humid environment, other cereals grown include bread wheat, barley and maize.

Durum wheat is used to produce the nation's staple food, pasta. Data from the Italian statistical office (ISTAT) indicate that in the past 20 years, 68% of national durum wheat production came from the southern regions of the country (ISTAT, n.d.). The regions Sicilia and Puglia alone produced 40% of Italy's output of durum wheat.

In Sicilia, for example, farmers often grow more than one durum wheat variety at a time, driven by a combination of heterogeneous agroecological conditions and end-use demand. The area is prone to drought and in some regions there is no irrigation. Some varieties provide higher protein content or preferred grain colour, characteristics that matter to food processors. Table 16.3 lists the varieties of durum wheat grow in the study regions from 1970 to 1993 by improvement status. Although the adoption of newer varieties (e.g. Ciccio, Gianni, Colosseo) is rising rapidly, some farmers' varieties (e.g. Russello, Timilia) are still in use. Newer varieties are typically of shorter stature.

Old improved varieties are still widely grown, including Adamello, Appulo, Capeiti, Simeto, Trinakria and Valnova. Some of these

Table 16.3. List of varieties grown in study regions from 1970 to 2000, southern Italy. (From Statistiche Agrarie, ISTAT (n.d.).)

Cultivars	Improvement status
Adamello	Old improved variety
Appulo	Old improved variety
Arcangelo	New variety
Balsamo	New variety
Capeiti	Old improved variety
Ciccio	New variety
Colosseo	New variety
Cosmodur	New variety
Creso	Old dwarf variety
Crispiero	New variety
Duilio	Old improved variety
Fenix	New variety
Fortore	New variety
Gianni	New variety
Grazia	Old improved variety
Iride	New variety
Messapia	Old improved variety
Norba	New variety
Nudura	New variety
Ofanto	New variety
Platani	New variety
Radioso	New variety
Russello	Landrace
Rusticano	New variety
Salentino	New variety
Simeto	Old improved variety
Svevo	New variety
Tavoliere	Old improved variety
Timilia	Landrace
Trinakria	Old improved variety
Tresor	Old improved variety
Valbelice	Old improved variety
Valnova	Old improved variety

Note: A new variety is typically of medium or short stature, and has been released during the past decade. An old improved variety is a tall variety released up to 50 years ago. A landrace is a farmers' variety.

Table 16.2. Cereals' share of agricultural land and durum wheat share of land in cereals, southern Italy 1970–1990. (From authors' calculation on ISTAT (n.d.) data.)

	% Cereals	% Durum wheat
Abruzzi	0.37	24.42
Molise	0.44	38.89
Campania	0.36	28.23
Puglia	0.55	46.62
Basilicata	0.54	35.13
Calabria	0.38	20.24
Sicilia	0.45	38.36
Sardegna	0.28	23.71

Table 16.4. Number of agricultural cooperatives in southern Italy, by region, by decade. (From Annuario dell' Agricoltura Italiana, INEA (n.d.).)

Year	Abruzzi Molise[a]	Campania	Puglia	Basilicata	Calabria	Sicilia	Sardegna
1951	8	34	85	31	36	155	86
1961	60	111	190	102	86	438	361
1971	194	430	547	146	192	1161	686

[a]Abruzzi and Molise was one single region up to 1973.

taller varieties have been grown for decades and farmers know their performance well. A number of the old improved varieties incorporate genetic material from farmers' varieties or improved varieties used in the 1920s (e.g. Cappelli).

In southern Italy, agricultural cooperatives have an important role in producing, processing and marketing durum wheat. After the 1950 agrarian reform, the agricultural sector in the south was partitioned into very small landholdings tenured by a multitude of different owners. Production cooperatives developed in order to overcome difficulties associated with this structural arrangement.

Table 16.4 shows the dramatic development and spread of agricultural cooperatives in southern Italy. For instance, in Campania the number of agricultural cooperatives went from 34 in 1951 to 430 in 1971. During the same time span in Sardegna the number of cooperatives passed from 86 to 686. In 1971 in Sicilia there were 1161 registered cooperatives. This upward trend continued steadily throughout the 1980s and 1990s. Figure 16.1 depicts the change in the number of cooperatives over the time span considered in the empirical analysis.

Cooperative members retain private property rights on their land, but have a common property regime for some fixed capital, such as threshing machines. Each cooperative can market the harvest as a monopsonist vis-à-vis

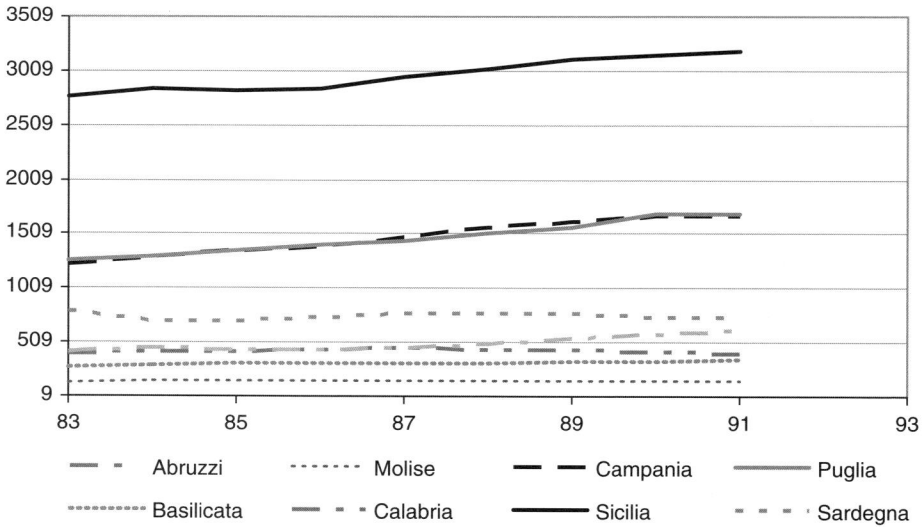

Fig. 16.1. Number of agricultural cooperatives by region, 1983–1991. (Adapted from Annuario dell' Agricoltura Italiana, INEA (n.d.).)

its members. In some cases, the cooperative mediates with respect to its members in a bargaining process with the agri-food industry; in others, the cooperative processes the crop. When cooperatives intermediate with the agri-food industry, they play a crucial role in determining the pay-off to members because the price paid by the industry to the farmers is determined by the cooperative's bargaining power. The bigger the cooperative, the bigger is its marketing and negotiating power. Potential for market power is amplified if one considers that because of tariffs in the EU and because of protection, the food industry is unable to buy cereals from non-EU countries.

Econometric Approach

Initially, estimation was conducted in two stages. To assess the effect of cooperatives on intracrop diversity of durum wheat and intercrop diversity of cereals, aggregate variety diversity was regressed against the density of cooperatives in the region. To test the effect of diversity on long-run productivity, a production function was estimated in the second stage. The predicted values from the first-stage regressions were used as explanatory variables in the production function, along with conventional inputs and a variable for weather conditions. Then the Arellano–Bond dynamic panel model was estimated to address the endogeneity of durum wheat diversity and productivity.

Data

Data were obtained from ISTAT, the Italian National Institute of Statistics, and the INEA, the National Institute for Agricultural Economics. The series are drawn from the Statistiche Agrarie and Annuario for the south of Italy (Abruzzi, Molise, Campania, Puglia, Basilicata, Calabria, Sicilia and Sardegna), including the years 1980–1993. Although the regional level of aggregation is driven primarily by the structure of the secondary data that were available, data on intracrop diversity are recorded only at the regional level.

This level of aggregation is also defined by the nature of the problems presented in this book. Understanding differences in levels of biodiversity, and the way that explanatory factors influence these levels as the scale of observation and analysis changes from farmer, to village, district, and region, is important for the design of conservation programmes. In southern Italy, regions are administrative units that implement and coordinate policy intervention in agriculture.

The definition of the variables used in our empirical analysis is reported in Table 16.5. The quantity of hard wheat produced is in tonnes per hectare. Pesticide applications per hectare and labour force participation are conventional inputs. Meteorological impact on productivity has been captured with the quantity of rainfall per year. The density of cooperatives expresses the role of cooperatives at the regional level. Density was calculated as the number of agricultural cooperatives in the region divided by the land allocated to agriculture in that region. The

Table 16.5. Variables' definition.

Variable	Definition
Durum wheat yield	Durum wheat output (t/ha), by region and year
Cereal yield	Cereals output (t/ha), by region and year
Pesticides	Pesticides use (100 kg/ha), by region and year
Rainfall	Rainfall (mm/year), by region and year
Labour	Labour units (no./ha), by region and year
Cooperative density	Number of cooperatives per hectare, by region and year
Intracrop diversity	Simpson index over durum wheat area allocated among varieties, by region and years
Intercrop diversity	Simpson index over cereals area allocated among crops, by region and year

majority of agricultural cooperatives (around 70%) are devoted to arable production, and most have been involved in durum wheat production, during the time period considered.

Many biodiversity indices have been developed in the general literature and a number of these are discussed in Chapter 1. In this chapter, the Simpson index of proportional abundance was used to measure both the intracrop diversity of durum wheat and the intercrop diversity of cereals. The Simpson index is 'heavily weighted towards the most abundant species in the sample while being less sensitive to species richness' (Magurran, 1988, p. 40). In the case of intercrop diversity, durum wheat is relatively abundant among cereals crops and the numbers of cereals grown are not large. In the case of intracrop diversity of durum wheat, however, the numbers of varieties are large and no single variety dominates for more than a short period of time.

Stage 1 estimation: crop biodiversity and cooperatives

The relationship of crop biodiversity to cooperative density was assumed to be linear:

$$D = a_0 + a_1 C + v_{it} \quad (16.1)$$

Stage 2 estimation: crop biodiversity and productivity

In agricultural productivity analysis, a range of mathematical representations of the production technology has been invoked (Mundlak, 2001). In this study a standard Cobb–Douglas production function has been applied. Along with a set of standard inputs and a control variable for rainfall, crop biodiversity was added separately as an explanatory variable.

Let $y = f(\mathbf{x})$ denote the production function, where y is quantity of durum wheat and $\mathbf{x}$ is a $n \times 1$ vector of inputs. In the single output case, the Cobb–Douglas production function is written as:

$$y = A \prod_{i=1}^{n} x_i^{\alpha_i} \text{ where } \alpha_i > 0, \forall i = 1, \ldots, n$$

By taking logarithms we have an expression that is linear in parameters,

$$\ln(y) = \alpha_0 + \alpha_i \Sigma_i \ln(x_i) \quad (16.2)$$

where $\alpha_0 = \ln(A)$

This implies that $((\partial Y/\partial x_i)/(Y/x_i)) = \alpha_i$. The estimated ith coefficient can be readily interpreted as the marginal productivity of the ith input. The Cobb–Douglas specification imposes unitary elasticity of substitution between inputs. In order to relax this property an interaction term was added. Since the fit of the model was not more robust in the more flexible case than it was with the standard specification, the assumption of unitary elasticity did not constrain the estimation.

Empirical analysis

The data set is a combination of a cross-sectional time series, suggesting that the use of panel analysis is appropriate. Panel data analysis improves reliability of estimates, and can control for individual heterogeneity and unobservable or missing values (Baltagi, 2001). Fixed and random effects eliminate problems arising from stochastic trends that are specific to a variable, but cannot eliminate those related to specific regions (Hsiao, 1986). In order to eliminate regional stochastic trends in the variables, a First Difference Estimator was used. Let Y_{it} be the dependent variable and X_{it} a set of explanatory variable, hence

$$Y_{it} = \mu_i + X_{it}\beta + v_{it}$$

Taking the first difference, the equation becomes:

$$\Delta Y_{it} = \beta \Delta X_{it} + \Delta v_{it} \quad (16.3)$$

Assuming that Δv_{it} are uncorrelated with ΔX_{it} Eq. (16.3) may be estimated by use of ordinary least squares (OLS). This transformation eliminates the individual effects (Baltagi, 2001) and reduces serial correlation. Moreover, if there are omitted integrated variables, the First Difference Estimator is consistent. This approach does induce residual autocorrelation. Estimated models should therefore be tested for autocorrelation.

Results

Table 16.6 reports the effects of cooperative concentration on the intraspecific diversity of durum wheat measured at the regional level in the southern Italy. The overall significance of the model is good. Cooperative concentration is positively and significantly correlated with aggregate variety diversity in durum wheat. In regions where cooperative organizations are denser, levels of biodiversity in durum wheat are also higher. The results were statistically weaker for the intercrop diversity analysis and both steps were affected by high autocorrelation. No conclusions could be drawn.

Table 16.7 reports the second-stage production function coefficients for the First Difference and Arellano–Bond estimators. In the First Difference model, predicted levels of durum wheat diversity are positively and significantly related to long-run productivity. Conventional inputs such as labour and pesticide show the expected positive signs and are both statistically significant.

Rainfall, although not statistically significant, has a positive impact on production. Although the overall fit of the model is good, heteroskedasticity was found in first and second stages (treated with White's standard errors) and autocorrelation was present. In the Arellano–Bond equation, the effect of intra-crop diversity remains strong but the sign on labour is no longer consistent with theory. None the less, the Sargan test supports the validity of the instrumentation, and since we fail to reject the hypothesis of no second-order correlation, the estimator is consistent. An extra lagged variable did not alter the results.

Conclusions

This chapter has considered the impact of cooperatives of production on the intracrop (intraspecific, or variety) diversity of durum wheat in regional crop productivity in southern Italy from 1983 to 1991. Given the aggregate nature of the data, first differencing techniques have been used to eliminate regional stochastic trends and improve the estimation. Findings demonstrate that in areas of Italy that are economically marginalized, such as southern Italy, cooperatives can play a role in maintaining variety diversity in a major crop. This finding likely reflects the role of cooperative production and marketing in highly differentiated industry for the food staple, durum wheat. Durum wheat varieties have characteristics that relate to quality in end-use, such as protein content, colour and grain moisture

Table 16.6. The effect of cooperative concentration on durum wheat diversity.

Variables	Coefficient	Standard error
Constant	0.4	0.28
Cooperative density	1.2*	0.22

R^2: 0.4; F-test : 47.9.
Significance: *1% with one-tailed test.

Table 16.7. Contribution of intracrop diversity and conventional inputs to durum wheat productivity.

	First Difference estimator		Arellano–Bond estimator	
Variables	Coefficient	Standard error	Coefficient	Standard error
Lagged dependent variable			0.018	0.177
Constant	−0.49*	0.12	−0.0218*	0.00378
Intracrop diversity (fitted)	1.2*	0.18	2.2**	1.092
Rain	0.0003	0.078	0.035	0.037
Pesticide	0.21*	0.092	0.0158527	0.0219
Labour	0.15**	0.071	−0.794*	0.133

Significance: * 1%; ** 5% with one-tailed test. First Difference: R^2: 0.73; F-test = 31.27*; Arellano–Bond: Sargan test: $chi^2(27)$ = 36.62, Prob > chi^2 = 0.1102; test that average autocovariance in residuals of order 1 is 0 (H0: no autocorrelation) yields z = −1.65 Pr > z = 0.0988; test that average autocovariance in residuals of order 2 is 0 (H0: no autocorrelation) yields z = 0.85 Pr > z = 0.3930.

content or humidity. These qualitative differences are related in turn to product differentiation. Consumer demand for a range of wheat-based food products drives the food processing industries to acquire several varieties of crops, each having a slightly different combination of properties. In Italy, the agri-food industry requires diversity in durum wheat and other cereals to satisfy diversified consumer demand for quality food. The same analysis was undertaken for intercrop diversity. However, the results were affected by a serious degree of autocorrelation and are statistically weak.

In the overall context of this book, it is important to recognize that this finding reflects the well-articulated (in market prices) consumer demand for food products found in advanced industrial economies with high incomes. The cost of the market infrastructure that supports this differentiation is borne by consumers, and in this case study, has the positive, unintended side effect of supporting regional diversity levels. In less industrialized agricultural economies, and countries with lower incomes, such market infrastructure does not yet exist and it would be costly to construct solely for the purposes of maintaining crop biodiversity on farms.

Furthermore, keeping crop biodiversity appears to positively affect long-run productivity, at least at the aggregate level. It is possible that agricultural cooperatives can be a viable way to internalize some of the economic and ecological benefits of biodiversity. In other words, the cooperative can 'internalize' the public good externalities of individual variety choice decisions, such as the genetic diversity that results on farms scattered across a crop-producing region. To test this hypothesis fully, relationships to yield stability or resilience would need investigation.

Implications

The policy implications of this case study are transparent. The conservation of durum wheat diversity in an important crop-producing region of southern Italy (and the EU, for that matter) is an increasing function of cooperative density. Policies that serve to enhance cooperative formation, reduce the cost of membership of cooperatives or the cost of coordination have encouraged the cultivation of a diversity set of durum wheat varieties over the time period studied. Also, the strong and positive marginal effect of variety diversity (including both landraces and modern varieties) on the long-run productivity of durum wheat is a salient finding. This case study contributes to the ongoing debate on biodiversity conservation by providing an empirical example from a high-income, developed European country situated in a Vavilov megadiversity area for wheat.

Acknowledgements

We would like to thank Antonino Bacarella, University of Palermo, Melinda Smale and John Hoddinott for useful comments.

References

Altieri, M. and Liebman, M. (1986) Insect, seed, and plant disease management in multiple cropping systems. In: Francis, C. (ed.) *Multiple Cropping Systems*. Macmillan, New York, pp. 183–217.

Arellano, M. and Bond, S. (1991) Some tests of specification for panel data: Monte Carlo evidence and an application to employment equations. *Review of Economic Studies* 58, 277–297.

Baltagi, B.H. (2001) *Econometric of Panel Data Analysis*. John Wiley, Chichester, UK.

Harlan, J.R. (1992) *Crops and Man*. American Society of Agronomy, Crop Science Society of America, Madison, Wisconsin.

Heisey, P.W., Smale, M., Byerlee, D. and Souza, E. (1997) Wheat rusts and the costs of genetic diversity in the Punjab of Pakistan. *American Journal of Agricultural Economics* 79, 726–737.

Hsiao, C. (1986) *Analysis of Panel Data*. Cambridge University Press, Cambridge, UK.

INEA (n.d.) *Annuario dell' Agricoltura Italiana*. INEA, Rome.

ISTAT (n.d.) *Annuario di Statistica Agraria*. ISTAT, Rome.

Just, R.E. and Pope, R.D. (1978) Stochastic representation of production functions and econometric implications. *Journal of Econometrics* 7, 67–86.

Magurran, A.E. (1988) *Ecological Diversity and Its Measurement*. Croom Helm, London.

Meng, E.C.H., Smale, M., Rozelle, S., Ruifa, H. and Huang, J. (2003) Wheat genetic diversity in China: measurement and cost. In: Rozelle, S. and Sumner, D.A. (eds) *Agricultural Trade and Policy in*

China: Issues, Analysis and Implications. Ashgate, Burlington, Vermont.

Mundlak, Y. (2001) Production and supply. In: Gardner, B.L. and Rausser, G.C. (eds) *Handbook of Agricultural Economics.* North Holland, Amsterdam.

Naeem, S., Thompson, L.J., Lawler, S.P., Lawton, J.H. and Woodfin, R.M. (1994) Declining biodiversity can affect the functioning of ecosystems. *Nature* 368, 734–737.

National Research Council (NRC) (1972) *Genetic Vulnerability of Major Crops.* National Academy of Sciences, Washington, DC.

North, C.N. (1991) Institutions. *The Journal of Economic Perspectives* 5(1), 97–112.

Priestley, R.H. and Bayles, R. (1980) Varietal diversification as a means of reducing the spread of cereal diseases in the United Kingdom. *Journal of National Institute of Agricultural Botany* 15, 205–214.

Qualset, C.O., McGuire, P.E. and Warburton, M.L. (1995) 'Agrobiodiversity': key to agricultural productivity. *California Agriculture* 49(6), 45–49.

Smale, M., Hartell, J., Heisey, P.W. and Senauer, B. (1998) The contribution of genetic resources and diversity to wheat production in the Punjab of Pakistan. *American Journal of Agricultural Economics* 80, 482–493.

Sumner, D.R., Doupnik, B. and Boosalis, M.G. (1981) Effects of tillage and multicropping on plant diseases. *Annual Review of Phytopathology* 19, 167–187.

Tilman, D. and Downing, J.A. (1994) Biodiversity and stability in grasslands. *Nature* 367, 363–365.

Tilman, G.D., Wedin, D. and Knops, J. (1996) Productivity and sustainability influenced by biodiversity in grassland ecosystems. *Nature* 379, 718–720.

Vavilov, N.I. (1951) The origin, variation, immunity, and breeding of cultivated plants. *Chronica Botanica*, 13, 1–366.

Widawsky, D. and Rozelle, S. (1998) Varietal diversity and yield variability in Chinese rice production. In: Smale, M. (ed.) *Farmers, Gene Banks, and Crop Breeding.* Kluwer Academic Publishers, Boston, Massachusetts, 159–187.

Wood, D. and Lenné, J.M. (eds) (1999) *Agrobiodiversity: Characterization, Utilization and Management.* CAB International, Wallingford, UK.

17 Scope, Limitations and Future Directions

M. Smale, L. Lipper and P. Koundouri*

Scope of Current Research

Dimensions of crop biodiversity

Concepts, theoretical principles and econometric approaches are interrelated throughout the chapters of this book, but generalizations are not so straightforward. One reason why is that although the range of empirical contexts represented is broad, the crops and countries studied were selected purposively. The selection of contexts reflects the joint decisions of the national and international scientists involved, as well as the research policy environment of the country. In other words, empirical research has been conducted in countries where at least some national stakeholders have recognized on-farm conservation of crop biodiversity as a policy issue.

Another feature that complicates generalization is that the studies themselves consist of in-depth research that is both location- and crop-specific. Although the conceptual variables defined by the underlying models are similar, the dependent and explanatory variables have been measured with survey instruments that are adapted to each farming system and crop context.

Table 17.1 assembles the 'dimensions of crop biodiversity' encompassed by the studies in this book: country, national income, farming system, crop, level or scale of observation and diversity concept measured. Countries are classified by group according to gross national income per capita, as listed by the World Bank 2004 Development Indicators. Five countries are low-income (Ethiopia, Uganda, Nepal, India and Uzbekistan); one is lower-middle income (Peru); two are upper-middle-income (Mexico and Hungary) and one is high-income (Italy). Two countries are classified as economies in transition from state-controlled to market-based, and Hungary is an accession state to the European Union (EU). The regions studied in Italy are classified as 'backward' or relatively poor and underdeveloped within the EU. Geographical area represented include North America (Mexico) and South America (Peru); Central Asia (Uzbekistan) and South Asia (Nepal and India); East Africa (Uganda) and the Horn of Africa (Ethiopia) and eastern and southern Europe (Hungary and Italy).

Most farming systems include both modern varieties and farmers' varieties or 'landraces', as the term is used in this book (see Chapter 1 for definitions), although in most instances they are dominated by farmers' varieties. Some, such as the *milpa* system of Mexico or home gardens in Uzbekistan and Hungary, are microecosystems. While the farming systems represented are generally found in comparatively remote areas with relatively low

---

* The final section in this chapter is based on the contributions of authors during a workshop sponsored by IPGRI, IFPRI and FAO: Bálint Balázs Györgyi Bela, Ekin Birol, Sam Benin, Romina Cavatassi, Evan Dennis, Salvatore Di Falco, George Dyer, Svetlana Edmeades, Ágnes Gyovai, Pablo Eyzaguirre, Devendra Gauchan, Berhanu Gebremedhin, Toby Hodgkin, Jarilkasin Ilyasov, Latha Nagarajan, Edilegnaw Wale, Paul Winters and Patricia Zambrano.

© CAB International 2006. *Valuing Crop Biodiversity: On-farm Genetic Resources and Economic Change* (ed. M. Smale)

Table 17.1. Dimensions of crop biodiversity analysed in book chapters.

Chapter	Country	Income group[a]	Farming system	Crop	Crop reproduction system	Unit of observation (level or scale)	Diversity concept[b]
4,6,11,14	Ethiopia	Low	Mixed modern and traditional	Cereals (maize, wheat, barley, teff, finger millet, pearl millet, sorghum); coffee; wheat and maize, multiple crops	Range of self- and cross-pollinating rates; vegetative	Household and plot; village; some regional variables	Intra- or intercrop
10	Nepal	Low	Focus on traditional	Rice	Highly self-pollinating	Household and plot; breeding programme; some ecosite variables	Intracrop
7	Uganda	Low	Mainly traditional	Highland banana	Vegetative	Household and plot; some village and regional variables	Intracrop
12	Uzbekistan	Low	Microecosystem; mixed modern and traditional	Fruit trees, grapes and nuts	Vegetative	Household and plot	Intra- and intercrop
13	India	Low	Mixed modern and traditional	Sorghum, pearl millet, finger millet, other minor millets	Range of self- and cross-pollinating rates	Village; some household variables; some district variables	Intra- and/or intercrop
9	Peru	Lower middle	Mixed modern and traditional	Potato	Vegetative	Household; some regional variables	Intracrop
3,8,15	Hungary	Upper middle	Microecosystem; mixed modern and traditional	Home gardens; maize and beans	All systems	Household and plot; settlement; some regional variables	Intra- and/or intercrop
2,5	Mexico	Upper middle	*Milpa* microecosystem	Maize only; maize, beans and squash	Highly cross-pollinating	Household and plot; some village and regional variables	Intra- and intercrop
16	Italy	High	Mixed modern and traditional	Durum wheat	Self-pollinating	Region	Intracrop

[a]The World Bank (2004) defines GNI per capita as 'the gross national income, converted to US dollars using the World Bank Atlas method, divided by the midyear population. Low-income economies had GNI per capita of US$735 or less in 2002; middle-income economies had more than US$735 but less than US$9076; lower-middle-income and upper-middle-income economies are separated at US$2935; high-income economies had US$9076 or more'.
[b]All chapters base the classification of varieties on farmer and/or breeder taxonomies. Diversity indices are spatial (for definitions see Chapter 1).

productivity potential, within each study context, market infrastructure, services and production environment vary by ecosite, village, settlement or region.

Studies have been undertaken in locations where the current state of scientific knowledge considers that crop biodiversity of global economic value remains in the fields of farmers. All survey sites are located in known centres of origin and/or diversity, although not always for the crops investigated: maize, rice, durum wheat, sorghum and millet, potato, highland banana, coffee, fruits and nuts. Hungary is a centre of origin for rye, but the Institute of Agrobotany found few landraces remaining on farms in previous collection missions. Hungary represents one of the more interesting cases from the standpoint of valuing crop biodiversity and economic change. A relatively rich nation undergoing fundamental structural changes in the economy, Hungary is situated within a conducive policy framework (the EU) that explicitly recognizes the multiple functions of agricultural landscapes and their economic value, yet has seed policies that may contribute to the process of genetic erosion (Chapter 15).

Crop reproduction systems range from highly cross-pollinating (maize and pearl millet) to highly self-pollinating (rice and wheat), including plants that are vegetatively reproduced in several ways. Potato tubers can serve as planting material or food. The planting material of a banana is a shoot from the parent plant and the fruit, although it contains seed that is not used for propagation. Fruit trees are primarily reproduced through clonal propagation, with some crops propagated through grafting scion wood on to rootstock. Propagation techniques vary across crops; apples are entirely propagated by grafting, grapes are rarely grafted and walnuts are often grown from seedlings. Bananas and other fruit trees are perennial crops, compared with potato and the other cereal crops studied. Smallholder farmers propagate coffee either from seedlings (especially improved coffee types) or by vegetative cuttings.

In most chapters of this book, the unit of observation and analysis is the household farm. The notion of the household farm includes the social unit of the household and its members, the physical unit represented by the land it cultivates or owns and the crop varieties as recognized by those who make crop production decisions. The economic unit includes those family members who reside elsewhere but remit cash income or transfers, as well as the non-farm activities of those who reside on the farm. In several analyses based on the household farm, variables measured at higher levels of aggregation have been introduced as explanatory factors that condition the decisions made by individual households but that households cannot individually influence. In three of the studies, dependent variables are themselves measured at the village or regional level and the village or region is the unit of observation and analysis. In a number of chapters, seed supply variables measured at the level of the village, breeding programmes or rural development programmes have been included as determinants of crop biodiversity on farms.

The diversity indices applied throughout the book are spatial indices adapted from the ecological literature. The definitions and relevance of these indices for social science analysis of crop biodiversity on farms are discussed in Chapter 1. Taxonomies for classifying crop varieties have been linked to or overlain with those of crop breeders where feasible, emphasizing differentiation within the typologies of modern and traditional.

Determinants of crop biodiversity

The household model of crop biodiversity on farms is derived from the theoretical concept of utility maximization in the presence of market imperfections for crop products, seed or labour (Singh *et al.*, 1986; de Janvry *et al.*, 1991). When the conditions for maximization are met, reduced form equations for the optimal choices of farmers can be expressed in terms of vectors of independent variables that consist of the characteristics of individual households, their farms and the markets in which they trade commodities or labour. Diversity metrics or indices can be constructed over observed, optimal choices, retaining the underlying structure of the reduced form equation. The crop biodiversity observed on a farm is expressed as the outcome or consequence of a choice rather than a choice in and of itself (Chapter 5). In the lexicon of impure public goods presented in Chapter 1, the outcome represents a public good externality associated with a private choice of seed types and crops.

The vectors of independent variables or characteristics can be interpreted in terms of any one of several vocabularies used to describe rural development processes. For example, household characteristics can be understood as a combination of social and demographic descriptors, or as indicators of human and financial capital. Farm characteristics are physical, environmental or agroecological features of the production unit. Except for the case of southern Italy, the farm technologies in this book are non-mechanized and constitute human labour, implements and, in some cases, animal traction, and land. Slopes, elevation, moisture conditions, soil quality and plot fragmentation are fixed land quality and farm physical descriptors. Market-related characteristics include distances to different types of markets that proxy for fixed transactions costs and physical impediments to participating in product, seed or labour markets. Household and market characteristics, as compared with farm characteristics, are those most amenable to public investments and interventions designed to promote development or sustainable management of crop biodiversity.

A summary of the statistical findings from econometric estimation of reduced form equations from chapter case studies is shown in Table 17.2.

Human capital

Across lower- and middle-income countries, the formal education of farm decision makers contributes positively to sustaining crop biodiversity, whether they be men or women. In particular, there is evidence that in some locations and crops, women's education and participation in crop production is associated with a greater number of varieties grown. Women's education and participation in farm production supports intracrop diversity in the Ethiopia cereals case, in the Nepalese case for several indicators of the rice diversity indices and in the Ugandan case when women are decision makers in banana production – that is, in countries where women's levels of educational attainment are on average less than completion of primary school. Schooling – and not only traditional knowledge – is associated with access to seed-related information for any seed type or genetic material.

In all cases but one, where age matters at all, households with higher levels of intracrop biodiversity have older decision makers. That case is for maize in Ethiopia, a newer crop for which modern varieties have been recently introduced. In all cases except Uganda, there is a positive correlation between the age of the household head and his or her farming experience. In the Ugandan study, when experience is adjusted for age, the effect of experience continues to be positive. In higher-income locations such as Mexico and Hungary the effect of age diminishes – elderly farmers cut back in terms of crop and variety diversification. Where ageing farm populations are not being replaced by younger generations of farmers, such as in higher-income countries with declining farm populations, traditional knowledge about crop genetic resources indeed could be lost; where they are being replaced, as in lower-income countries that still retain large farm populations, public investments may need to be undertaken to ensure the continuity of local knowledge.

The quantity and quality of family labour, and family participation in crop production, often bear strong, positive associations with crop biodiversity levels on farms. In challenging production environments with ox-drawn or labour-intensive technology, like in the highlands of northern Ethiopia, greater involvement of men as compared with women tends to be associated with intercrop diversification, perhaps because more physical labour is required to prepare land for multiple cereal crops. The magnitude of the farm labour effect is also strong, however, for crop variety diversity in the multi-crop *milpa* system of Mexico, the rice systems of Nepal and home gardens in Hungary. These are labour-intensive farming systems, where diversification requires even heavier investments of labour. Combined with the education and experience findings, it is evident that cultivating diverse crops and varieties requires higher quality labour, or some specialization in labour that is related to knowledge – a point underscored in the chapter about the effects of migration on the *milpa* system, and one that has repeatedly emerged in the project findings from Nepal.

Table 17.2. Determinants of crop biodiversity on household farms, by case study.

Country (chapter)	Household						Characteristics of farm					
	Age or experience, household head	Education, household head	Women's education or partici-pation	On-farm labour, family size	Other income, transfers, migration	Wealth	Farm size	Good quality land, moisture	Elevation, slope	Number of plots, fragments	Markets	Seed supply, including modern varieties
Ethiopia (6)												
Intercrop	0	0	−	+	0	+	+	0	0	+	0	
Intracrop	−,+	+	+	+	+,−	+,−	+	−,+	+,−	−	+,−	0
Ethiopia (4)	0	0				−	0			+	−	
Ethiopia (14)	0	0		0	0	+	+	+	+	+	+,−	0,+
Uganda (7)	+	+	+ª			+	0	−	−	+	+,−	+
Nepal (10)	+	+	+,−	+		+		+		+	+,−	
Peru (9)	0	0		0	−	+	+(−)	−	+	+	−	
Uzbekistan (12)	+(−)	+		0	+	+	0					+
Mexico (5)	0			+	+,−	0	+		+	0	+,−ᵇ	+
Hungary (8)												
Intercrop	0	0		0	0	+,−	+	+			−	
Intracrop	+(−)			+		+,−	+	−			0	

Note: + indicates statistically significant, positive direction of effect on coefficient of variable in econometric regression; − indicates negative effect; +,− means both directions of effects observed for different equations; (−) shows that second-order effect is decreasing; 0 indicates no effect; blank indicates that the factor was not measured or was not relevant to the study.
ªEffect if banana production decision maker is a woman.
ᵇIn particular, labour markets.

Off-farm income and migration

Rising opportunity costs for farm family members in countries undergoing rapid economic change may lead to less diversity within cropping systems, other factors held constant, although the evidence provided in the chapters of this book is mixed. Income from regional employment, permanent migration and participation in social networks that facilitate migration to the USA have a detrimental effect on diversity in the *milpa* (maize, beans, squash) system in the Sierra Norte de Puebla, Mexico, offsetting the positive impacts achieved through cash earned in temporary migration. On the other hand, off-farm employment of family members supports the diversity of fruit and nut trees in the backyard gardens of Samarqand, Uzbekistan. In northern Ethiopian highlands, the relationship of transfers, gifts and remittances to the richness and evenness of varieties grown differs by crop. In Hungary, no relationship was apparent.

Assets

The message concerning wealth is relatively more uniform. In almost all case studies conducted in lower-income countries, the relationship between crop biodiversity levels observed on farms and assets, denominated in terms of livestock, land or consumer durables, is strong and positive. Wealthier households are those that maintain a greater number of crops and varieties, which are more evenly distributed. As overall national income rises, the effects of asset ownership become more ambiguous. Like the finding concerning human capital, this finding reminds us that in poorer communities, possessing more generally has other ramifications – such as access to seeds and related information, as well as more resources to cultivate a range of crops and varieties with different soil, moisture and management regimes. In higher-income countries, having more assets means specialization or leaving agriculture entirely. An exception in terms of the sign of the estimated coefficient is coffee in Ethiopia, where farmers with a lower ratio of assets to consumption requirements demand more coffee attributes, most of which are yield-related. Wale and Mburu explain that these households are more vulnerable to fluctuations in cash income earned from coffee sales.

Physical conditions on the farm

Although farm size doubles as an indicator of wealth, in those studies where spatial diversity indices are used to measure crop biodiversity on farms, the extent of land area on the farm is a factor that controls for scale. The literature suggests that the probability of encountering an additional species or subspecies rises with the geographical scale of analysis. The consistency of this effect is evident across all income levels and crops. The Peruvian example provides an additional piece of information – as land areas farmed rise, the positive effect of an additional unit of area diminishes. There is only so much diversification that a farm household demands or is capable of managing.

Physical and agroecological determinants are also crucial to crop and variety diversification on individual farms, consistent with scientific literature about plant population genetics and biogeography. Conflicting signs in Table 17.2 reflect different farming systems and empirical proxies, although in all cases, the block of physical features shapes the crop biodiversity observed on farms. Where measured, higher numbers of plots and fragments bore an almost universally positive relationship to inter- and intracrop diversity. Similarly, more diversity was generally found at higher elevations with more variable slopes and land quality. The Mexico and Peru cases, which build on the earlier work where some of these hypotheses were initially tested (Brush *et al.*, 1992; Bellon and Taylor, 1993), confirm earlier findings. In the case of cereal crops in the Ethiopian highlands, slope, erosion, fertility and irrigation were independent of the diversity among crops grown by farmers, while the direction of their effect on infraspecific diversity depended on the crop; in eastern Ethiopia, considering all crops, higher elevation and good farming conditions contributed positively to interspecific diversity. This second finding is consistent with the notion that having access to 'more' (more fertile land) lends itself to diversification in an environment where production diversification remains an important strategy for managing risk. In contrast, in Uganda, the higher elevation, higher rainfall areas of Uganda specialize in production of particular banana types for the commercial market; in Peru, a predominance of fertile black soils also implied specialization in fewer potato varieties.

These results have two fundamental implications for sustainable levels of crop biodiversity on farms and economic change, each related to the propositions advanced in Chapter 1. First, as long as there are harsh production environments where markets function imperfectly, there will be rural households that depend very much on the diversity of the materials they grow for the goods they consume, they will not be able to substitute farm production with goods purchased on the market and a range of crops and varieties will be necessary to ensure the family food supply through home-produced goods. As a consequence, these locations will also be those where supporting sustainable management of diversity will cost least in terms of public investments or effective subsidies.

Product and seed markets

Yet, this does not necessarily mean that those who maintain crop biodiversity need be 'left out' of the process of economic development. With respect to the development of markets, the case studies presented in this book extend those of previous literature, but raise more questions than they answer.

The working hypothesis in the literature reviewed in Chapter 1 and that advanced above suggests that market development provides disincentives to maintaining crop biodiversity. Indeed, market isolation almost always has the expected positive effect on crop biodiversity in the case studies of this book. None the less, the relationship of market development and commercialization to crop biodiversity appears more complex when specific aspects of markets, other than sheer isolation from physical infrastructure or road density, are disengaged. Market participation as a product seller enhances the range of endemic banana varieties grown in Uganda, while participating as a product buyer has the opposite effect. In the hillsides of Ethiopia, different types of markets or road access seem to influence the richness (numbers) of varieties grown in opposing ways. Cooperative marketing supports durum wheat diversity in an economically marginalized area of southern Italy.

Seed supply through markets sometimes enhances and sometimes detracts from crop biodiversity. Greater numbers of distinct varieties available in a village are associated with richer and more evenly distributed banana landraces on farms in Uganda. Access to a combination of official and unofficial seed supply institutions, including the bazaar, national plant breeding institute and other village social networks is one determinant of the total diversity of fruit varieties in home gardens of Samarqand, Uzbekistan. In Nepal, local grain markets clearly provide incentives to grow landraces with aromatic quality, but not those with coarse grains. Seed volumes traded through local weekly markets contribute to greater diversity in minor millet landraces grown in villages of southern India; larger quantities of seed traded through dealers, regardless of identity, contribute to a wider range of pearl millet varieties grown. Unexpectedly, seed supply interventions through disaster relief and extension programmes, including the introduction of modern varieties, do not appear to diminish the richness or evenness of potatoes in Peru or crop diversity in eastern Ethiopia. This topic requires much more study before conclusions can be drawn, however.

Villages, settlements and regions

Within the same region of a country, determinants of inter- and intracrop diversity are highly location-specific, as illustrated by the Ethiopia, Peru, India, Nepal and Hungary case studies. Regional fixed effects are typically pronounced, and data support both region-specific levels of diversity and distinct marginal effects of explanatory variables. In the India study (Chapter 13) and in one of the chapters about cereal crops in Ethiopia (Chapter 12), the unit of observation and analysis was the village. That is, both dependent and explanatory variables were tabulated at the village level.

In Amhara, as well as in the more environmentally degraded region of Tigray, villages with households that are better off in terms of human and financial capital have higher levels of inter- and intracrop diversity. The influence of fixed transaction costs differs by region, depending also on whether they involve distance from the village to a major road or district markets. If other factors are held constant, villages with more extensive eroded land tend to grow more cereal crops that are more evenly distributed across the cultivated landscape.

Literacy levels in the farming community and overall access to oxen and credit affect intracrop diversity positively across cereal crops, in some instances with a large magnitude. Agroecological features and market infrastructure bear both positive and negative coefficients, according to crop. Location of a village in the region of Tigray augments the number of both barley and finger millet varieties per village by more than one, decreasing the number of maize varieties by nearly one and the richness of sorghum varieties by over one. The introduction of modern varieties of maize has added to intracrop diversity in villages of Tigray. Maize is a relatively new crop in Ethiopia and less of it is grown in that region. The introduction of varieties of bread wheat has no appreciable effect one way or another.

Among communities (*panchayats*, containing multiple villages) in southern India, district fixed factors alone explain most of the variation in levels of millet intercrop diversity. The density of roads in the community lessens the dominance of the most widely grown variety of pearl millet by providing a wider range of improved varieties, but the opposite is the case for sorghum, and by a very large magnitude. The greater the proportion of village women involved in farming the greater is the diversity of sorghum and pearl millet varieties. In contrast with the findings in Table 17.2, wealthier villages in southern India generally appear to have less intracrop diversity in millets, although those with higher land values grow more diverse minor millets. In this arid zone with limited irrigation, larger rain-fed areas in communities imply more richness in pearl millet and sorghum varieties and less dominance by any single variety.

Seed system parameters were introduced in the Peru, southern India and eastern Ethiopia case studies. In the millet-based systems of southern India, seed system factors significantly affect the level of variety diversity in almost all regressions. Higher seed replacement ratios in a community suggest higher equilibrium levels of farmer demand for seed, given seed system supply. The average seed replacement ratio in a village is positively correlated with the spatial richness and relative abundance of varieties of major and minor millets in villages of Andhra Pradesh and Karnataka. Greater seed volumes traded through local weekly markets enhance the diversity of minor millet varieties, and those traded by dealers are significant for pearl millet diversity, a crop that is highly cross-pollinating and for which hybrids have been developed. In the Peru study, the introduction of a new, late-blight-resistant variety was not associated with less potato diversity. The eastern Ethiopia study demonstrates that a supply intervention promoting the seed of a minor cereal influenced its proportional abundance on farms relative to other crops, but not the total number of crops grown.

The one regional study in the book (southern Italy) is an analysis based on the partial productivity analytical framework rather than the household farm model, and is consequently based on secondary data. In this economically marginalized area of a rich, industrialized country, cooperative production and marketing positively affect the intracrop diversity of durum wheat, a food staple.

Policy Implications

Private and public value

The chapters of this book have fundamental implications for the way we look at the private and public value of crop genetic resources. Two chapters apply stated preference approaches to examine the private value farmers themselves associate with the non-market benefits of crop biodiversity. Positive values are evident among small-scale, traditional farmers in environmentally sensitive areas of Hungary for several components of agrobiodiversity, including the richness of crops and varieties, the genetic diversity contributed by local landraces and integrated livestock and crop production (Chapter 3). Yet, the predictions of economic principles about the value of agrobiodiversity to farmers are confirmed, even among regions within this relatively rich nation. Farmers in the less productive, most remote regions of this high-income country value agrobiodiversity the most. As the settlements in which farmers reside develop and the physical infrastructure of their markets becomes denser, they rely less on their home-produced goods for food and the value they ascribe to agrobiodiversity on their farms diminishes.

Results that reflect economic changes at the margin are of no use in describing corners and jumps in decision making – that is, what occurs when it is not feasible or optimal for farmers to engage in an activity at all, or what happens to them when the changes they face are abrupt and relatively large. For instance, it is likely that despite the structural changes that may occur with Hungary's accession to the EU, some isolated regions will continue to be disfavoured agroecologically and economically. As a part of Hungary's development strategy and the EU's policy of multi-functional agriculture, other social goals might be addressed by policies that would support more sustainable agriculture in the sites already targeted for biodiversity conservation and land-extensive (labour-intensive) agriculture.

Dyer (Chapter 2) too questions the relevance of static comparisons of marginal value to predicting the costs and benefits of on-farm conservation. He contends that not only do these values vary among households, but also they are jointly determined with the decision making process. Instead, the key question to him is how farmers respond to policy-induced abrupt income or price changes by choosing among competing crops, varieties and economic activities. In the context of his research, the external 'shock' is the North American Free Trade Agreement (NAFTA). The evidence that non-market benefits of maize production continue to be great is that the supply response to NAFTA has not been what was expected – remarkably, maize supply has remained above the 1990 level even in the rainfed areas where maize landraces dominate and semi-subsistence farmers have not benefited from subsidies. He finds that responses to both maize price and income changes depend clearly on the type of grower and household characteristics – supporting the viewpoint that marginal values are endogenously determined. In other words, while maize landraces in Mexico are known to be of global value, clearly they are also of private value to the farmers who grow them – even in this upper middle-income country.

The crux of least cost conservation, as the concept was explained in Chapter 1, is to identify the factors that increase the likelihood that farmers will find privately valuable what is also publicly valuable. In the case of crop genetic resources, the non-market, public good benefits are embodied in the seed, for which the costs and benefits could be more easily measured on markets if markets were performing adequately. The chapters about Mexico illustrate these points. However, not all landraces are equally valuable. Gauchan et al. (Chapter 10) use three proxies for the public value of rice landraces based on stated preferences of rice breeders and geneticists who are familiar with them. They then identify the predictors that farmers choose to grow landraces that also belong to the choice sets of breeders and conservators – that is, a coincidence in public and private value. Perhaps the single largest determinant is location in the hillside ecosite of Nepal. Within that location, however, it is the better-off households with more labour, more assets, more land and more rice area that are

Table 17.3. Determinants of crop biodiversity in villages, settlements or regions, by case study.

Country (chapter)	Household				Physical characteristics of the farm		Market infrastructure	Seed supply, including modern varieties	Cooperative density
	Education, literacy	Men as proportion of on-farm labour	Assets, access to credit, land or oxen	Off-farm labour	Good-quality land; moisture	Elevation			
Ethiopia (11)									
Intercrop	+		+		–	0	+,–		
Intracrop	+		+		+,–	+,–	+,–	+,0	
India (13)									
Intracrop	+,–	–	–	+,–	+		+,–	+	
Italy (16)									+

Note: + indicates statistically significant, generally positive direction of effect on coefficient of variable in econometric regression; – indicates negative; 0 indicates no effect; blank indicates that the factor was not measured or was not relevant to the study.

most likely to grow landraces thought to be of value to future crop improvement.

The controlled, highly articulated and differentiated markets for which they produce, combined with a challenging production environment and a historical endowment of local wheat diversity, contribute to productivity gains from intracrop diversification in the regions of southern Italy. Farmers earn additional revenues, the region earns a revenue share and Italy recoups a national revenue share in the EU through this effect. In this industrialized economy, there is no trade-off between revenues and diversification, or revenues and intracrop diversity of durum wheat.

Crop biodiversity on farms and economic change

Many of the case study findings suggest that factors associated with economic development may not, in the short-term, detract from intracrop and in particular intercrop diversity on farms. Education of men and women almost uniformly has a positive effect. In some marginal environments, the introduction of modern varieties broadens the range of materials grown rather than replacing it. Investments in different types of market infrastructure may have offsetting effects. Asset accumulation enhances rather than detracts from crop biodiversity in most of these studies.

On the other hand, those farmers currently maintaining crop biodiversity are generally older, and it is evident that diversification in any form is most often associated with relatively labour-intense production. The negative impact of long-term, international migration is highlighted by the Mexico case. In Peru, potato diversity declines with a rapid uptake by farmers of a labour-intensive, but profitable alternative – dairy farming.

These findings underscore an essential point: there will often be better ways to relieve poverty than through either the introduction of crop varieties or their diversification. Supporting crop genetic diversity conservation is not, in general, a way out of poverty – unless it is linked to an income-earning activity. Growing a stable food crop is not likely to be highly remunerative in a subsistence-oriented farming system, unless, as in the case of durum wheat in southern Italy, highly differentiated, commercial markets can be developed. Even in such cases, there are hefty public costs associated with the creation of this infrastructure and strong consumer demand is one prerequisite for their success.

Conservation objectives

Trade-offs were hypothesized between conservation objectives, but in fact few were found in the context of these chapters. The three diversity indices applied in most chapters of this book express different diversity concepts, or conservation goals: richness of crops or varieties, evenness or proportional abundance and relative abundance or dominance (Chapter 1). Benin *et al.* (Chapter 6) found no apparent trade-offs between policies that would enhance one type of diversity (richness) versus another (evenness) at the household level in the northern Ethiopian highlands; nor did Gebremedhin *et al.* (Chapter 11) at the village level – either for inter- or intracrop diversity of cereals. No offsetting effects are found for richness or equitability of highland banana varieties or use groups at the farm level in Uganda (Chapter 7), or for potato diversity in Peru (Chapter 9).

Gauchan *et al.* (Chapter 10) explore other trade-offs associated with an array of conservation objectives. With richness, evenness and dominance indices, which are metrics constructed over varieties or crops, conservation goals are related to the numbers, evenness or equitability of varieties grown in communities without regard to the nature of the varieties. A second type of trade-off involves differences in landraces targeted for conservation, according to the criteria established by rice geneticists (rarity, heterogeneity, adaptability). The findings reveal few trade-offs in either case, although some interventions may more effectively support the cultivation of rare landraces, such as those related to marketing the grain of landraces.

Trade-offs in policy impact across crops is pronounced. Programmes designed to encourage intraspecific diversity in one cereal crop might have the opposite effect on another crop (Chapters 6 and 11), while those supporting one component of agrobiodiversity might reduce the chances that another is sustained (Chapter 3).

Conservation and equity

Statistical profiles of households most likely to sustain crop biodiversity suggest social equity consequences that may be associated with launching conservation programmes. In Hungary, targeting the households most likely to maintain crop biodiversity at least cost is equivalent to targeting the poor, or relatively disadvantaged rural populations. Although most farmers on the hillsides of Nepal may be ranked as poor by global standards, targeting the households relatively more likely to maintain valuable landraces in those locations is by no means equivalent to targeting the poor. It is the better-off households with more labour, more assets, more land and more rice area that grow socially valuable landraces. In this nation with very low per capita income, sustaining diversity does not necessarily imply that farmers are kept in poverty. Local conservation initiatives might have greater probabilities of success, in fact, when *not* working with the most poor, unless it is to provide with access to genetic materials or related resources.

Women's education and participation, where measured, appear to relate positively to intracrop, or variety diversity. This finding is consistent with hypotheses from the literature, relating in some case to the gender division of labour (managing seed stocks along with food stocks), and in others to the importance of the crop in family subsistence and women's responsibility in food preparation and consumption.

Research Advances and Limitations

Published research that applies economics methods to investigate the value farmers themselves place on agrobiodiversity is sparse (see Chapter 18 for the bibliography). Most published research involves economic theory and detailed ethnobotanical or anthropological case studies. The chapters in this volume, and the original field studies from which they were drawn, contribute both in breadth and depth to that literature. Authors have consistently sought to ground their research in both theoretical principles and farmers' circumstances, although each has met challenges in addressing the topic of this book. Approaches and tools from several fields of economics have been combined in an attempt to gain fuller scientific comprehension and greater policy relevance. Fields include agricultural economics, environmental economics and institutional economics, although the three analytical approaches have not yet been integrated analytically. The authors' assessments of progress and limitations are summarized next.

Revealed preferences analysis based on the household model

Strictly speaking, the household model of on-farm diversity reveals the constrained preferences of farmers for crops and seed types. Linking social and economic factors to agricultural diversity on farms requires a theoretical model and an econometric approach that enables the testing of nested and multiple hypotheses as well as flexible formulations of similar hypotheses. The household model of on-farm diversity achieves both. The reduced form estimation permits both joint tests of hypotheses related to the separability of production and consumption decisions and individual tests of hypotheses concerning specific policy variables, such as public education and transactions costs. In addition, the dependent variable can be formulated in terms of any proposed diversity metric that best captures the concepts the researcher seeks to investigate. For instance, to investigate policy trade-offs in terms of conservation goals, the effects of the same set of explanatory variables (measurements taken on farm, household and market characteristics) on different diversity metrics were tested.

Here, diversity metrics have been adapted from indices of spatial diversity employed in the ecological and crop science literature. Units summarized by each scalar metric are counts or shares of crop varieties, as farmers, taxonomists or plant breeders understand them. More sophisticated indices, in terms of either mathematics or genetics, can also be constructed using molecular data (see Chapter 1 and Meng *et al.*, 1998, for an overview). In general, however, the more sophisticated the index, the more it is removed from the choices farmers make and the more costly it is to obtain in a large cross-sectional data set. Such indices communicate more to geneticists employed in plant breeding programmes or gene banks, or to conservationists, than they do to farmers and development practitioners. Relating farmer-managed units to those used by scientists is a continuing challenge.

The statistical and economic underpinning of the approach means that the econometric output can be understood in terms of predictions. Stratification of the sample captures large, discrete differences in indicators of economic change. On-farm diversity levels and their sensitivity to changes in explanatory factors can be predicted; farmers most likely to maintain diversity can be profiled.

In this way, in centres of crop diversity where public benefits are known to be relatively high, policy or intervention packages can be conceptualized in terms of a least-cost concept. That is, programme designers can potentially use the information to identify the farmers most likely to maintain diversity because they value it most. Among these farmers, costs of public intervention would be least. If these locations are found in centres of crop diversity where scientific knowledge confirms that public benefits are likely also to be high, conservation will achieve the highest total net benefits. This notion parallels that advanced by Krutilla (1967).

The model of the agricultural household is a suitable theoretical context in which to study crop biodiversity and economic change because it makes no assumptions about profit maximization and market function. At the same time, empirically, it contributes little without the contributions of past empirical and theoretical work on modelling the adoption of modern varieties. The approaches presented in the chapters of Part III are generally built from both strands of literature, although in many empirical settings, more emphasis could and perhaps should be made on the role of modern varieties within systems to understand better the conditions under which they support farmer income and complement local genetic materials.

The analysis of the role of modern varieties, as well as of specific rural development interventions, is impeded by econometric challenges related to simultaneity in censored variable systems, and multiple layers of selection or participation bias. Both the hypotheses related to policy interventions and depiction of these interventions at the farm level need fuller articulation.

Another methodological limitation of the household model of on-farm diversity relates to reduced form as compared with structural estimation, although this has been a matter of debate in applied agricultural economics for some time. The comparative statics of the reduced form are ambiguous for the non-separable case of the model. In the specific applications of this book, the dependent variables do not directly measure optimal choices but are metrics over optimal choices. Confronted with the difficulty in accounting either for permanent income in cross-sectional household databases or the endogeneity of labour allocation decisions, authors have in most cases defined exogenous income as receipts in a previous time period or remittances, gifts and transfers.

A practical limitation of the approach used so far is that the nature of market failure remains a mystery. As authors begin to disentangle specific components of markets in their chapters, the fundamental hypothesis that market isolation drives on-farm conservation appears less and less informative. Understanding the role of seed systems, and particularly supply interventions, is critical for those involved in efforts to raise productivity without sacrificing crop biodiversity.

While the information provided through detailed case studies of this type is enlightening when programme interventions are already envisaged, as in the cases of Hungary and Nepal, these studies are costly to implement and burdensome for respondents. Repeatedly, authors found a high degree of location specificity in findings, which suggests that there are few economies of scale to be achieved in conducting this type of research – or that problems of this type cannot be fixed with generalized solutions.

Questions of geographical 'scale' or 'level' of analysis were treated in several chapters through mixing variables measured at the household farm, village, settlement or community levels. For analysis to generate useful information for programme design, it is essential to have prior knowledge about whether the conservation goal is to sustain crop biodiversity levels for the average household, among targeted households, or at the level of a larger social and biological unit. Crop biodiversity levels might be adequately maintained at the village level by only a few farmers, or at the regional level, by only a few villages. Diversity metrics, conceptual approaches and variable measurement must be appropriately adapted to the level of observation and analysis. Analysis at the household level does not provide sufficient information about diversity in larger biological units, even when explanatory economic variables measured in larger units can be introduced into the equation. Moreover, variation across communities may be more important for programme design than variation within any single community. As the scale of programme intervention becomes more removed from the individual farmer, diversity metrics that are more removed from the choices of individual farmers will probably also be more appropriate where they are feasible to implement. In other words, molecular analyses might be suitable for some situations in which scientific research already suggests that public value is high and sampling can be effectively designed.

Stated preferences analysis

Contingent valuation has been applied extensively to value rare and endangered animal species, habitats and landscapes, and has been especially pertinent to assessment of conservation policy. One reason why it has not been widely employed to value agricultural biodiversity is that, even if provided with details, respondents find it challenging to value unfamiliar species or complex processes such as ecosystem functions and traditional management processes for crop and livestock types in centres of origin and diversity (Birol, 2004).

Two recent advances in environmental valuation, the choice experiment and a contingent behaviour approach, were applied in this book. Published literature contains few cases of the application of these approaches to valuing the biodiversity of domesticated crops or livestock. The first provides a monetary measure of the value people assign to a change in the provision of a non-market good. The second estimates the impact of a hypothetical change in order to predict the effect of a policy change (e.g. tax, increase in prices, possible market creation).

The choice experiment method provides four pieces of information about the values of environmental goods that may be of use in a policy context: (i) which attributes are significant determinants of the values people place on environmental goods; (ii) the implied ranking of these attributes among the relevant population(s); (iii) the value of changing more than one of the attributes at once; and (iv) as an extension of this, the total economic value of an environmental asset.

As a result of its choice format, the choice experiment method has several distinct advantages compared with contingent valuation. Respondents may be more comfortable with decisions among choice sets than with direct questions concerning willingness-to-pay (WTP) or willingness-to-accept (WTA) compensation. Choice sets are like menus, or options, that can be portrayed or illustrated in ways that are relatively easy for respondents to conceptualize. Also, the strategic bias of stating an extreme monetary value to get a point across is minimized with the choice experiment method since the prices of the goods are defined implicitly within the choice sets. Other types of respondent bias that are associated with contingent valuation are also eliminated.

The flexibility of stated preference and its compatibility with contingent valuation and revealed preference methods of valuation suggest that it will become a popular method of eliciting environmental preferences. Recent advances in the stated preference method include incorporating uncertainty in the choice models, including dynamic elements (state dependence and serial correlation), incorporating non-choice alternatives and a variety of experimental design and model validation issues. These are not well addressed in existing contingent valuation and revealed preferences approaches.

Stated preference models have a long history in marketing and transport literature, and are generally well accepted as methods for eliciting consumer responses. They also seem to be well suited to addressing questions that have troubled economists for some time, such as difficulties in estimating benefit transfers. If an activity can be broken down into its attribute components, and if models can be appropriately 'segmented' to account for different types of users, the stated-preference approach may provide a broad-enough treatment of responses to allow for more accurate benefit transfer calculations. These techniques will undoubtedly become more widely used in the valuation of environmental amenities and in the economics literature in general.

Like those based on the household farm model of on-farm diversity, the approaches share the essential drawback that they require intensive, primary data collection. In the case of the choice experiment, the apparent simplicity of the survey instruments relative to household surveys disguises the complexity involved in data manipulation. Moreover, as in any household survey, the design of the survey instrument, as well as respondent comprehension of the concepts, is of utmost importance. As in the case of household surveys, measurement error in operational variables may be great. Ideally, instrument design should in both cases be preceded by informal surveys and some participant observation. The instrument itself should be pre-tested. Moreover, any hypothetical approach has the weakness that it seeks to measure the consequences of an event that has not transpired. This weakness can be minimized by proper design and interview practice.

Institutional analysis

Institutions, ranging from local norms of access and exchange to seed markets, national breeding programmes and international proprietary regimes for plant genetic resources, are the purveyors (conduits) of the public goods embodied in seed. Although they have been treated as exogenous variables in a number of ways throughout the book, applications of institutional analysis per se have been few. Yet, the opportunities for contributions from this field are substantial: in alternative approaches to valuation, in comprehending access the farmers have to crop genetic resources and in enabling stakeholders in local, national, and international policy to formulate their own solutions.

Contemporary institutionalism views the exercise of valuation as a social process of forming preferences, so that research methods should be applied in order to understand and make room for alternative types of valuation. Institutional analysis is also a means for linking the decisions of individual farm households to crop biodiversity observed at more aggregated levels of analysis, such as the identification of seed supply channels and actors.

Stakeholder analysis aims at identifying key actors or stakeholders of a system or a problem under examination. Mapping and stakeholder analysis situates households within the context that proscribes their behaviour and that they themselves can influence. These facilitate understanding of barriers in access to seed as well as to related information. The textual analysis presented by Bela *et al.* (Chapter 15) illustrates the dissonance of vocabularies and views that even well informed stakeholders often hold. Such analyses may also contribute to the process of articulating strategies to resolve conflicts. Policies act on institutions by changing rules. By understanding institutions better, more effective policies for on-farm conservation can be developed.

Future Research Directions

At the household level, perhaps the most promising research direction in terms of methodology would involve merging of stated and revealed preference approaches. Since both choice experiment and farm household data analysis are based on random utility theory and the data are from the same farm families, they are combined to get a richer data set and to take advantage of the relative strengths of different types of data. Both stated and revealed preference methods have advantages and drawbacks. Stated preference methods are criticized because of their hypothetical nature and the fact that actual behaviour is not observed; revealed preference methods suffer from collinearity among attributes and other modelling shortcomings. Combining the two is expected to increase the statistical efficiency of results and lend greater validity. There are also good arguments for embarking on institutional analysis as a precursor to analyses of stated and revealed preferences, and for comparing qualitative and quantitative findings.

In addition, the roles of production and consumption risk are relevant to stated preference formulation. In general, additional applications of stated preference methods are needed in order to assess the advantages and disadvantages of the research tool in poorer countries with less literate populations. Intertemporal, or dynamic, aspects should be considered in the household farm model or in a production function framework (as long as prices are endogenous) – both in terms of model structure and measures of cropping system resilience rather than crop biodiversity levels. Multi-output technologies, and interactions with other components of agrobiodiversity, such as livestock, probably underlie some of the results reported here, despite the fact that they were not explicitly treated.

Still at the household level, future research directions in terms of topics include the effects of crop biodiversity on other aspects of household welfare, such as nutritional values, and intra-household modelling of gender-related differences in valuation and management of crop genetic resources. Economic models of intra-household decision making have not been applied yet in this body of empirical research. Gender-disaggregated data permitted the testing of hypotheses in several chapters of this book, but more research is needed to better articulate the role of gender in models of the type presented here.

Although the practical interest of farmers underlies our perspective in this book, chapters have emphasized choices of crops and crop varieties rather than livestock. Research on the value of livestock genetic resources and their diversity has recently emerged (Drucker *et al.*, 2001), with some congruence in applied methods and tools. In many chapters of this book, livestock assets are used as indicators of wealth, and occasionally the suitability of a variety for feed or fodder is used to explain its cultivation. The private value of mixed livestock and crop production on small farms has been estimated in one chapter. In none of the chapters are livestock numbers or races modelled as choices, separately or simultaneously with the choice of crops or varieties.

The authors have concluded that, in parallel with continued advances in valuation methodologies, future research should seek to link household modelling with higher levels and scales of observation and analysis. There are compelling arguments that stakeholder analysis should precede formal modelling given the policy sensitivity and communications challenges encountered in proposing and implementing local conservation initiatives. The paradigm of institutional environmental economics offers a constructive way to begin bridging scales or levels of observation and analysis.

One entry point for examination of crop biodiversity at larger geographical scales is the local seed system, although this type of economic analysis will also require advances in terms of conceptual and theoretical frameworks. Some tentative definitions and concepts are found in Part IV of this book. The analysis of seed systems naturally moves the research agenda towards several of the unresolved and important issues that have emerged from this set of studies, including consideration of institutions and how they influence individual and group behaviour; the role of markets in the seed system and the incentives or disincentives they provide to farmers for maintaining high levels of crop biodiversity and the need for analysis at higher geographical scales of analysis. In addition, interventions in seed supply are an important part of agricultural development programmes. Focusing more analytical attention on the role of local markets in seed systems, the relationship of local markets to other seed system institutions and the impact of the seed system institutions on farmers' access to genetic materials is critical. By analysing seed system interventions, more can be learned about the possible trade-offs and synergies between agricultural development and *in situ* conservation.

As noted in the first chapter, the social dilemma of on-farm conservation stems from the mixed good properties of the crop genetic resources embodied in seed. A social dilemma implies that there are no simple policy solutions. Situations where policies can promote both rural development and the conservation of crop biodiversity on farms may be uncommon. As concepts, conservation is often perceived as static, while development is dynamic. Thinking of conservation as a process by which farmers are better equipped to maintain useful crop genetic resources during economic change is perhaps more appropriate. Social gains might be achieved if, through a better comprehension of the complex of seed system institutions that affect the local supply of seed and farmer access to diverse genetic materials, we can identify practical entry points to support both conservation and rural development.

References

Bellon, M.R. and Taylor, J.E. (1993) 'Folk' soil taxonomy and the partial adoption of new seed varieties. *Economic Development and Cultural Change* 41, 763–786.

Birol, E. (2004) Valuing agricultural biodiversity on home gardens in Hungary: an application of stated and revealed preference methods, PhD thesis, University College London, University of London, London.

Brush, S.B., Taylor, J.E. and Bellon, M.R. (1992) Technology adoption and biological diversity in Andean potato agriculture. *Journal of Development Economics* 39, 365–387.

de Janvry, A., Fafchamps, M. and Sadoulet, E. (1991) Peasant household behaviour with missing markets – some paradoxes explained. *Economic Journal* 409, 1400–1417.

Drucker, A.G., Gómez, V. and Anderson, S. (2001) The economic valuation of farm animal genetic resources: a survey of available methods. *Ecological Economics* 36, 1–18.

Krutilla, J.V. (1967) Conservation reconsidered. *The American Economic Review* 4, 777–786.

Meng, E.C.H., Smale, M., Bellon, M.R. and Grimanelli, D. (1998) Definition and measurements of crop diversity for economic analysis. In: Smale, M. (ed.) *Farmers, Gene Banks and Crop Breeding: Economic Analyses of Diversity in Wheat, Maize, and Rice.* Kluwer Academic Publishers, Boston, Massachusetts, pp. 19–31.

Singh, I., Squire, L. and Strauss, J. (1986) A survey of agricultural household models: recent findings and policy implications. *The World Bank Economic Review* No. T I-1.

World Bank (2004) World Bank Development Indicators 2004. CD-ROM.

18 An Annotated Bibliography of Applied Economics Studies about Crop Biodiversity *In Situ* (On Farms)

P. Zambrano and M. Smale

The purpose of this bibliography is to provide researchers with a comprehensive list of the applied economics literature about *in situ* (on-farm) conservation of crops as of December 2004. This effort was initiated several years ago by Amanda King and Pablo Eyzaguirre at the International Plant Genetic Resources Institute (IPGRI), and updated at the International Food Policy Research Institute (IFPRI) by the authors. The first list of references, entitled 'Economics of Crop Genetic Resource Diversity and Conservation: A Selected Bibliography of Related Literature', was made available by IFPRI and IPGRI on their websites. The scope of the current web posting also includes *ex situ* conservation and animal genetic resources.

The first step in building the bibliography in this chapter was to extract all references from the existing IPGRI–IFPRI bibliography that related to *in situ* (on-farm) conservation of cultivated plants. To further expand its coverage, many other sources were consulted, ranging from the lists of references in relevant papers to electronic databases.

Although the general literature on *in situ* crop biodiversity appears initially to be abundant, when only peer-reviewed articles and published books or reports are considered, the studies that apply economics principles or methods are limited. For example, CAB Direct (an electronic database available for subscribers at http://www.cabdirect.org/) has at least 257 references in its database related to *in situ* conservation of crops. Nevertheless, a review of these 257 references shows that very few of them conduct their analysis by using economics methods.

The following references use economics principles or methods to analyse the conservation of crop genetic resources on farms. Either the abstracts have been drawn directly from those written by the authors or they have been newly drafted because there was no reference to the methods employed in the original abstracts. Where the abstracts have been reprinted, permission has been provided by the publisher.

1. Bellon, M.R. (2004) Conceptualizing interventions to support on-farm genetic resource conservation. *World Development* 32, 159–172.

Abstract: Ongoing public investment in the establishment and maintenance of gene banks around the world needs to be complemented with interventions that support on-farm conservation. To develop effective interventions a proper diagnosis of the causes of on-farm loss of biodiversity should be made. The author identifies specific supply and demand factors that influence the decision of farmers to maintain diverse crops and illustrates his points with a case study of an on-farm conservation project in the Central Valleys of Oaxaca, Mexico. His main conclusions are that if loss of biodiversity is supply-driven, interventions should focus on decreasing the costs of accessing crop diversity, but if the loss is demand-driven, interventions should aim to increase the value of crop diversity for farmers or decrease the farm-level opportunity costs of maintaining it.

2. Benin, S., Smale, M., Pender, J., Gebremedhin, B. and Ehui, S. (2004) The economic determinants of cereal crop diversity on farms in the Ethiopian Highlands. *Agricultural Economics* 31, 197–208.

Abstract: In less favoured areas such as the highlands of Ethiopia, farmers manage risk through land allocation to crops and varieties since they cannot depend on market mechanisms to cope. They also grow traditional varieties that are genetically diverse and have potential social value. Supporting the maintenance of crop and variety diversity in such locations can address both the current needs of farmers and future needs of society, though it entails numerous policy challenges. We estimate a model of crop and variety choice in a theoretical framework of the farm household model to compare the determinants of crop and variety diversity, revealing some of these policy considerations. Farm physical features and household characteristics such as wealth and labour stocks have large and significant effects on both the diversity among and within cereal crops, varying among crops. Policies designed to encourage variety diversity in one cereal crop may have opposing effects in another crop. Trade-offs between development-related factors and diversity in this resource-poor system are not evident, however. Market-related variables and population density have ambiguous effects. Education positively influences cereal crop diversity. Growing modern varieties of maize or wheat does not detract from the richness or evenness of these cereals on household farms.

Printed with permission from Elsevier B.V. © 2004

3. Birol, E., Smale, M. and Gyovai, A. (2004) Agri-environmental policies in a transitional economy: the value of agricultural biodiversity in Hungarian home gardens. EPTD Discussion Paper – Environment and Production Technology Division, International Food Policy Research Institute No. 117. International Food Policy Research Institute, Washington, DC.

Abstract: Much of the agricultural biodiversity remaining *in situ* today is found on the semi-subsistence farms of poorer countries and the small-scale farms or home gardens of more industrialized nations. The traditional small farms of Hungary are labelled *home gardens* as a reflection of their institutional identity during the collectivization period: homesteads managed with family labour, they continue to serve essential food security and diet quality functions during economic transition. Home gardens contain relatively high levels of several components of agricultural biodiversity. The role of home gardens in the agri-environmental programme that is now being formulated by Hungary and the European Union has not been elucidated, although the stated goal of these policies is to support multi-functional agriculture. This study estimates the private value that Hungarian farmers assign to home gardens and their biodiversity attributes, and indicates how such information might be used in designing least-cost mechanisms to support their maintenance as part of the national agri-environmental programme.

© *IFPRI*

4. Birol, E. (2004) Valuing agricultural biodiversity on home gardens in Hungary: an application of stated and revealed preference methods. PhD thesis, University College London, University of London, London.

Abstract: This thesis contributes to the economics of conservation of agricultural biodiversity on farm with a case study on traditional Hungarian home gardens. The aims of the thesis are to (i) measure the private values of home gardens and components of agricultural biodiversity found in them that accrue to farm families who manage them and (ii) investigate the effects of household, market, agroecological, cultural and economic factors on farm families' demand for and supply of agricultural biodiversity in their home gardens. Data on farm families' revealed and stated preferences for agricultural biodiversity in home gardens are collected from 323 farm households in 22 communities across three regions of Hungary, with an original farm household survey and an original, choice experiment. Data are analysed with theoretical and empirical models from agricultural and environmental economics literature to identify those farm families, communities and regions that attach the highest values to agricultural biodiversity and that are most likely to conserve home gardens with high levels of agricultural biodiversity. The results disclose that the most economically and environmentally marginalized communities in the country are most likely to sustain and attach the highest values to home gardens and their attributes. Within these communities, farm families that are larger, have elderly decision makers, lower income levels and home gardens with unfavourable production conditions tend to conserve higher levels of, and attach the highest values to, agricultural biodiversity in home gardens. The findings of the thesis may assist the national policy makers in designing efficient and cost-effective agri-environmental policies for conservation of Hungary's agricultural biodiversity and cultural heritage.

The dissertation *citations and abstracts contained here are published with permission of ProQuest Information and Learning. Further reproduction is prohibited without permission.*

5. Brush, S.B., Taylor, J.E. and Bellon, M.R. (1992) Technology adoption and biological diversity in Andean potato agriculture. *Journal of Development Economics* 39, 365–387.

Abstract: The effects of the introduction of high-yield varieties are researched using household survey data from two Andean valleys of eastern Peru. These two regions are characterized as being in different stages of the adoption process. A simultaneous equation model is used to estimate adoption of improved varieties and the effects of adoption on the diversity of potato landraces. Econometric results show that although biological diversity declines as the area in improved varieties increases, considerable diversity remains and the displacement of landraces by modern varieties depends on the stage of adoption.

6. Brush, S.B. and Meng, E. (1998) Farmers' valuation and conservation of crop genetic resources. *Genetic Resources and Crop Evolution* 45, 139–150.

Abstract: This paper focuses on the value of landraces (traditional and local crop varieties) to farmers in centres of agricultural diversity. Additional information on factors contributing to the private value that farmers assign to landraces may help to identify a strategy for ensuring the conservation of the crop genetic resources (CGRs) that are embodied in landraces while at the same time minimizing the costs. Economic and ethnobotanical approaches for examining the value of landraces complement one another. A formal economic approach establishes a framework for quantitative analysis while ethnobotanical methods provide qualitative data for assessing the likelihood that particular farmers or farm sectors will maintain landraces. This research synthesizes the two approaches in order to examine farmer selection of local wheat landraces in relation to that of modern varieties in three provinces in western Turkey. Multiple farmer concerns (e.g. yield, risk, quality), environmental heterogeneity and missing markets contribute to the persistence of landraces. Household characteristics informing variety choice will also affect the household's perceptions of the importance and value of landraces.

© *2004 Kluwer Academic Publishers, with kind permission of Springer Science and Business Media.*

7. Brush, S.B. and Meng, E. (1998) The value of wheat genetic resources to farmers in Turkey. In: Evenson, R.E., Santaniello, V. and Gollin, D. (eds) *Agricultural Values of Plant Genetic Resources.* CAB International, Wallingford, UK, pp. 97–113.

Abstract: Landraces as other genetic resources have a unique valuation problem because they have both a social and a private value. Farmers who choose to conserve and use landraces do so because they benefit from their consumption and production. Any social valuation attempt should take this into account as private value will ultimately determine the actual supply of landraces. This chapter proposes a valuation approach, focusing on the private value farmers assign to landraces, and uses wheat data from a household survey covering three provinces in Turkey, a centre of domestication and diversity for wheat. The econometric analysis confirms the existence of multiple factors affecting plot-level varietal selection decisions. Landraces have a positive private value for farmers derived from multiple factors, and as long as this happens they will continue to cultivate them.

8. Cromwell, E. and Oosterhout, S.V. (1999) On-farm conservation of crop diversity: policy and institutional lessons from Zimbabwe. In: Brush, S.B. (ed.) *Genes in the Field: On-farm Conservation of Crop Diversity*. IDRC/IPGRI/Lewis Publishers, Boca Raton, Florida, pp. 217–238.

Abstract: There has been little research on the economic, sociocultural and environmental variables influencing farmers' attitudes towards maintaining on-farm conservation. Using previous overall and regional research, as well as participatory rural appraisals, the authors designed a comprehensive questionnaire that included relevant economic, sociocultural and environmental variables. The research was carried out in Mutokp and Mudzi districts in Zimbabwe. The regions were chosen because they were characterized as being rich in crop diversity but were threatened by intense livelihood pressures. Multiple regression analysis was used to determine factors influencing number of crops grown on-farm, number of varieties grown and farm area allocated to all non-hybrid maize cereals. One important result of this analysis is that there is no single set of variables that determines on-farm crop diversity, but rather a complex combination of variables.

© 2000 IDRC and IPGRI

9. Dalton, T.J. (2003) A household hedonic model of rice traits: economic values from farmers in West Africa. *Agricultural Economics* 31, 149–159.

Abstract: New crop varieties often have been promoted in developing countries based upon superior yield vis-à-vis locally available varieties. This research presents a hedonic price model for upland rice by drawing upon the input characteristics and consumer good characteristics model literature. Model specification tests determine that a combination of production and consumption characteristics best explains the willingness to pay for new upland rice varieties. The household model specification determined that five traits explain the willingness to pay for new rice varieties: plant cycle length, plant height, grain colour, elongation/swelling and tenderness. Yield was not a significant explanatory variable of the willingness to pay for seed. The implications of this model are twofold. First, varietal development and promotion must include postharvest characteristics in addition to production traits when determining which varieties to promote for official release. Second, non-yield production characteristics such as plant height and cycle length are significant factors in producers' assessments of the value of a new variety. Overall, this paper provides an alternative explanation for limited adoption of modern upland rice varieties in West Africa: varietal evaluation programmes have focused too narrowly on yield evaluation and have not promoted varieties with superior non-yield characteristics than locally available varieties.

© *2003, with permission from Elsevier.*

10. De Ponti, T. (2004) Combining on-farm PGR conservation and rural development. Clash or synergy? MSc dissertation, Wageningen University, Wageningen, The Netherlands.

Abstract: Although the premise of many of the interventions is that on-farm PGR conservation is beneficial to farmers' livelihoods as well as to PGR conservation, no empirical evidence in support of this assumption could be found in the literature. Considering that Bhutan's unique development philosophy – in which conserving the integrity of its natural, cultural and spiritual heritage is the corner stone of all of its policies – likely offers the most beneficial policy environment to be found in any country in the world for combining on-farm PGR conservation with rural development, it was decided to determine whether, and if so how on-farm PGR conservation of rice – Bhutan's most important staple crop – is compatible with rural development in rice-based cropping systems in Bhutan. It is argued that one can only determine whether on-farm PGR conservation and rural development are compatible or even synergistic when it is made explicit by all involved institutions which unit of diversity (e.g. pure landraces/traditional varieties, or genes or a dynamic system with a broad but fluid genetic base) is to be conserved and at which geographical scale of diversity (e.g. within each agroecological zone, or at the national or global level). To answer the compatibility question, farmer group discussions and group exercises, as well as short individual farmer interviews, were held both in a locality with high exposure to formal research and extension and in a locality with a low exposure. Further, extensive interviews were held with staff of the RNRRCs (research centres), the extension services, the National Biodiversity Centre, an external coordinating NGO and different divisions of the Ministry of Agriculture. The situation in the research localities revealed that despite the faith that researchers and policy makers generally have in de facto conservation – the continued cultivation of traditional varieties without any government intervention – a lot of varieties have already been discarded as a result of farmers' efforts to improve their livelihoods. Further it revealed that, with the exception of one 'ancestral' variety in each locality, variety choice is governed by how well a variety meets farmers' use requirements and growing conditions, rather than traditional varieties having an intrinsic value.

The dissertation citations and abstracts contained here are published with permission of ProQuest Information and Learning. Further reproduction is prohibited without permission.

11. Di Falco, S. and Perrings, C. (2002) Cooperative production and intraspecies crop genetic diversity: the case of Durum wheat in southern Italy. Paper presented at the first BIOECON Workshop on Property Right Mechanisms for Biodiversity Conservation. International Plant Genetic Resources Institute (IPGRI), 30–31 May 2002. Rome.

Abstract: In the standard resource economics literature, the private solution is always suggested as solution to resource degradation as opposed to a common property regime, since the latter is considered as an open access situation. In a dynamic setting, Larson and Bromley (1990) showed that this result does not hold if the common property regime is not free access. In this paper we apply a simple impure public good model in order to show that an agricultural cooperative (assuming homogeneity) or a system of agricultural cooperatives might act as a centralized decision maker, effectively promoting genetic diversity conservation on farms through land allocation among cultivars. This is an application example based on data from southern Italy.

Published with permission from the authors.

12. Di Falco, S. (2003) Crop genetic diversity, agroecosystem production and the stability of farm income. PhD dissertation, Environment Department, University of York, York, UK.

Abstract: Crop genetic resources are the raw materials for crop breeding, pest resistance, productivity, stability and future agronomic improvements. In the last decade in the agricultural and resource economics literature a number of studies on farm conservation of crop genetic diversity have been published. These studies can be categorized under two main strands. The first strand focuses on the contribution of crop genetic diversity to the mean and variance of agricultural productivity and of farm income. The second strand offers both theoretical and empirical investigations of the determinants of loss of crop genetic diversity, mainly in developing countries. These studies have found that production risk is an important driving force behind conservation of crop genetic diversity and loss of crop genetic diversity increases with market integration. This thesis contributes to both strands of the literature by presenting a theoretical and empirical investigation of the phenomenon. The impacts of the Common Agricultural Policy and other institutions on crop diversity conservation are also analysed.

The dissertation citations and abstracts contained here are published with permission of ProQuest Information and Learning. Further reproduction is prohibited without permission.

13. Dyer Leal, G.A. (2002) The cost of *in situ* conservation of maize landraces in the Sierra Norte de Puebla, Mexico. PhD dissertation, University of California at Davis, California.

Abstract: The integration of rural markets and economic development are expected to hinder *in situ* conservation of landraces in the long run, raising conservation costs. The integration of maize markets under the North American Free Trade Agreement (NAFTA) raised concerns for *in situ* conservation of Mexican maize landraces. Although 7 years into NAFTA, rain-fed maize growers have not reacted as expected to falling maize prices and advocates believe that low prices are an imminent threat to landrace conservation. Current economic explanations for the resilience of maize agriculture in Mexico, based on transaction costs, suggest that a downturn in production can be expected if prices decrease further. This study examines the threat to maize conservation in the *milpa* system in Zoatecpan, a village in the Sierra Norte de Puebla, and assesses the potential cost of an *in situ* conservation programme. Although the market value of maize has dropped following NAFTA, it appears that factors other than transaction costs have kept the shadow value of *milpa* above its opportunity cost. Responses to price and income changes, estimated using the contingent valuation approach, show that the elasticity of household supply is positive for price increases, on average, but is nil for price decreases. Analysis of individual household responses suggests that different factors, including liquidity constraints and non-market benefits, influence production decisions. Village-wide responses to policy and market shocks, simulated using a computable general equilibrium model, suggest that price decreases promote a shift from commercial to subsistence maize production and an increase in varietal diversity within household. It is concluded that NAFTA does not pose a threat to *in situ* conservation of maize landraces in the region.

The dissertation citations and abstracts contained here are published with permission of ProQuest Information and Learning. Further reproduction is prohibited without permission.

14. Franks, J.R. (1999) *In situ* conservation of plant genetic resources for food and agriculture: a UK perspective. *Land Use Policy* 16, 81–91.

Abstract: The value of plant genetic resources for food and agriculture (PGRFA) is discussed and the contribution of the UK's agri-environmental schemes to the conservation of these genetic resources is reviewed. It is concluded that the UK's agri-environmental conservation schemes do not prioritize the conservation of genetic diversity of wild relatives of agricultural crops. Surveys of the distribution of genetic variation are required so that PGRFA can be safeguarded by incremental amendments to existing conservation schemes, by adopting new schemes and by altering the contract between the conservation body and farmers to allow farmers to contract as groups rather than as individuals.

Reprinted with permission from Elsevier, © 1999.

15. Gauchan, D. and Smale, M. (2003) Choosing the 'right' tools to assess the economic costs and benefits of growing landraces: an example from Bara district, Central Terai, Nepal. *Plant Genetic Resources Newsletter* 134, 18–25.

Abstract: Economists often use marginal analysis based on partial budget as a tool for estimating the economic returns farmers might expect from using (or choosing not to use) a new practice. However, caution must be exercised when applying this tool in semi-commercial agriculture and especially in analysing the costs and benefits of growing landraces. In semi-commercial agriculture, incomplete markets cause the effective input and output prices actually faced by farmers to diverge within a band defined by producer and consumer prices. In addition, markets may be partially absent for landraces or market prices may fail to reflect their distinctive attributes. Here, we illustrate and expand these points with an analysis that compares the costs and benefits of growing landraces instead of modern varieties in Nepal, a centre of rice diversity. We also suggest other types of economic tools that may be used in assessing the costs and benefits of growing landraces and in addressing issues related to design, implementation and monitoring of projects to conserve crop biodiversity on farms.

© *2003 IPGRI*

16. Gauchan, D. (2004) Conserving crop genetic resources on-farm: the case of rice in Nepal. PhD dissertation, University of Birmingham, Birmingham, UK.

Abstract: Conservation of crop genetic resources is essential to meeting livelihood needs of many small farmers in developing countries and to providing future options for crop improvement. This thesis about on-farm conservation of crop genetic diversity in Nepal analyses the rice variety choices of farmers and plant breeders as well as the policy and market incentives that influence these choices. A sample survey of farm households and a key informant survey of plant breeders, market traders and other stakeholders in crop genetic resource systems were conducted. A farm household variety choice model based on microeconomic theory was developed, which was then tested econometrically to identify the factors that affect farmers' variety choices as well as those choices recognized by the plant breeders. In addition, institutional and market analyses were carried out to understand the market-based and policy-induced (dis)incentives that influence conservation of rice diversity on farms. The findings indicated that the current policy environment and market-based incentives favour modern varieties although bulk of the farmers continue to grow landraces for their livelihood. Market-based incentives also favour a certain group of aromatic landraces, which are more likely to be grown and maintained by relatively better off farmers. Econometric analysis revealed that the factors that influence on-farm rice diversity are market distance, agroecological heterogeneity and adult family labour working on farm. Households and farm plots that are located farther away from market centres and those who own and cultivate heterogeneous lands are more likely to maintain rice diversity as well as to grow socially valued landraces based on breeders' criteria of diversity, rarity and adaptability. Based on the predicted probabilities, the location and profile of farmers that have high likelihood of maintaining socially valued rice diversity were identified. Finally, issues are raised on the development goals, incentives and equity implications of the findings for designing on-farm conservation programmes.

The dissertation *citations and abstracts contained here are published with permission of ProQuest Information and Learning. Further reproduction is prohibited without permission.*

17. Gauchan, D., Smale, M. and Chaudhary, P. (2005) Market-based incentives for conserving diversity on farms: the case of rice landraces in central Tarai, Nepal. *Genetic Resources and Crop Evolution* (in press).

Abstract: Market-based incentives are one means of encouraging farmers to grow landraces that are also of social value, thereby contributing to the conservation of crop genetic diversity on farms and are in principle, the cheapest. This study uses a participatory market systems approach supplemented by baseline data from an ongoing project to analyse markets for rice landraces and modern varieties in Nepal. Nepal is located in the area of origin and diversity for Asian rice. With the exception of traditional Basmati rice (which is of high aromatic quality), most rice landraces are traded through small-scale informal channels. Traders earn higher profits handling modern varieties rather than landraces, with the exception of Basmati, which competes with modern varieties. The superior consumption qualities of Basmati are valued in markets, but conserving these landraces may not have great social value. Furthermore, farmers who grow Basmati are clearly better off than those who do not. These findings raise questions about the role of market-based incentives for conserving landraces on farms and the costs entailed in establishing a structure to generate them, and about efficiency versus equity considerations in the design of conservation programmes.

© *2004, Kluwer Academic Publishers, with kind permission of Springer Science and Business Media.*

18. Gollin, D. and Evenson, R.E. (1998) An application of hedonic pricing methods to value rice genetic resources in India. In: Evenson, R.E., Gollin, D. and Santaniello, V. (eds) *Agricultural Values of Plant Genetic Resources*. CAB International, Wallingford, UK, pp. 139–150.

Abstract: A pedigree analysis was undertaken for the 306 rice varieties released by Indian breeders during 1965–1986. This enabled a quantitative description of varieties in terms of year of release, releasing institution, characteristics emphasized, parent and grandparent combinations, number of landraces in the pedigree, generations from landrace materials and crosses of landrace material. Hedonic price evaluation was undertaken. This involved a statistical regression relating a measure of varietal improvement in farmers' fields to factors expected to cause or produce varietal improvement. Findings of an economic interpretation of the results are discussed.

19. Heal, G., Walker, B., Levin, S., Arrow, K., Dasgupta, P., Ehrlich, P., Maler, K.-G., Kautsky, N., Lubchenco, J., Schneider, S. and Starrett, D. (2004) Genetic diversity and interdependent crop choices in agriculture. *Resource and Energy Economics* 26, 175–184.

Abstract: The extent of genetic diversity in food crops is important as it affects the risk of attack by pathogens. A drop in diversity increases this risk. Farmers may not take this into account when making crop choices, leading to what from a social perspective is an inadequate level of diversity.

Reprinted with permission from Elsevier, © *2004.*

20. Heisey, P.W., Smale, M., Byerlee, D. and Souza, E. (1997) Wheat rusts and the costs of genetic diversity in the Punjab of Pakistan. *American Journal of Agricultural Economics* 79, 726–737.

Abstract: The theory of impure public goods is used to demonstrate why farmers may not grow wheat cultivars with the socially desirable level of rust resistance. First, they may grow cultivars that are high yielding although susceptible to rust. Second, many farmers may grow cultivars with a similar genetic basis of resistance. Expected rust losses can be reduced by (i) more diversified genetic background in released wheat cultivars; (ii) greater spatial diversity in planted cultivars or (iii) use of a temporally changing list of cultivars known to be rust-resistant. Yield trade-offs associated with these policies illustrate potential costs of increasing genetic diversity.

With permission from Blackwell Publishing.

21. Mayer, E. and Glave M. (1999) Alguito para ganar (a little something to earn): profits and losses in peasant economies. *American Ethnologist* 26, 344–369.

Abstract: We explore various ways in which small-scale peasants in the highlands of Peru conceptualize the everyday concept of profit in the contemporary context of neoliberalism. Through a process of approximation, we use the results of a survey of potato fields in two comparable valleys in Peru to clarify the differences between a strict business accounting procedure to establish profits or losses and the procedure that peasants use to evaluate the profitability of cash crops. We suggest that peasants evaluate profits or losses of cash crop in terms of a simple cash-out and cash-in flow. We indicate that this kind of calculus carries an implicit subsidy that permits market participation but provides little or no long-run benefit under prevailing productivity conditions and price levels. We also look at how farmers evaluate the status of their subsistence crops by showing that they ignore important cash expenses that are necessary to produce them. Finally, we describe accounting procedures characteristic of Andean peasants to understand how they monitor resource flow in their household-based farms. Analysis of the data leads us to question the 'subsistence first' model of peasants economies and to posit and interdependent relationship between subsistence and commercial sectors in which money plays an important but perverse role as it cycles through to the market and the household.

Taken from: http://www.grade.org.pe/asp/brw_pub11. asp?id =135

22. Meng, E.C.H. (1997) Land allocation decisions and *in situ* conservation of crop genetic resources: the case of wheat landraces in Turkey. PhD dissertation, University of California at Davis, California.

Abstract: This study contributes to the ongoing discussion of the feasibility of *in situ* conservation of crop genetic resources by developing a linkage between a household-level analysis of farmer incentives to cultivate traditional varieties and the diversity outcomes observed in the household for those varieties. The availability of both household-level socio-economic data and scientifically measured diversity data from the same households in an area of wheat diversity in Turkey permits the empirical application of the model. Estimation of the model of diversity outcomes suggests that diversity observed in the household is shaped primarily by the household's choice of variety, rather than its management of the variety once the crop has been planted in the field. Household risk attitudes, the agroecological conditions on the household farm and access of the household to markets were found to be significant factors in the household's varietal choice decision. Market-related factors, such as district-level market development and the relative prices between modern and traditional varieties, were particularly important in the household's decision to cultivate traditional varieties. Because the decision to cultivate traditional varieties appears to be the most important determining factor of household levels of diversity, an effective public-policy approach to maintaining the existing diversity level at the least cost is likely to consist of targeting the households with the highest *ex ante* probabilities of cultivating traditional varieties. An examination of the diversity held by a subset of households with a probability above 95% of cultivating traditional varieties showed these households, concentrated in three of the six surveyed districts, accounted for almost all of the named landraces in the survey. Findings also suggest that price policies specifically targeting traditional varieties and market development focusing on the consumption characteristics associated with traditional varieties may be the most effective means of encouraging their cultivation in the future.

The dissertation *citations and abstracts contained here are published with permission of ProQuest Information and Learning. Further reproduction is prohibited without permission.*

23. Meng, E.C.H., Taylor, J.E. and Brush, S.B. (1998) Implications for the conservation of wheat landraces in Turkey from a household model of varietal choice. In: Smale, M. (ed.) *Farmers, Gene Banks and Crop Breeding: Economic Analyses of Diversity in Wheat, Maize, and Rice.* Kluwer Academic Publishers, Boston, Massachusetts, pp. 127–143.

Abstract: Despite the potential advantages of conserving traditional varieties on farm, there is a gap between this de facto conservation and the establishment of a viable, long-term framework for on-farm conservation. This study presents concrete steps for monitoring, predicting and developing potential mechanisms to encourage farmers' *in situ* conservation. This chapter discusses findings from a behavioural model, analysing the incentive influencing a household's decision to grow traditional wheat varieties in three major wheat-producing provinces in Turkey. The model also estimates factors affecting the diversity outcomes observed for these varieties and tests for the linkages between choice varieties and diversity outcomes. Results confirm the role of multiple factors affecting the household's plot-level choice of variety. Different policy interventions might be required depending on the goal such policies seek. If the goal is to maintain morphological diversity in the wheat populations cultivated by farmers the additional information provided by the diversity estimation is helpful. The similarity in the range of variation maintained in a given landrace by both high- and low-probability households implies that this group of households requires the minimum number of external incentives for the de facto conservation.

© 1998 *Kluwer (extracted from different parts of article), with kind permission of Springer Science and Business Media.*

24. Meng, E.C.H., Smale, M., Rozelle, S.D., Hu, R. and Huang, J. (2003) Wheat genetic diversity in China: measurement and cost. In: Rozelle, S.D. and Sumner, D.A. (eds) *Agricultural Trade and Policy in China: Issues, Analysis and Implications.* Ashgate, Burlington, Vermont, pp. 251–267.

Abstract: In this chapter, recently developed statistical methods for classifying crop populations and indices of spatial diversity adapted from the ecology literature are used to measure wheat genetic diversity in seven major wheat-producing provinces in China. These diversity indices are then linked to economic decisions through the estimation of a cost function for wheat, using panel data on input and output prices, expenditures, environmental conditions and government interventions from 1982 to 1995. By using this approach, the marginal economic cost (or benefit) of wheat genetic diversity and its effect on input allocations are examined. Although econometric results indicate that evenness in morphological groups is a positive factor in overall costs per hectare of wheat production, the relationship of morphologically represented diversity to specific input use carries potentially important cost-saving implications. If the influx of new sources for pest and disease resistance has simultaneously resulted in increased levels of measured diversity, interaction with other required production inputs may have also changed. Diversity may thus contribute to a more efficient use of inputs, such as pesticides, which otherwise would have been required for a similar level of production stability.

© *Ashgate. Reprinted with permission.*

25. Morris, M.L. and Heisey, P.W. (1998) Achieving desirable levels of crop diversity in farmers' fields: factors affecting the production and use of commercial seed. In: Smale, M. (ed.) *Farmers, Gene Banks and Crop Breeding: Economic Analyses of Diversity in Wheat, Maize, and Rice.* Kluwer Academic Publishers, Boston, Massachusetts, pp. 217–238.

Abstract: This chapter examines crop diversity in rice and maize, focusing on spatial, temporal and genealogical dimensions of the adoption and diffusion of modern varieties (MVs) in commercial production systems. After measuring and examining the limitations of temporal, latent and spatial diversity, the authors recognize that a solid conclusion regarding trends in the level of diversity found within sets of MVs will be difficult to establish unless better measures of crop diversity are developed. The demand for seed of modern varieties is also examined. As the expected profitability is the main factor driving the demand for MVs in commercial cropping systems, farmers may in fact be demanding the same MVs, which could lead to overall reduction in crop diversity. This situation might lead to a 'social trap' as all individuals are acting in their own interest but the results are undesirable for the group as a whole, which in this case is the risk of a catastrophic disease. Policy recommendations to avoid this situation are evaluated.

26. Ninan, K.N. and Sathyapalan, J. (2003) The economics of biodiversity conservation – a study in a coffee growing region of India. Contributed paper selected for presentation at the 25th International Conference of Agricultural Economists, 16–22 August 2003, Durban, South Africa.

Abstract: This paper analyses the economics of biodiversity conservation in the context of a tropical ecosystem in India, where coffee is the main competitor for land use. Using primary data covering a cross-section of coffee growers, the study notes that the opportunity costs of biodiversity conservation in terms of coffee benefits forgone are quite high. Even after including external costs due to wild life damages and defensive expenditures to protect against wild life, the NPVs and IRRs from coffee for all land holding groups were high. The study notes that the external costs accounted for between 7% and 15% of the total discounted costs of coffee cultivation, and smaller holdings proportionately incurred higher external costs as compared with larger holdings. The study also notes high transaction costs incurred by the growers to claim compensation for wild life damages. Notwithstanding these disincentives, the study notes that the local community were willing to pay in terms of time for participatory biodiversity conservation, and they preferred a decentralized government institution for this purpose.

© *2003 by Authors, permission to publish granted.*

27. Shaxon, L. and Tauer, L.W. (1992) Intercropping and diversity: an economic analysis of cropping patterns on smallholder farmers in Malawi. *Experimental Agriculture* 28, 211–228.

Abstract: The diversity of cropping patterns on smallholder farms in southern Malaysia was analysed using a framework that explicitly incorporates the extent of intercropping in each field. Diversity is defined as relative abundance of each crop in the overall cropping pattern. Six indices of diversity were constructed for 208 farms and used in a model of welfare maximization farm household to examine the reasons for diversity in cropping patterns; multiple regression techniques were used to determine the effect of different household characteristics in diversity. The results suggest that an increase in labour availability over the production period is associated with a more diverse cropping pattern. Landholding size also influences diversity, which rises to a maximum and then falls as the area per capita increases. Farmers who grow a non-food cash crop (tobacco) have more diverse cropping patterns than those who do not. As diversity increases farmers use intercrop patterns that are more substitute than additive.

© *Cambridge University Press.*

28. Smale, M., Hartell, J., Heisey, P.W. and Senauer, B. (1998) The contribution of genetic resources and diversity to wheat production in the Punjab of Pakistan. *American Journal of Agricultural Economics* 80, 482–493.

Abstract: Recent criticisms of the green revolution in wheat concern the effects of their popularity on crop diversity and the consequences for productivity and conservation. A Just–Pope production function is used to test the relationship of genetic resource and diversity variables to mean and variance of wheat yields in the Punjab of Pakistan. In irrigated areas, greater area concentration among varieties is associated with higher mean yields. In rain-fed districts, genealogical variables are associated positively with mean yield and negatively with yield variance. Further research is needed to overcome data limitations, capture biological relationships more accurately and specify a fuller decision making model.

© *1998 Blackwell Publishers, permission granted.*

29. Smale, M. and Bellon, M.R. (1999) A conceptual framework for valuing on-farm genetic resources. In: Wood, D. and Lenné, J.M. (eds) *Agrobiodiversity: Characterization, Utilization and Management*. CAB International, Wallingford, UK, pp. 387–408.

Abstract: While recognizing the ethical difficulties involved in assigning values to biodiversity, the authors propose a general approach for identifying which crop population to conserve on-farm and *ex situ*. The authors propose an economic model that depicts farmers' incentives to grow varieties that are identified as key genetic resources. A decision making model for agricultural households is formulated, using maize farming in Mexico as an example. The challenge is to identify varieties that are both attractive to farmers and contribute to future flexibility of the genetic resource system.

30. Smale, M., Bellon, M.R. and Aguirre Gómez, J.A. (2001) Maize diversity, variety attributes, and farmers' choices in southeastern Guanajuato, Mexico. *Economic Development and Cultural Change* 50, 201–225.

Abstract: This paper examines farmers' demand for varieties of maize landraces by applying a choice model in which variety attributes and features of the region of production determine area shares allocated among varieties. The relationship between farmers' demand for variety and the genetic diversity of maize landraces in the farmers' communities is also investigated. Data are based on a survey of 160 farm households in 21 communities in Guanajuato, in Bajio, Mexico, conducted from August 1995 to January 1996. It is argued that the area allocation among varieties of maize landraces in the study area is determined not by the utility of the varieties themselves but by the attributes they provide. It is also suggested that area allocation decisions of individual farmers contribute to an impure public attribute, namely, maize genetic diversity in the community.

© *The University of Chicago Press*.

31. Smale, M., Meng, E., Brennan, J.P. and Hu, R. (2003) Determinants of spatial diversity in modern wheat: examples from Australia and China. *Agricultural Economics* 28, 13–26.

Abstract: The spatial distribution of modern varieties, and the genes they embody, has economic value because it affects crop productivity from year to year. Since farmers choose varieties based on observable traits rather than the genes they cannot see, a first step in understanding the spatial distribution of genes is to better understand the determinants of the spatial distribution of varieties. In this paper, we have constructed spatial diversity indices from area distributions of modern wheat varieties in Australia and China. We hypothesize that factors explaining variation in these indices are related to farmers' demand for traits and the supply of varieties, given physical features of the production environment. We test these hypotheses using reduced form equations for three concepts of spatial diversity, richness, abundance and evenness, using Zellner's seemingly unrelated regression (SUR). Spatial diversity indicators and analyses of this type, if more fully developed and targeted to address specific policy issues, may assist in monitoring crop genetic diversity or 'refuge' targets associated with the diffusion of some genetically modified crops.

Reprinted with permission from Elsevier, © *2003 Elsevier Science B.V. All rights reserved.*

32. Smale, M., Bellon, M.R., Aguirre Gómez, J.A., Manuel Rosas, I., Mendoza, J., Solano, A.M., Martínez, R., Ramírez, A. and Berthaud, J. (2003) The economic costs and benefits of a participatory project to conserve maize landraces on farms in Oaxaca, Mexico. *Agricultural Economics* 29, 265–275.

Abstract: Conventional methods were used to assess the benefits and costs of a project (during 1999–2002) whose purpose was to test whether participatory crop improvement can encourage Mexican farmers to continue growing maize landraces by enhancing their current use value. Findings suggest that farmers as a group earned a high benefit–cost ratio from participating, although from the perspective of the private investor the returns were low. The project also generated social benefits, but these are difficult (and costly) to measure. There was a gender bias in both participation and benefits distributions, although there is some evidence of a welfare transfer to maize-deficit households. Application of other valuation approaches is necessary in order to assess both the private and the social benefits of similar projects.

© *2003 Elsevier Science B.V. All rights reserved.*

33. Smale, M., Bellon, M.R., Jarvis, D. and Sthapit, B. (2004) Economic concepts for designing policies to conserve crop genetic resources on farms. *Genetic Resources and Crop Evolution* 51, 121–135.

Abstract: The future food supply of all societies depends on the exploitation of genetic recombination and allelic diversity for crop improvement, and many of the world's farmers depend directly on the harvests of the genetic diversity they sow for food and fodder as well as the next season's seed. On-farm conservation is an important component of the global strategy to conserve crop genetic resources, although the structure of costs and benefits of on-farm conservation differ from those associated with *ex situ* conservation in gene banks. A fundamental problem that affects the design of policies to encourage on-farm conservation is that crop genetic diversity is an impure public good, meaning that it has both private and public economic attributes. This concept is defined and made operational in order to assist practitioners in identifying (i) least-cost sites for on-farm conservation and (ii) the types of policy instruments that might be appropriate for supporting conservation once a site has been located. Published findings regarding prospects for on-farm conservation as economies develop are summarized and empirical examples of suitable policies to support farmers' decisions are placed in the context of economics principles.

© *2004 Kluwer Academic Publishers, with kind permission of Springer Science and Business Media.*

34. Stonehouse, D.P. (1999) Economic evaluation of on-farm conservation practices in the Great Lakes region of North America. *Environmetrics* 10, 505–520.

Abstract: Empirical research, divided into three alternative types of conservation practices, confirmed that two conservation crops and riparian buffer strips provide net costs to farmers, and that conservation tillage was not profitable under all circumstances. Research showed that riparian buffer strips and conservation tillage could be economically beneficial to society as a whole. This raised the question as to whether and to what extent society, as economic gainers, should offer compensation to farmers as economic losers. It was shown that not all conservation practices that result in reduced erosion will lead to decreased sediment and phosphorous loadings into watercourses, that not all reduced sediment and phosphorus loadings lead to improved water quality and that even where an improvement to water quality in chemical, biological, physical and aesthetic terms can be obtained, the costs to society for achieving the improvement may exceed the economic benefits. It was concluded that such outcomes readily promote disagreements between environmentalists, ecologists and socio-economists.

© *2004 John Wiley & Sons. Reproduced with permission.*

35. Van Dusen, E. (2000) *In situ* conservation of crop genetic resources in the Mexican *milpa* system. PhD thesis, University of California at Davis, California.

Abstract: This dissertation focuses on the theoretical modelling and empirical testing of household motivations for the *in situ* conservation of crop genetic resources (CGRs). An original household survey is used to test whether the household diversity outcomes are different for the cropping system as a whole, for the principal crop, maize, or for the secondary crops, beans and squash. Agroecological characteristics and market characteristics are found to significantly affect the levels of diversity maintained by households. A review of the economic literature relevant to modelling *in situ* conservation is presented. A theoretical model is developed in which a household's decision to plant a *milpa* variety is linked to household, agroecological and market variables. A household farm model appropriate to CGR conservation is presented, and extended to the case of missing markets. The agricultural ecology of the Sierra Norte de Puebla is described, as well as the principal CGR in the *milpa* system. The empirical methodology uses a Poisson regression, for the total number of crop varieties and for each crop group separately. The econometric work is extended to a hurdle model for sample selection and a SUR model utilizing a Shannon diversity index as a linear measure of diversity. The results from the regressions of household-level diversity show that a range of household, village, environmental and market conditions affect the diversity outcomes.

Market integration, measured by distance to a regional market, use of hired labour and international migration, was found to negatively affect diversity outcomes. Agroecological conditions, measured by the number of plots, plots with different slopes and the high-altitude region, were found to positively increase household diversity outcomes. The econometric findings were different for the combined *milpa* system than for the individual crops, and individual crops were affected by different factors. The principal crop, maize, seems mainly affected by the agroecological characteristics, while the levels of market integration are found to affect the minor crops, beans and squash. Conclusions are presented on the links between this study and conservation planning issues, and possible directions for future research are discussed.

The dissertation *citations and abstracts contained here are published with permission of ProQuest Information and Learning. Further reproduction is prohibited without permission.*

36. Van Dusen, E. and Taylor J.E. (2005) Missing markets and crop diversity: evidence from Mexico. *Environment and Development Economics* 10, 513–531.

Abstract: Recent microeconomic studies of *in situ* conservation of crop diversity focus on competition between modern and traditional varieties of major food crops. Our paper offers a different crop system approach and a limited dependent variable econometric technique to model *in situ* conservation of both intra- and infraspecies crop diversity in a context of heterogeneous ecological and market environments, using unique household-farm data from Mexico. Our findings reject separability and indicate that market integration significantly reduces crop diversity. They underline the importance of studying diversity in the context of larger cropping systems and economic environments.

© *2005 Cambridge University Press.*

37. Wale, E. (2004) The economics of on-farm conservation of crop diversity in Ethiopia: incentives, attribute preferences, and opportunity costs of maintaining local varieties of crops. PhD thesis, University of Bonn, Center for Development Research (ZEF), Bonn, Germany.

Abstract: The principal objectives of the study were to examine the farm household-related contextual factors motivating farmers to diversify on local varieties, study farmers' variety attribute preferences and examine their demand for local varieties and quantify the opportunity costs of growing local varieties and to analyse the contextual factors affecting opportunity costs. To address these objectives, the study uses household survey data from Ethiopia concerning coffee, sorghum and wheat. It examines the above objectives using a variety of microeconomic theories (including the characteristic model, the random utility theory, theory of impure public goods and the theory of joint production) and econometric techniques, including Poisson regression, multinomial logit and switching regression.

The dissertation *citations and abstracts contained here are published with permission of ProQuest Information and Learning. Further reproduction is prohibited without permission.*

38. Wezel, A. and Bender S. (2003) Plant species diversity of homegardens of Cuba and its significance for household food supply. *Agroforestry Systems* 57, 39–49.

Abstract: The cultivation of different plants in homegardens for self-sufficiency has a long tradition in Cuba, but knowledge about homegardens in Cuba is small. To analyse this more deeply, cultivated plants of 31 homegardens were surveyed in three villages in eastern Cuba in 2001. Two of the study villages were located in a humid area with an annual precipitation of about 2200 mm. The third village was situated in a semiarid area with about 450 mm precipitation. The plants studied in the homegardens included those for human consumption such as fruits, vegetables, tubers and cereals as well as spices and medicinal plants. In total, 101 different plant species were found with an average number of 18–24 species per homegarden for the three villages. A broad range of species was found in all villages, because irrigation is used under semiarid conditions, which leads to a relative high similarity in species composition between the villages. But, also differences due to the climatic situation became evident, particularly with the medicinal plants. In general, homegarden production provided a broad and diverse basis for self-sufficiency of the households.

Although homegarden production showed to be only a small source of income, it is particularly important because of low-paid outside work and minimal food provision of the state.

© *Kluwer Academic Publishers, with kind permission of Springer Science and Business Media.*

39. Wossink, G.A.A. and Wenum, J.H. (2003) Biodiversity conservation by farmers: analysis of actual and contingent participation. *European Review of Agricultural Economics* 30, 461–485.

Abstract: This paper examines actual and contingent participation by Dutch arable farmers in biodiversity conservation programmes. Probit and Tobit modelling were used to analyse the effect of farm and farmer characteristics and farmers' attitudes on participation. The optimal bid offer was derived from a referendum contingent valuation (CV) survey (with 250 respondents) for a proposed field margin programme. The results indicate that actual and contingent participation are better explained by the production environment and by familiarity with conservation programmes than by farmer characteristics or field characteristics. Contingent participation was significantly affected by farmers' perceptions of weed risks. The CV experiment suggested that up to 60% participation might be achieved with appropriate bid offers.

© *2003 Oxford University Press and the Foundation for the European Review of Agricultural Economics.*

Index

Note: page numbers in *italics* refer to figures, tables and boxes

abundance 9
adoption, partial 10
agricultural product demand 37
Agriculture and Rural Development Operational Programme (ARDOP) 257
Amhara region (Ethiopia) 180, 185–189
Andhra Pradesh (India) 213–214
attribute count index of diversity 54–55
attribute index, coffee 58
autarky, millet 224, *226*

bananas
    area planted 112
    attributes 110, 113–114
    biology 104
    bunch size 115
    consumption attributes 114
    cultivars 103–104, 106, 108, *109*, *110*
        demand 107
        diversity 113–114, 117
    dependent variables 109
    diseases 105, 112, 114, 117
        resistance 115
    diversity 97–98, *99–102*, 102–103
    econometric estimation 109–110, *111*, 112–113, 117
    endemic 103, 108
    farm characteristics 114
    genetics 102–105
    hybrids 104–105
    independent variables 109–110, *111*, 112
    labour 107
    markets 106, 112–113, 115–116, 117
    non-endemic 103
    pests 105, 114, 117
    planting material 114, 117
        replacement 105, 106
    rainfall 115
    selling 115–116
    site description 107–108
    taxonomy *99–102*, 102–105
    traits 103–104
    Uganda 97–117
    use group diversity indices 109, *110*, 115, *116*
Bara (Nepal) *164*, *165*, *171*
barley 82, 84, *86*, 184–185
    intracrop diversity *91*, *187*
bazaars 201, *202*
    seed source 200, *202*
beans
    haricot 235, 239
    seed market 260–261
    seed systems 259
Berger–Parker index of dominance 9, 109, *156*, 168
    cereal crop diversity 182, 188
    intercrop diversity 240
    potato diversity 157
    seed systems 247
biodiversity, agricultural
    conservation 34
    dependent variables 134–135
    development 35–36
    explanatory variables 134–135
    farmer demand 119–143
    home gardens 119–143
    Hungary 32–44, 119–143
    least cost options 43
    milpa system of Mexico 63–76

biodiversity, agricultural *continued*
    private value 37
    settlement characteristics 37
    wheat 270–278
    see also crop biodiversity
Biological-base Tender 257
Bulungur (Uzbekistan) 195–209

Cajamarca (Peru) 147, 148–152
capital, potato diversity 157
censored least absolute deviations (CLAD) 85, 181
cereal crops
    adoption 92, *93*
    cross-pollinating 84
    cultivation 82–84
    dependent variables 85–86, 181–182
    diversity 78, 81, 177–191
        index 85
        trade-offs 92, 94, 189–190
    econometric approach 180–182, *183*, 184
        estimation 84–85
    Ethiopia 78–94, 177–191
    household farms 81
    independent variables 182, *183*, 184
    intercrop diversity 79, *88*, 89, *90*, 185–186, 190
    intracrop diversity 79, 85, *88*, 89–92, 186–190
    Italy 273
    labour 90
    land
        area 87
        distribution 182
    landraces 85–86, 184
    markets 87, 89, 92, 184, 185
    modern varieties 92, *93*
    ox ownership 182, *183*
    policy implications 189–191
    regression models 180–181
    self-pollinating 84
    trade-offs in diversity 92, 94, 189–190
    variety names 85–86
change, response to 20–22
characteristics models 53–54
chat, intercrop diversity 239
chaykhana 202, 203
choice experiment 37–39, *40*, 292, 293
Cobb–Douglas production function 276
coffee
    attribute count 58
        index 59, 60
    attributes 51, 53, 54–55, 56, *59*
        index 58
    characteristics models 53–54
    choice decisions 51–53
    consumption attributes 55
    crop diversity index 49
    dependent variables 56
    diversity in Ethiopia 48–60
    econometric methods 58, 59–60
    explanatory variables 56–58
    farm household demand 53–54
    genetic diversity 50
    genetic resources 48
    household production 57–58
    human capital variables 58
    improved types 51–53
    indigenous 51–53
    labour 57–58
    land allocation 57
    markets 56–57, 58
    production
        attributes 55
        in Ethiopia 49–50
    propagation 51
    risks 55–56
    rural development interventions 60
    smallholding production 50
    tree replacement 51
    types 54–55
    utility 53
    varieties 49
    variety choices 54
collectivization 255
community groups 287
    Uzbekistan 194
    variation across 292
compensation
    farmer willingness to accept 37
    maize growers in Mexico 20
conditional logit model 41–42, 43
conservation
    attitudes towards 265
    equity 290
    landraces 28
    least cost 288
    objectives 289–290
    on-farm 3
        benefit–cost ratios 5–6
        landraces 18–19
        opportunity costs 19
    plant genetic resources 266–267
    programme design 140–143
    trade-offs 289–290
consumption attributes, coffee 55
contingent behaviour analysis 21, 292
contingent valuation 21, 292
Convention on Biodiversity (CBD) 2
cooperatives, agricultural 270–278, 286
    density 275–276, *277*
    functions 271–272
    southern Italy 274–275
credit access for households 154, 157, 182, *183*, 184
crop(s)
    cross-pollination 282

diversity index of coffee 49
evenness of planting 248
farmer motivation for growing 244–245
genetic diversity 38, 146–147, 177–178
    Uzbekistan 194–195
genetic erosion 63
genetic resources 5, 234
home gardens *123–125, 126–129, 129–131*
multiple varieties 64–65
perennial tree 197–198
productivity in Ethiopia 79
reproduction systems 282
risks 66, 78
self-pollination 282
'service' groups *245*
social trade-offs 178
spatial diversity 239
utilization patterns 234–249
valuation 17–19
variety
    choices 54
    home gardens 38
    value of local 2–3
crop biodiversity 2, 3–4, 5
agricultural household 291
bananas 97–98, *99–102*, 102–105
cereals 78
determinants 282–283, *284*, 285–287
dimensions 280, *281*, 282
economic change 289
economic development 286
farms 289
genetic 137–138
    maintenance 272
home gardens
    genetic 137–138
    variety 135–137
household farms 81–82
Hungarian family farms 251–267
intercrop 79, *88*, 89, *90*, 185–186, 190
    Ethiopia 236, 239–249, 285–286
    seed systems 248
intracrop 79, 85, *88*, 89–92, 186–190
    households 283
location specificity 291
metrics 7–10
milpa system 66, 67, 73, 75
reduced form 291
rural development impact 148
seed system 237
social costs–benefits 7
sustainable levels 286
trade-offs 92, 94, 189–190
Uzbekistan 197–199
variety 135–137
wheat productivity 278
crop-area shares 9–10

cultivars 3–4
cultivation of cereal crops 82–84

dependent variables, coffee 56
dichotomous choice, double-bounded 21
displacement hypothesis 10, 147
diversity
    metrics 7–10
    outcome 66, 68
    regression equations 85
    trade-offs 6–7
diversity index 8, 9, *156*, 168, 282
    attribute count 54–55
    cereal crops 85
    milpa system 70
    scalar 85
    wheat 272
dominance index 9

econometric methods, coffee 58, 59–60
economic change, crop biodiversity 289
economic development, crop biodiversity 286
economic value 4–5
    crop biodiversity 1
education 287
    households 171, 182, *183*, 189
    women 290
environmental valuation 292
    home gardens 36–37
environmentally sensitive areas (ESAs), Hungary 33,
    37, *40*, 41, 43, 120
equilibrium effects, maize 28
equity in conservation 290
Ethiopia
    cereal crops 78–94, 177–191
    coffee
        diversity 48–60
        production 49–50
    crop productivity 79
    highlands of north 178–179
    household farms 80–81
    households 179–180
    intercrop crop biodiversity 285–286
    markets 179–180
    peasant associations 82, 87, 237–239
    rural development 79, 190–191
    seed supply 233–249
    soil loss 78–79
European Union, Hungary in 34, 143, 256, 288
explanatory variables, coffee 56–58

farm characteristics
    cereal diversity 184, 189
    location 229

farm characteristics *continued*
   milpa system 71
   rice diversity 170
farmer field schools (FFS) programme (Peru) 152, 154–155, 159
farmers
   access to markets 35
   ageing 255
   decision making 252
   elderly 261, 283
   market-oriented cultivation 263
   motivation for growing crop 244–245
   response to change 20–22
   risk attitudes 55–56
   seed systems 261, *262*, 263
   small and food production 35
farmers' cultivars *see* landraces
farming, social status 266
farms
   crop biodiversity 289
   physical conditions 285–286
farms, family in Hungary 251–267
   historical patterns 255
   institutional economics 252–253
   stakeholder analysis 253
   stakeholder interviews 253–254
farms, household 63–76
   cereal crop diversity 81
   crop biodiversity 81–82
   Ethiopia 80–81
   model 66–67
   Uzbekistan 196–199
   see also home gardens
fertilizer use 263
fields, home gardens 121, 123, 126, 129
First Difference Estimator 276–277
folk varieties 3
food consumption, self-security 42
food market index 43
food security, home gardens 33, 34
fruit and nut production in Uzbekistan 192–209
   agricultural information 201–205
   crops 198
   diversity 198
   econometric analysis 205–208
   planting material 201
   regression analysis 205–208
   seed systems 199–200, 201, *202*
fruit trees 129, 197–198

garden plots, Uzbekistan 192–209
gardens, closed 123, 126
genetic resources
   potential use 18
   price data 1
grapes 197–198

diversity 198
vines 199
green revolution 1–2
guzar *202*, 204, 205

Hararghe Catholic Secretariat (HCS) 233–249
Hararghe region (Ethiopia) 234
haricot beans
   intercrop diversity 239
   seed system 235
hashar 201, *202*, 203, 204
Heckman's two-step estimation procedure 84–85
herbicide use 263
holidays, Uzbekistan 201
home gardens, Hungarian 32, 33–44, 255
   agricultural biodiversity 119–143
      least cost options 43
   attributes *39*
   choice experiments 37–39, *40*
   conservation programme design 140–143
   crops *123–125*, *126–129*, *129–131*
      varieties 38
   dependent variables 134–135
   econometric analysis 41–44, 134–140
   economic importance 33–34
   environmental valuation 36–37
   explanatory variables 134–135
   farmer management of diversity 120–143
   fields 121, 123, 126, 129
   food consumption 42
      household 38
   households 131–133
   landraces *138*, 261, 263
      cultivation 142–143
   livestock production 139–140
   markets 121–122, 135
   microecosystems 135
   orchards 129
   organic produce 123, 140
   physical characteristics 135
   seed systems 261, *262*, 263
   settlements 39, *40*
      characteristics 41–42
   site description 39, *40*, 41
   soil quality 137
home gardens, Uzbekistan 192–209
households
   agricultural 291
   allocation of garden plots 194
   assets 169–170, 285
   banana consumption 105, 107, 112
      decision making 110, 112
   banana production 106, 107
   cereal crops 87
      intercrop diversity 89, *90*
      intracrop diversity 89–90, *91*, 92

characteristics 87, *171*, 182, *183*, 186, 189
    millet farming 225
coffee production 57–58
conservation of rice varieties 171–175
consumption 105, 107, 132–133
    decisions 80, 166–167, 236–237
credit access 154, 157, 182, *183*, 184
crop biodiversity model 282–283, *284*
crop variety choices 54
decision making 213, 236, 283
education 171, 182, *183*, 189
Ethiopia 179–180
farm 107
food consumption 38
fruit and nut diversity 206, 209
Hungarian 131–133
income 154
    change 26
intercrop diversity 236, 239–249
    cereal crops 89, *90*
intracrop diversity 283
    cereal crops 89–90, *91*, 92
labour 171, 173
land endowment 65, 66
landrace cultivation 142–143
maize management implications 27
market constraints 65, 80
migrants 76
milpa system 73
non-tradability constraint 106
on-farm diversity 236, 237
participation in social groups 209
potato production 152–155
production constraints 133
production decisions 80, 166–167, 236–237
profiles 142
revealed preferences analysis 290–292
rice-growing 165–175
seed supply 237–239
seed system impact 240–241
self-sufficiency 133
unobserved shadow price 66
utility 66, 105
    of crops 132
Uzbekistan 196–199
wealth
    index 196–197
    indicators 135
human capital 283
human capital variables
    coffee production 58
    rice growing 171
Hungary
    agricultural biodiversity 32–44, 119–143
        farmer demand 119–143
    economic change 43

environmentally sensitive areas 33, 37, *40*, 41, 43, 120
European Union 34, 143, 256, 288
family farms 251–267
    Legislative Institutions 257
    market infrastructure 43
    Market Support System 258
    plant breeding 259–260
    policy regime 255–257
    seed system 259–261, *262*, 263

income, off-farm 64, 66
    cereal crops 90
    diversification 148
    Hungary 255
    migration 285
    Nepal 170
    Peru 147–148, 154, 157, 159–160
India, southern 211–230
institutional analysis 293
institutional economics 252, 266
    environmental 252–253
institutions
    fruit and nut diversity 194
    Uzbekistan 194, 200, 201, *202*, 203–205, 208, 209
    wheat production 271
intensification, agricultural 89
International Treaty on Plant Genetic Resources for Agriculture (2004) 2
inverse Mills ratio 85, 181, 189
Italy
    agricultural cooperatives 274–275
    regions 272–273
    wheat 270–278

Karnataka (India) 213–214
Kaski (Nepal) 163, *164*, *165*, *171*

labour
    bananas 107
    cereal crops 90
    coffee production 57–58
    family 70–71, 72, 73–74, 283
    intensity 72, 73–74
    markets 68
    migration impact 76
    milpa system 65, 70–71, 72, 73–74
    rice diversity 171, 173
    rural markets 63–64
Lancaster theory of consumer choice 53, 55
land
    degradation 184
    distribution 182

landraces 3, 4, 10–11, 280
    cereal crops 85–86, 184
    conservation 18–19, 28
    crop genetic diversity 38
    cultivation 142–143
    definition 263
    diversity equation 11
    home gardens *138*, 261, 263
    maize 18, 26, 28, 35
    market-oriented cultivation 263
    millet 213–223
    organic produce 263, 265, 266
    potato 35
    protection 265
    replacement by improved varieties 146–147
    rice 35–36, 162–163, 171–173
        choice 166–167
    sorghum 242–243
    terminology 264
    value 288
    wheat 273–274
livestock production
    Ethiopia 179
    home gardens 139–140
    Peru 150–151
logit model, conditional 41–42, 43

mahalla 201, *202*, 203
maize 84
    area grown 24–25
    biodiversity 18
    cross-pollination 282
    equilibrium effects 28
    Ethiopia 82, 83, 84, 86, 184–185
    growing in Mexico 22–28, 76
    intercrop diversity 239
    intracrop diversity *91, 187*
    land demand 27
    land use 24–25
    landraces 18, 26, 28, 35
    liberalization in Mexico 20
    marginal value 28
    market value *27*
    markets 26
    price change 24, 25, 26–27, 28
    production 19–20
        costs 26, *27*
    protection 19–20
    sample survey form 31
    seed industry 261
    seed market 260–261
    seed systems 259
    subsistence growing 24, 28
    supply 288
    types 24, 27
    varieties

        in Hungary 260
        modern 189
        old 264
Margalef index 9, 84, 109
marginal value 28
Market Support System, Hungary 258
markets
    access to 35, 148
    bananas 106, 112–113, 115–116, 117
    cereal crops 87, 89, 92, 184, 185
    coffee 56–57, 58
    constraints on household 65, 80
    cooperative 286
    Ethiopia 179–180
    failures 21–22
    home gardens 121–122, 135
    infrastructure 43
    labour 68
    maize 26
    millet 214, 225, 229
    milpa system 65, 68, 71–72
    potatoes 157
    products 286
    rice 166–167, 170, 171–172, 173
    rural areas 65
    seeds 259, 286
        supply 286
    Uzbekistan 198
    wheat 277–278
Mexico
    maize growing 22–28, 76
    migration 75–76
    milpa system 63–76
microecosystems 135, 280
migration 64, 66, 73, 285
    Hungary 255
    labour 68
    Mexico 75–76
    Peru 151, 152–153
    temporary 74–75
milk production, Peru 147–148, 149–151, 154, 159
millet 82, 83, 84, 184–185
    autarky 224, *226*
    biodiversity 214–215, *216*
    cropping choices 215
    diversity variation 227, *228*, 229
    econometric analysis 225, 227, *228*, 229
    farmers' cultivars 213–223
    hybrids 230
    intracrop diversity 188
    markets 214, 225, 229
    planting material 212
    replacement rate 216–218, *219, 220*, 227,
        229, 230
    seed
        saving 215
        sources 216–218, *219*

supply channels 220, *221*, 222–224, *225–226*, 229, 230
    trading 224
    transactions *221, 222*
  seed dealers 223–224
  seed lots 216–218, *219, 220*
  seed system 212–214, 229
  southern India 211–230
  supply transactions 218
  transfer rate 216–218, *219*
  types 214
  varieties *216*
milpa system, Mexico 63–76
  altitude 73
  crop biodiversity 66, 67, 73
  diversity index 70
  econometric methods 73
  farm characteristics 71
  genetic diversity 64
  households 73
  labour 65, 70–71, 72, 73–74
  markets 65, 68, 71–72
  multi-cropping systems 64
  multiple varieties 64–65

NAFTA programme 20
National Agri-Environmental Programme (NAEP, Hungary) 34, 37, 43, 143, 256–257
National Rural Development Plan (NRDP, Hungary) 257
natural resources management 194–195
Nature Conservation Act (1996, Hungary) 255
nature protection, Hungarian 255–256
Nepal, rice diversity 162–175
non-governmental organizations (NGOs) 259
North American Free Trade Agreement (NAFTA) 288
nut production *see* fruit and nut production in Uzbekistan

orchards 197–198
  home gardens 129
organic produce
  home gardens 123, 140
  landraces 263, 265, 266
ox ownership 182, *183*

peasant associations, Ethiopia 82, 87, 237–239
Peru
  livestock production 150–151
  migration 151, 152–153
  milk production 147–148, 149–151, 154, 159
  potato diversity 146–160

rural development 147, 157
pesticide use 263
  wheat 275
plant breeders 163
plant breeding, Hungary 259–260
plant genetic diversity 3
  crops 38
plant genetic resources
  conservation 266–267
  legal framework 263–264
  private value 287–289
  protection strategy 265, 266
  public value 287–289
  sustainable utilization 233
plantains 103, 104
Poisson regression model 58, 60
  cereal crops 181
  fruit growing 205, *207*
  milpa system 73, *74*, 75
  potato diversity 157
  rice 170
  seed systems 247
policy trade-offs 289–290
pollination systems 282
population density 42, 184
  cereal crops 186
  index 43
potatoes
  altitude of cultivation 153, 157, 159
  cultivation area 151
  diversity 147
  econometric approach 155–157
  harvests 159
  landraces 35
  management 151–152
  markets 157
  modern variety introduction 147
  Peru 146–160
  planting 151
  plots 157, 159
  production intensification 159
  propagation 152
  seed 152, 154
  varieties 152, *156*, 157
poverty 289
price data, genetic resources 1
PROCAMP programme 20
product markets 286
production
  attributes of coffee 55
  possibility frontier 65, 66
productivity 6–7
profit maximization 80
public goods 5
  theory of impure 271
pudrat 202, 203

rainfall, banana growing 115
random utility 36
    coffee attributes 55
regions 286–287
regression analysis
    fruit growing 205–208
    seed systems 246, *247*
    see also Poisson regression model
revealed preferences analysis 290–292
rice
    agroecosystems 164, *165*
    conservation trade-offs 171–172
    diversity
        conservation 174
        trade-offs 170–171
    econometric methods 167–170
    farm characteristics 169
    genetic resources 163
    human capital variables 171
    labour 170, 171
    land
        allocation 167
        types 164–165
    landraces 35–36, 162–163, 171
        choice 166–167
    markets 166–167, 170, 171–172, 173
    Nepal 162–175
    plot dispersal 170
    production decision makers 166
    regression models 170
risks
    coffee 55–56
    crops 66, 78
rural development
    Ethiopia 79, 190–191
    interventions in coffee growing 60
    Peru 147, 148, 157
    potato diversity 147, 157, 159
rural poverty 193–194

Samarqand (Uzbekistan) 195, *196*
seed(s)
    accessibility 234
    availability 234
    certification 256
    exchange 213
    farm-saved 264
    informal trade 265
    markets 259, 286
    registration 256
    supply
        households 237–239
        markets 286
    trade 265
    utilization patterns 233–234
Seed Act (Hungary) 257

seed systems
    beans 259
    characteristics of farmers 240
    crop diversity 237
    crop genetic resources 234
    crop utilization patterns 234–249
    econometric analysis 246–249
    farmer 261, *262*, 263
    formal *262*, 263
    home gardens 261, *262*, 263
    Hungary 259–261, *262*, 263
    intercrop diversity 248–249
    maize 259
    millet 212–214
    parameters 287
    sorghum 242–245, *246*, 249
    stakeholders 263–266
    wheat 242–245, *246*, 249
self-sufficiency of households 133
selsoviet 200
settlement(s) 286–287
    characteristics of home gardens in Hungary 39, *40*, 41–42
    development index 42–43
    population density 42
    unemployment 42
shandies 214, 224, 230
    traders *226*
Shannon index 9, 84, 109, 117, *156*
    cereal crop diversity 182
    fruit diversity 198
    potato diversity 157
    seed systems 247, 248
shirkat 200
shock
    external 288
    hypothetical 24
Sierra Norte de Puebla (Mexico) 22–28, 69–76
Simpson index
    fruit diversity 198
    intercrop diversity 239–240
    wheat 272, 276
smallholdings, coffee production 50
social dilemmas 5
social networks in Uzbekistan 194
soil loss in Ethiopia 78–79
soil microorganism diversity 140
soil quality
    cereal diversity 184
    home gardens 137
solkbak 202, 203
sorghum 82, 83, 184–185, 214, 215
    biodiversity 229
    breeding 235–236
    intercrop diversity 239, 244
    intracrop diversity *188*
    landraces 243–244

replacement rate *220*
seed lots 218, *219*
seed purchase 246
seed supply channels *222*
seed systems 235, 242–245, *246*, 249
service groups *245*
Soviet Union 192, 193
Special Accession Programme for Agriculture and Rural Development (SAPARD) 34, 143
species diversity 3
stakeholders 257, 258
    analysis 253, 293
    interviews 253–254
    plant genetic resource protection strategy 266
    seed system 263–266
stated preference analysis 292–293
subsistence growing, maize 24, 28
sustainable utilization of plant genetic resources 233

teff 82, 83, 84, 86, 184–185
    intracrop diversity *91*
Terai lowlands (Nepal) 163, *164*
Tigray (Ethiopia) 178–179, 180, 185–189

Uganda, bananas 97–117
unemployment, settlement characteristics 42
urbanization index 43
Urgut (Uzbekistan) 195–209
utility theory 73, 105
Uzbekistan 192–209
    holidays 201
    institutions 194, 200, 201, *202*, 203–205, 208, 209
    markets 198

valuation, contingent 21
varieties 3–4
    named 8–9
vegetable production, Uzbekistan 193, 194
villages 286–287
    cereal diversity 177–178, 184, 185
    characteristics 184

weddings 201, *202*
wheat 84, *86*, 184–185
    agricultural cooperatives 275–276
    breeding 235
    crop biodiversity 276
    crop productivity 270–278
    diversity 270–278
        intercrop 239, 244, 275
        intracrop *91, 187*, 275, *277*
    durum 82, 83, 273, 275, 277–278
    improved varieties 235
    landrace 273–274
    markets 277–278
    modern varieties 189, *244*
    pesticide use 275
    production function estimation 277
    productivity 276, 278
    renewed seed 246
    seed purchase 246
    seed systems 235, 242–245, *246*, 249
    service groups *245*
    southern Italy 270–278
women's education 290
work brigade *202*, 203

Zoatecpan (Mexico) 22–28

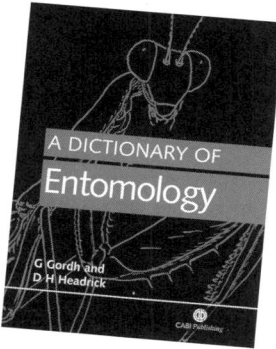